HUMORAL CONTROL

OF GROWTH AND DIFFERENTIATION

VOLUME II
Nonvertebrate Neuroendocrinology and Aging

CONTRIBUTORS

D. Bellamy

Hans Bode

John Tyler Bonner

Max M. Burger

William H. Darden, Jr.

Albert S. Gordon

George M. Happ

William J. Kuhns

Joseph LoBue

V. A. Maslennikova

D. Mertz

Robert Turner

Fleur L. Strand

George Weinbaum

HUMORAL CONTROL
OF GROWTH
AND DIFFERENTIATION

VOLUME II

Nonvertebrate Neuroendocrinology and Aging

Edited by

JOSEPH LOBUE

*Department of Biology
Graduate School of
Arts and Science
New York University
Washington Square Campus
New York, New York*

ALBERT S. GORDON

*Department of Biology
Graduate School of
Arts and Science
New York University
Washington Square Campus
New York, New York*

1973

ACADEMIC PRESS   *New York and London*

A Subsidiary of Harcourt Brace Jovanovich, Publishers

ACADEMIC PRESS, INC.
111 Fifth Avenue, New York, New York 10003

United Kingdom Edition published by
ACADEMIC PRESS, INC. (LONDON) LTD.
24/28 Oval Road, London NW1

Library of Congress Cataloging in Publication Data

LoBue, Joseph.
     Humoral control of growth and differentiation.

     Includes bibliographies.
     CONTENTS: 1.   Vertebrate regulatory factors.–
2.   Some nonvertebrate humoral principles.
     1.   Hormones.   2.   Cell proliferation.   3.   Cell
differentiation.   4.   Cellular control mechanisms.
I   Gordon, Albert, joint author.   II.   Title.
[DNLM: 1.   Cell differentiation.   2.   Cells–
Growth & development.   3.   Growth substances.
4.   Hormones–Physiology.   5.   Homeostasis.
6.   Vertebrates. WK515 L799h]
QH604.L6   574.3   73-2067
ISBN 0–12–453802–9 (v. 2)

PRINTED IN THE UNITED STATES OF AMERICA

CONTENTS

LIST OF CONTRIBUTORS   ix

PREFACE   xi

CONTENTS OF VOLUME I   xiii

I SOME NONVERTEBRATE HUMORAL PRINCIPLES

1. Hormonal Control of Insect Diapause
V. A. Maslennikova

I. Introduction   3

II. The Hormonal Control of Growth and Morphogenesis   4

III. Ecological and Physiological Features of Diapause   8

IV. Hormonal Control of Pupal and Larval Diapause   11

V. Hormonal Control of Imaginal Diapause   18

VI. Hormonal Control of Embryonic Diapause   21

VII. Bifactorial Principle of Diapause Hormonal Control   26

References   29

2. Humoral Influences in Hydra Development
Hans Bode

I. Introduction   35

II. Growth   36

III. Head Regeneration   40

IV. Cell Differentiation   46

V. Conclusion   53

References   55

3. Aggregation Factors of Marine Sponges

William J. Kuhns, George Weinbaum, Robert Turner, and Max M. Burger

I.	Introduction	59
II.	Sponge Differentiation and Development	60
III.	Sponge Cell Aggregation and Aggregating Factor (AF)	62
IV.	Characterization and Properties of Sponge Aggregating Materials	66
V.	Studies on *Microciona* Sponge Receptor Sites Reactive with AF	69
VI.	Immunological Studies of Sponge Cell–Cell Interactions	71
VII.	Sponge Aggregation: Summary	75
VIII.	Conclusion	76
	References	77

4. Hormones in Social Amoebae

John Tyler Bonner

I.	Hormones which Orient Morphogenetic Movements	84
II.	Hormones which Affect Pattern	88
III.	Hormones which Affect Differentiation	90
IV.	Hormones which Affect Spore Germination	92
V.	Discussion	92
	References	97

II PLANT GROWTH AND DIFFERENTIATION

5. Hormonal Control of Sexuality in Algae

William H. Darden, Jr.

I.	Sexual Differentiation in Algae	102
II.	Control of Differentiation in *Volvox*	104
	References	117

6. Hormonal Control of Root Growth

D. Mertz

I.	Introduction	121
II.	Root Organization	122
III.	Nutritional Requirements of Root Growth	125
IV.	Development of the Hormone Concept	127
V.	Vascular Differentiation	137
VI.	Concluding Remarks	142
	References	143

III ADDITIONAL ASPECTS

7. Chemical Signals between Animals: Allomones and Pheromones

George M. Happ

I. Introduction	150
II. Characteristics of Chemical Signals	151
III. Methods of Study	154
IV. Allomones	155
V. Pheromones	163
VI. Exocrines and Endocrines	178
References	181

8. Endocrineurology

Fleur L. Strand

I. Introduction	191
II. Effects of Hormones on Neurogenesis	192
III. Effects of Hormones on the Mature Central Nervous System	202
IV. Effects of ACTH and Adrenal Cortical Hormones on Peripheral Nerve	210
V. Conclusions	212
References	213

9. Aging as a Process

D. Bellamy

I. Definitions	219
II. Homeostasis and Aging	226
III. Aging at the Tissue Level	256
IV. Evolution and Aging	269
V. General Conclusions	271
References	271

10. General Summary — 281

Joseph LoBue and Albert S. Gordon

AUTHOR INDEX	295
SUBJECT INDEX	312

LIST OF CONTRIBUTORS

Numbers in parentheses indicate the pages on which the authors' contributions begin.

D. BELLAMY (219), Department of Zoology, University College, Cardiff, Wales, Great Britain

HANS BODE (35), Department of Developmental and Cell Biology, School of Biological Sciences, University of California, Irving, California

JOHN TYLER BONNER (81), Department of Biology, Princeton University, Princeton, New Jersey

MAX M. BURGER (59), Department of Biochemistry, Biozentrum der Universität, Basel, Switzerland

WILLIAM H. DARDEN, JR. (101), Department of Biology, University of Alabama, University, Alabama

ALBERT S. GORDON (281), Department of Biology, Graduate School of Arts and Science, New York University, Washington Square Campus, New York, New York

GEORGE M. HAPP* (150), Department of Biology, New York University, Bronx, New York

WILLIAM J. KUHNS (59), Department of Pathology, New York University School of Medicine, Bellevue Hospital, New York, New York

* Present address: Department of Zoology and Entomology, Colorado State University, Fort Collins, Colorado.

JOSEPH LOBUE (281), Department of Biology, Graduate School of Arts and Science, New York University, Washington Square Campus, New York, New York

V. A. MASLENNIKOVA (3), Entomological Laboratory, Biological Research Institute, Leningrad State University, Leningrad, U.S.S.R.

D. MERTZ (121), Division of Biological Sciences, University of Missouri-Columbia, Columbia, Missouri

ROBERT TURNER (59), Department of Biochemistry, Biozentrum der Universität, Basel, Switzerland

FLEUR L. STRAND (191), Department of Biology, New York University, Washington Square Campus, New York, New York

GEORGE WEINBAUM (59), Albert Einstein Medical Center Research Laboratories, Philadelphia, Pennsylvania

PREFACE

The purpose of this volume remains the same as that of the first, namely, to introduce the reader to the intriguing and complex subject of humoral regulation of growth and differentiation. As in Volume I, emphasis is placed on those aspects of hormonal control that have generally been neglected in standard texts on endocrine physiology. Similarly, considerable editoral selection of the subject matter included in this volume was also made necessary by the obvious fact that not all facets of this broad field could possibly be covered. Thus the ultimate contents of this second book were determined using the same criteria as for Volume I. In particular, topics for inclusion had to be associated with a sufficient body of literature to allow a substantive, critical, up-to-date review to be conducted. The plan of Volume II is somewhat different from that of Volume I in that the framework is considerably more extensive, encompassing as it does nonvertebrates, neuroendocrinology, and aging. Specifically, Parts I and II of the book relate exclusively to nonvertebrate hormones. Thus, Part I takes the reader by rapid descent from the Insecta in which diapause and "diapause hormone" are considered through the Coelenterata (differentiation in Hydra) and Porifera (aggregation in sponges) to the curious and remarkable protist being, the social amoeba, for which humoral regulation of aggregation, spacing, and differentiation are considered. The name "amoeba" encouraged us to include this subject in company with animals although clearly its true taxonomic status and evolutionary affinities are not certain. Part II considers humoral control in two extreme groups, moving abruptly from a discussion of regulation of sexual differentiation in the plantlike Protista (the green alga *Volvox*) to chemical control of root growth and development. Neuroendocrinology is treated in Part I, in which neurosecretions are seen as an important part of both the insect's and Hydra's hormonal control mechanism, and also in Part III, in which endocrine effects on nerve growth and function ("endocrineurology") are discussed. Also in Part III is an extensive evaluation of the process of aging as it occurs in many diverse organisms. Consideration of the hormonal aspects of the subject as well as many other facets of this enigmatic phenomena are included. In addition, some space has been alloted to an examination of humoral controls as they are exerted among individuals, both vertebrate and invertebrate, through the mediation of allomones and pheromones. Thus, this second volume is really intended to be a "sampler" of this exciting and important general field. We hope it will serve this purpose and be of real interest and value.

Once again, we are pleased to express our gratitude to Mrs. Blanche Ciotti for her expert secretarial help. Thanks are also due to a wife, Catherine LoBue, a young son (Joseph), and daughter (Ellen). Catherine was an invaluable aid in the mechanical aspects of subject indexing while Joseph and Ellen lent much needed moral support. Sincere gratitude is also extended to the staff of Academic Press for their friendly advice and cooperation during all phases of production.

Joseph LoBue
Albert S. Gordon

CONTENTS OF VOLUME I

VERTEBRATE REGULATORY FACTORS

I Chalones

Chalone Control Systems
  William S. Bullough

II Blood Cell Formation and Release

Erythropoietin: The Humoral Regulator of Erythropoiesis
  *Albert S. Gordon, Esmail D. Zanjani, Anthony S. Gidari, and
  Robert A. Kuna*

Humoral Regulation of Neutrophil Production and Release
  Edward F. Schultz, David M. Lapin, and Joseph LoBue

Humoral Regulation of Eosinophil Production and Release
  Natalie S. Cohen, Joseph LoBue, and Albert S. Gordon

The Colony Stimulating Factor (CSF)
  Donald Metcalf

Humoral Regulation of Thrombocytopoiesis
  T. T. Odell, Jr.

Humoral Aspects of Blood Cell Dyscrasias
  T. N. Frederickson and P. F. Goetinck

Regulation of Erythropoiesis in Lower Vertebrates
  *Esmail D. Zanjani, Albert S. Gordon, Anthony S. Gidari, and
  Robert A. Kuna*

Humoral Regulation of Lymphocyte Growth *in Vitro*

   Klaus Havemann, Manfred Schmidt, and Arnold D. Rubin

Possible Feedback Inhibition of Leukemic Cell Growth: Kinetics of Shay Chloroleukemia Grown in Diffusion Chambers and Intraperitoneally in Rodents

   Philip Ferris, Joseph LoBue, and Albert S. Gordon

III   Humoral Control of Organs and Tissue Growth

The Nerve Growth Factor

   Ruth Hogue Angeletti, Pietro U. Angeletti, and Rita Levi-Montalcini

Humoral Aspects of Liver Regeneration

   Frederick F. Becker

Renal Growth Factor

   Ronald A. Malt

Hormonal Influence on Skeletal Growth and Regeneration

   Edgar A. Tonna

Trauma and Tumor Growth with Special Emphasis on Wound Stress and "Wound Hormones"

   Philip Ferris, Norman Molomut, and Joseph LoBue

General Summary

   Joseph LoBue and Albert S. Gordon

Author Index-Subject Index

I

SOME NONVERTEBRATE HUMORAL PRINCIPLES

1

HORMONAL CONTROL OF INSECT DIAPAUSE

*V. A. Maslennikova**

I. Introduction 3
II. The Hormonal Control of Growth and Morphogenesis 4
    A. Activation Hormone 5
    B. Molting Hormone or Ecdysone 6
    C. Juvenile Hormone 6
III. Ecological and Physiological Features of Diapause 8
IV. Hormonal Control of Pupal and Larval Diapause 11
    A. Pupal Diapause 11
    B. Larval Diapause 16
V. Hormonal Control of Imaginal Diapause 18
VI. Hormonal Control of Embryonic Diapause 21
VII. Bifactorial Principle of Diapause Hormonal Control 26
    References 29

I. INTRODUCTION

Insect diapause is a temporary arrest of growth, development, or reproduction arising as a result of profound changes in metabolism. Diapause as a remarkable ecological and physiological adaptation for survival under various unfavorable conditions has long been of concern to ecologists and physiologists.

Ecological factors controlling diapause have been thoroughly investigated (Andrewartha, 1952; Beck, 1962, 1968; Danilevsky, 1961; Lees,

* This paper was prepared for the memorial readings devoted to my teacher Professor A. S. Danilevsky.

1955, 1968; Müller, 1965, 1970; Wilde de, 1962). Many investigations have been devoted to the intrinsic physiological mechanisms regulating diapause (Danilevsky *et al.*, 1970; Gersch, 1964; Goryshin and Tyshchenko, 1968; Fukuda, 1963; Harvey, 1962; Lees, 1955, 1956, 1961, 1968; Morohoshi, 1959, 1969a, b, c, d e; Novák, 1966; Schneiderman, 1957; Wilde de, 1961; Williams, 1952).

The occurrence of diapause at larval, pupal, and imaginal stages is considered to be due to the deficiency of hormones controlling growth and metamorphosis. Embryonic diapause, though, is conditioned by the influence of a specific diapause hormone produced by the subesophageal ganglion. The majority of experimental data leading to these conclusions are sufficiently convincing. However, one cannot think these conclusions fully reveal the nature of diapause, which in fact is considerably complicated.

In this chapter attention is focused on some results which do not agree with common concepts of hormonal control of insect diapause. An attempt has been made to give a general scheme of hormonal control based on the interaction of hormones inhibiting and activating morphogenesis. This concept was first introduced by H. E. Hinton (1953) but, unfortunately, was not accepted to any great extent at that time as it was not sufficiently supported by experimental evidence. In our opinion the absence of a reliable general hypothesis of hormonal control of diapause hinders future investigations in this field; however important data accumulated recently make it possible to revise the commonly accepted concepts and to advance new ones.

II. THE HORMONAL CONTROL OF GROWTH AND MORPHOGENESIS

This paper deals mainly with the hormonal control of insect diapause, but a brief examination of the hormonal control of growth and morphogenesis is appropriate, as the same neurohormonal systems are involved, in different ways, in the hormonal regulation of both events.

Ontogenesis in holometabolous insects includes several larval instars and a pupal one. The transition from one instar to another is carried out through molt, which is the most striking feature of insect ontogenesis.

It is well known (Gersch, 1964; Gilbert, 1964; Karlson, 1967; Novák, 1966, 1969, 1970; Tamarina, 1966; Wigglesworth, 1954, 1966, 1967, 1970) that three hormones [activation hormone (AH), molting hormone (MH) or ecdysone, and juvenile hormone (JH)] take part in the regulation of insect morphogenesis, interacting in a complicated way.

A. Activation Hormone

Activation hormone is a true neurohormone and plays a paramount part in growth and development processes. However, its influence in all these processes is indirect. AH is necessary for the activation of two other endocrine organs: prothoracic glands, which are a source of MH, and corpora allata, which are a source of JH. AH activates these endocrines, but the final effect depends on the ratio of the titers MH and JH in the insect. Besides, AH probably activates some other glands and influences some vital functions of the organism. The absence of AH reduces the level of general metabolism (Gersch, 1968; Novák, 1966, 1969).

The neurosecretory cells of the brain are a source of AH. There are about six groups of such cells in the brains of different insects (Herman and Gilbert, 1965; Hinks, 1971; Kind, 1968a, b; Panov, 1962a, b, 1964, 1969, 1971; Panov and Kind, 1963; Raabe, 1964, 1965; 1971). In each half of the brain there are three medial (M_1, M_2, M_3) and three lateral (L_1, L_2, L_3) groups comprised of neurosecretory cells. These groups and types of neurosecretory cells remain considerably constant in the different insects. Axons of medial and lateral cells by the nerves of corpori cardiaci I and II reach corpora cardiaca and a part of them reach even the corpora allata. It is not certain which neurosecretory cells are responsible for AH production, although the A cells of the medial group of the protocerebrum are supposed to be a possible source of AH in Lepidoptera (Panov, 1969).

A detailed biochemical analysis of the brain hormone performed by Yamazaki and Kobayashi (1969) induced them to conclude that this hormone is a glycoproteid with a molecular weight of about 20,000.

The investigation of brain hormone showed that it is comprised of two neurohormones, C and D, which later were divided into C_1 and C_2 and D_1 and D_2 (Gersch, 1961, 1968; Gersch and Berger, 1962; Gersch and Richter, 1963; Gersch et al., 1963, 1964). It was also found that only the neurohormone D_1 possesses the property of AH. The AH injection even at extremely low concentration of 10^{-11} gm/1 ml Ringer's solution stimulates the pupation in *Calliphora* and *Lymanthria* (Gersch, 1964).

B. Molting Hormone or Ecdysone

Molting hormone is produced by prothoracic gland cells and their homologs which were found in all major insect groups (Gersch, 1964; Novák, 1966). The extirpation of these glands makes the molt and metamorphosis impossible. Implantation of these glands or injection of MH into the isolated insect abdomen induce apolysis.

It is considered that MH acts as a gene-activating agent. Its primary influence on the organism is the stimulation of synthesis of all RNA types (Clever, 1964; Gilbert, 1964; Kroeger, 1963; Novák, 1969; Wigglesworth, 1970). Hypodermis and ectodermal tissues are found to be the main target for MH. Ecdysis is the final effect of the MH action. In case the prothoracic glands are stimulated by AH, the MH titer in insect determines the occurrence of either molt or metamorphosis. It has been proved that MH can induce molt even in adults, which normally can not molt.

Cristalline ecdysone was obtained in 1954 by Butenandt and Karlson. Later it was divided into two fractions: α-ecdysone and β-ecdysone (Karlson, 1957). α-Ecdysone had higher activity: 0.0075 μg of this substance induced the pupation in *Calliphora erythrocephala* larvae. Probably the number of ecdysone active fractions is still larger (Burdette and Bullock, 1963). The chemical formula of ecdysone is $C_{27}H_{44}O_6$. It is a steroid, belonging to α,β-unsaturated ketones.

C. Juvenile Hormone

Juvenile hormone, secreted by corpora allata, influences development in a more complicated manner as compared to ecdysone. The effects of JH on insect development are very diverse, therefore there are contradictory opinions as to the mechanism of its action. JH determines the trend of the metamorphosis, in particular the type of arising cuticle. Sufficient concentration of JH in unmature insects facilitates retaining larval pecularities during molting and prevents imaginal differentiation. After corpora allata extirpation the next molt turns into the pupal molt in holometabolous insects or the imaginal molt in hemimetabolous insects. On the contrary, the implantation of corpora allata results in extraordinary larval molt in last instar larvae. The pupae can undergo a second pupal molting.

Undoubtedly, JH can be considered as one of the most important morphogenetic hormones, as it may lead to the changes in morphogenesis during any period of ontogenesis and to the appearance of abnormal states. The important role of JH in morphogenetic processes has been shown in numerous papers (Gersch, 1964; Gilbert, 1964; Gilbert and Schneiderman, 1960, 1961; Meyer, 1965; Novák, 1966, 1969; Pflügfelder, 1958; Pridantseva *et al.*, 1971; Tamarina, 1966; Wigglesworth, 1966, 1967, 1970; Williams, 1958).

In the mature adult, JH produces a gonadotropic effect, i.e., it stimulates ovogenesis in females and the development of accessory glands in males. JH also influences the prothoracic gland function, the level of general metabolism, polymorphism, the change of coloration and many other developmental patterns. Therefore, it is not surprising that some authors attribute these various effects of JH to the action of two or more hormones secreted by corpora allata.

JH belongs chemically to the class of terpenoids. Williams (1956a) was the first to obtained the JH active extract from *Hyalophora cecropia*. The injection of this extract into various Lepidoptera pupae caused the same result as the implantation of corpora allata. Purified JH was first obtained in 1965 by Röller *et al.* (1965), and in 1969 the method of hormone purification was considerably improved (Röller *et al.*, 1969). Chemical and physical analysis of JH properties led Röller *et al.* (1967) to view this structure as:

Such a structural formula is possible for eight JH geometric isomers. All those isomers were synthesized, and their biological activity was tested on *Tenebrio molitor* (Röller and Dahm, 1968). One of them, the so called synthetic JH, was identical in its properties to the natural one and its activity was about 5000 "*Tenebrio* units" in 1 μg.

Some peculiarities of the interaction of three main metamorphosis hormones should be noted. Temporal reduction of metabolism and a decrease in the hormone production was observed in majority of insects during the ecdysis period, i.e., during chitinous cuticle molting. The level of concentration of the three hormones in hemolymph drops below the threshold of their effectiveness at the beginning of each instar. The production of these hormones is resumed after molting and resumption of feeding. AH is secreted first. It activates the prothoracic glands to secrete MH and the corpora allata to secrete JH. MH causes all ectodermal cells to divide and to differentiate which leads to molting at the next instar. The presence of JH, which determines larval peculiarity of the next instar, is necessary for the larval molting. When MH stimulates the larval cells to grow and molt, high titer of JH forces them to secrete the larval cuticle. In the case of low titer of this hormone the same cells secrete the pupal cuticle. If JH is absent, the hypodermal cells can secrete imaginal cuticle even without the pupal molt.

The alternative differentiation of epidermal cells, which is determined by different ratio of MH and JH, has been convincingly demonstrated (Gilbert, 1964; Wigglesworth, 1966, 1970). Thus the insect growth and morphogenesis during ontogenesis are controlled by the balance between the two main hormonal factors, MH and JH. As was noted above, AH influences the insect development only through the stimulation of production of two other hormones, and the final morphogenetic effect depends on their action.

III. ECOLOGICAL AND PHYSIOLOGICAL FEATURES OF DIAPAUSE

Growth and morphogenetic processes in insects as well as in all poikilothermal animals can occur only within a very limited temperature range (Andrewartha and Birch, 1954; Danilevsky, 1949, 1961; Lees, 1955; Ushatinskaya, 1957). Insects are unable to endure low temperatures for a long time during growth and morphogenesis. They can withstand temperatures and other season conditions unfavorable for development in a particular physiological state, i.e., quiescence or other diapause forms. This chapter deals only with the true diapause, other forms of dormancy are not discussed here.

Diapause, as was noted above, is a peculiar physiological state which is characterized, as a rule, by long interruption of growth, morphogenesis, or reproduction, and is a result of profound changes in general metabolism. Not all insects possess diapause. The capacity to enter diapause is a hereditary characteristic. According to it all insects are subdivided into univoltine (or monovoltine), polyvoltine with a facultative diapause, and polyvoltine without a diapause (or homodynamic) species. Univoltine species are the species which have a diapause at a definite developmental stage in each generation, i.e., they have obligatory diapause. Such insects produce only one generation a year. Polyvoltine nondiapausing insects develop without diapause through the year and their diapause cannot be induced by any environmental change within the norm. Polyvoltine species with a facultative diapause may develop either with or without a diapause, depending on external conditions. Their facultative diapause is easily controlled by environmental factors such as temperature, photoperiod, and food supply. The species with facultative diapause have two different possibilities for seasonal development, which is a genetical norm for them, whereas the homodynamic species normally develop without diapause, and the univoltine species have an obligatory diapause in each generation. Expression of this genetic pattern in the species with facultative diapause is controlled both by intrinsic neuroendocrine factors and by external ecological conditions influencing the insect during its ontogenesis.

Seasonal photoperiodism or seasonal changes of day length is the main factor among other ecological factors inducing diapause in the life cycle. Numerous investigations (Beck, 1968; Danilevsky, 1961; Danilevsky *et al.*, 1968, 1970; Lees, 1955, 1968; Wilde de, 1962) have shown that photoperiodism is the determinating factor, whereas other factors (temperature, food, humidity) are additional regulators.

A valuable contribution to the study of ecological and physiological regulation of insect diapause and, in particular, the role of photoperiodism was made by A. S. Danilevsky and his colleagues.[*] Their experiments on over 100 insects and mites with various seasonal cycles prove the universality of photoperiodic regulation of diapause and make it possible to classify the main types of photoperiodic reaction (Fig. 1).

Photoperiodic information passing through a corresponding oscillator link of the nervous system tunes the activity of neuroendocrine system in accordance with a received signal (Bünning and Joerrens, 1959; Danilevsky *et al.*, 1970; Emme, 1967; Highham, 1968; Lees, 1960, 1963; Truman and Riddiford, 1970; Tyshchenko, 1966; Wilde de, 1961). Thus, the information about external ecological conditions "is translated into hormonal language" with the help of which the realization of the

[*] I thank the staff of the Entomology Laboratory organized by Professor A. S. Danilevsky who made it possible for me to conduct my investigations.

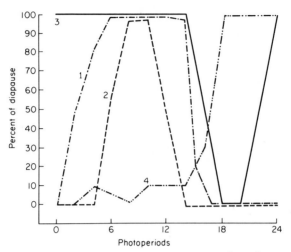

Fig. 1. The types of insect photoperiodic reaction: (1) long-day type in northern *Pieris brassicae;* (2) long-day type in southern *Pieris brassicae;* (3) intermediate type in *Euproctis similis* (Geispitz, 1953); (4) short-day type in *Stenocranus minutus* (Müller, 1958).

genetic information is carried out. In full accordance with ecological program, insects either enter diapause or develop without it. Thus, regular seasonal alternation of diapausing and nondiapausing generations as well as morphogenesis are regulated by hormones.

The changes in the hormonal system considerably influence respiratory, protein, lipid, and water metabolism, and fermentative activity (Andrewartha, 1952; Chefurka and Williams, 1952; Harvey, 1962; Gersch, 1964; Lees, 1956, 1961; Kurland and Schneiderman, 1959; Novák, 1966; Schneiderman and Williams, 1953; Sláma, 1960; 1964a, b). This complex of physiological alterations makes it possible for diapausing insects to resist the conditions which are lethal for nondiapausing ones (Danilevsky, 1961; Gersch, 1964; Lees, 1955, 1956; Novák, 1966; Ushatinskaya, 1957).

It is known that the diapausing insects can withstand prolonged exposure to low temperature (0°–10°C), can endure deep chilling (down to —30° to —40°C), and some can even withstand tissue freezing. Besides that, diapausing stages possess wide complex resistance to many other unfavorable conditions, i.e., to dryness, high temperature, poisoning, and starvation (Andrewartha and Birch, 1954; Geispitz and Orlovskaya, 1971; Danilevsky, 1949, 1961; Kozanchikov, 1938, 1948; Lozina-Lozinsky, 1963; Ushatinskaya, 1957).

Insects do not enter diapause suddenly. Diapause is determined by prolonged external influences acting on the prediapause instars. Gradual quantitative metabolic changes are especially illustrative in *Bombyx mori*. It diapauses at the egg stage, but this diapause is determined by the incubation conditions of all stages of the maternal generation. Gradual metabolic changes leading to the production of diapause or nondiapause eggs are seen during these stages.

As is well known, the insect diapause may appear at any ontogenetic stage, but it is strictly fixed for each species. There are embryonic, larval, pupal, and imaginal types of diapause. As each of the diapause types has its own peculiarity, the mechanism of hormonal regulation should be examined individually. It seems justified to begin the review of the data on the various diapause types with the examination of the pupal diapause which has a simpler mechanism.

IV. HORMONAL CONTROL OF PUPAL AND LARVAL DIAPAUSE

A. Pupal Diapause

The physiology of pupal diapause was more extensively studied in the giant silkworm *Hyalophora cecropia* by Williams and co-workers (Schneiderman *et al.*, 1953; Schneiderman and Williams, 1952, 1953, 1954a, b, 1955; Williams, 1946, 1947, 1948, 1952, 1956b). *Hyalophora cecropia* is a univoltine species, i.e., immediately after pupal molt each generation enters diapause, which continues for about 5 months at 25°C. But exposure to low temperature of about +5°C shortens its duration down to 1.5 months.

Williams (1946) showed that if a diapausing pupa is joined to a chilled pupa, both members of the pair in this parabiotic experiment resume development. In this case the factor circulating in the hemolymph of the chilled pupa penetrates into the unchilled one and induces imaginal development in both insects.

The experiments with a diapausing pupa transsected behind the thoracic segments demonstrated that immediately after the implantation of chilled brain into the anterior half of the pupa, the recipient can also resume the development (Williams, 1948). But the similar implantation into the posterior pupal half cannot induce imaginal differentiation. Such an isolated abdomen is capable of developing only after simultaneous implantation both of the chilled brain and the prothoracic glands. Thus, the main control of diapause in this species is carried out by brain and prothoracic glands, i.e., by the same neuroendocrine organs which regulate metamorphosis in insects that develop without a diapause.

On the basis of experiments on *H. cecropia* and on other Lepidoptera, Williams has concluded that the pupal diapause is a result of hypofunction of endocrine system controlling the development of the nondiapausing insects. The absence of brain and prothoracic gland activity is considered to be the direct physiological cause of a diapause. Special investigation concerning the significance of corpora allata, subesophageal ganglion, thoracic ganglions, and gonads by implanting them in the chilled pupae of *H. cecropia* did not reveal any inhibition of morphogenesis in recipients (Williams, 1947). Therefore, according to Williams's concept, the inhibitors do not participate in hormonal control of pupal diapause.

However, the data on the physiology of facultative pupal diapause in *Pieris brassicae* confirm the interference of inhibitors in the diapause control. The northern populations of *P. brassicae* develop without diapause within a temperature range of 12° to 25°C and under photoperiod with over 16 hours of light per day, but they enter diapause under shorter photoperiods. This pupal diapause continues at least for 5 months, as in *H. cecropia*. The chilling at the temperatures of about 0°C is necessary to terminate their diapause. Reactivation at higher temperatures (above 12°C) results in abnormal imaginal development (Danilevsky, 1950).

The main endocrines controlling diapause in *P. brassicae,* as in *H. cecropia,* are brain and prothoracic glands. However, there are some differences in the succession of hormonal events during the diapause induction in these species (Belozerov, 1962; Claret, 1966; Maslennikova, 1961, 1968, 1970, 1971).

The brain stimulation of prothoracic glands in *P. brassicae* takes place during prepupal stages. Therefore brain extirpation in 1-day nondiapausing pupae of *P. brassicae* does not avert their imaginal differentiation. In some cases brain and subesophageal ganglion extirpation in diapausing pupae also does not prevent the imaginal development after chilling (Maslennikova, 1970). If the diapause of *P. brassicae* is determined by short-day regimes with 6–12 hours of light, such extirpation results, as it does in *H. cecropia,* in the formation of permanent pupae, which cannot develop further without additional hormonal stimuli (Fig. 2,B). Some operated brainless specimens complete their imaginal development under critical photoperiodic conditions with 14–16 hours of light. The prothoracic glands were responsible for the development resumption after chilling during these experiments. The possibility to activate brainless pupae of *P. brassicae* and *Antheraea polyphemus* (McDaniel and Beery, 1967) by chilling proves that prothoracic glands are under special

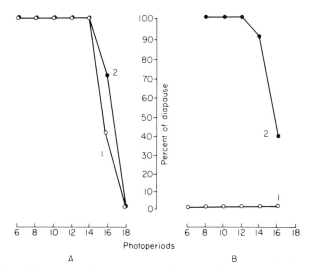

Photoperiods

A   B

Fig. 2. The influence of the extirpation of neuroendocrine complex on induction (A) and termination (B) of pupal diapause in *Pieris brassicae*. (A) The induction of diapause at 18°C: (1) nonoperated pupae; (2) extirpation of brain, corpora allata, corpora cardiaca, and subesophageal ganglion immediately after pupation. (B) The termination of diapause at 10°C and constant illumination: (1) nonoperated pupae; (2) pupae without brain, corpora allata, corpora cardiaca, and subesophageal ganglion, which were extirpated immediately after pupation.

inhibited state during a long diapause period. Only thus do they remain nonresponsive to their own trigger factor which has been present in the diapausing pupae before the operation. Long chilling restores their temporarily lost responsiveness. In our opinion, the obtained data prove the occurrence of specific inhibition processes in the pupal diapause control in these species.

The participation of inhibiting factor in the brain and prothoracic gland inactivation of *P. brassicae* was demonstrated by using the peculiar method of biological testing (Maslennikova, 1968). Chalcide *Pteromalus puparum* (pupal parasite of *P. brassicae*), which is very sensitive to even slight hormonal balance alterations in the host, was taken as a subject for the biotesting. The effect of inhibition of the parasite development in the head halves of the ligaturated diapausing pupae was obtained when the head and abdomen fragments of these pupae were parasitized separately (Fig. 3A). A stronger tendency to diapause in the test-object in the head, but not in the abdomenal fragments of the diapause pupae, cannot be explained by the hormonal deficiency. It is undoubtedly evidence of the active blocking of the neuroendocrine

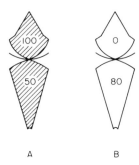

A   B

Fig. 3. The influence of hormonal balance of *Pieris brassicae* pupae, ligatured immediately after pupation, on diapause induction in test-object *Pteromalus puparum*. (A) Percent of diapausing *P. puparum*, which developed in diapausing host pupae at 18°C and 16 hours of light per day. (B) Percent of diapausing *P. puparum*, which developed in nondiapausing host pupae at 23°C and 12 hours light per day.

centers in both participants by some inhibition sources. Other events are observed during the analogous experiment with nondiapausing pupae. The stimulation in the development of the test object occurs in the head fragments of nondiapausing *P. brassicae* pupae. It may be easily explained by high brain and prothoracic gland activity in the host. But a stronger tendency to diapause is actually connected with the absence of the hormones stimulating morphogenesis in the non-diapausing pupa abdomen. All the above proves that the endocrine system inactivation during the diapause induction is not a passive process, but an active one. *Pieris brassicae* brain and prothoracic glands become inactive under the influence of specific diapause factor. This factor seems to be of hormonal nature as the test object gets the inhibition information while feeding on the host tissues.

It is interesting to note a number of facts which show a potential capacity of a diapause brain to secrete. The reimplantation of the *A. polyphemus* brain and the implantation of the brain of the diapausing donor may break the diapause in this species, whereas the intact dia-pause brain remains inactive due to some blocking mechanism. It be-comes hormonally competent only after its injury (McDaniel and Barry, 1967). The extirpation of the brain together with the corpora allata, corpora cardiaca and subesophageal ganglion in the *P. brassicae* diapaus-ing pupae does not activate the operated insects (Fig. 2A). However, there is some activation in the diapausing pupae if only the subesopha-geal ganglion is removed, but the injured brain is left (Fig. 4). These

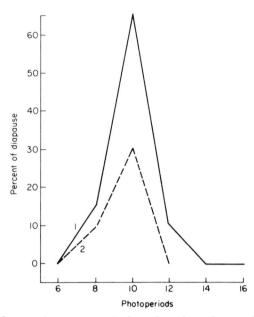

Fig. 4. The influence of extirpation of subesophageal ganglion on diapause induction in southern *Pieris brassicae* at 25°C. (1) Nonoperated pupae. (2) Pupae without subesophageal ganglion, which was extirpated immediately after pupation.

facts confirm the presence of some mechanisms which keep the diapause brain inactive.

The histological data on neurosecretion also demonstrate the existence of some blocking mechanisms. Brain neurosecretory cells in *P. brassicae* diapausing pupae are filled with great quantities of neurosecrete, which seems to have no exit into hemolymph (Kind, 1968a).

The following facts are very important for understanding diapause physiology. It is possible to induce a diapause of varying intensity in *P. brassicae* (Fig. 2B) and as a consequence of varying duration (Fig. 5). The pupae, whose diapause was determined under 10- to 12-hour photoperiod, had a longer period of dormancy under similar reactivation conditions. The duration of the diapause is shorter if it is determined under longer and shorter photoperiods. We have all grounds to suppose that these differences are due to the inhibition processes, because the information about physiological and ecological conditions during diapause induction is preserved until resumption of development. The inhibiting factors seem switched off the brain–prothoracic gland system

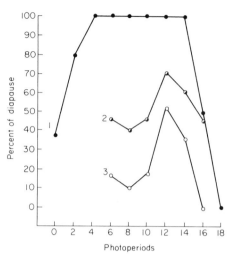

Fig. 5. The dependence of diapause duration in *Pieris brassicae* pupae upon photoperiodic conditions during diapause induction at 25°C. (1) The photoperiodic induction of diapause in *P. brassicae,* which was used for cold reactivation at 10°C and constant illumination. (2) Retention of diapause after 5 month chilling. (3) Retention of diapause after 6 month chilling.

for various time periods, the duration of which was being determined by hormone concentration at the diapause induction.

B. Larval Diapause

Hormonal mechanisms of the larval and pupal diapause are very much alike, but the former has its own peculiarities. Brain and prothoracic glands undoubtedly participate in the larvae diapause regulation (Beck and Alexander, 1964a,b; Church, 1955; Fukaya and Mitsuhashi, 1957, 1958; Kind, 1968a,b; Mitsuhashi and Fukaya, 1960; Novák, 1957; Rehm, 1952; Sláma, 1959, a, b, c; Waku, 1960).

The implantation of the brain of the postdiapause *Cephus cinctus* larvae into the abdomen of the diapausing ones activates the latter. The same result is obtained at the parabiosis with the postdiapause larvae (Church, 1955). The injection of α-ecdysone, ceasterone, ecdysterone, and inocosterone causes diapause termination and the pupation in the diapausing *Nomia melanderi* prepupae and *Megachile rotundata* larvae (Hsiao and Hsiao, 1969). However, injection of ecdysone into the diapausing *Ostrinia nubilalis* larvae does not interrupt their diapause, although apolysis occurs. In this case the physiological tissue insensitivity hampers the diapause breaking (Beck and Shane,

1969). The same physiological reason is a decisive factor during implantation and parabiosis in *C. cinctus*. The activation of the diapausing partner at parabiosis is effective in most cases if the partner was exposed to chilling for some time (Church, 1955).

Corpora allata and corpora cardiaca alongside with brain and prothoracic gland are involved in the larval diapause control in some species. Thus, the diapause of *Chilo supressalis* may be interrupted by brain and prothoracic gland implantation. However, the normal larval development is observed only after the removal of the corpora allata and corpora cardiaca, which seem to take an active part in the diapause maintenance in this species. The corpora allata in *Ch. supressalis* and in *O. nubilalis* remain active throughout the entire diapause period, but their activity gradually decreases at the diapause termination (Fukaya and Mitsuhashi, 1957, 1958; Mitsuhashi and Fukaya, 1960). The diapause does not terminate as long as the corpora allata are highly active. On the contrary, the brain and prothoracic gland activity in *Ch. supressalis* during diapause is low. The corpora allata secretion activity intensification during diapause was observed in *Plodia interpunctella* (Waku, 1960) and *Phragmatobia fuliginosa* (Kind, 1968a).

The special case of the diapause regulation and reactivation was observed in *O. nubilalis* in which neurosecretory cells of proctodaeum walls take part in the regulation processes (Beck and Alexander, 1964a, b; Beck et al., 1965). These neurosecretory cells produce specific hormone (proctodon) which stimulates the brain neuroendrocrines. The diapause brain implantation into the diapausing larvae causes recipient pupation (Beck and Alexander, 1964a; Cloutier et al., 1962). However, the implanted diapause brain is capable of being activated only under the proctodon influence. The ligature, applied at the seventh abdominal segment prevents the proctodon penetration into the frontal larval fragment and hinders the activation of both the intact and implanted brain.

Undoubtedly, the larval and pupal diapause is connected with the reduction in activity of the main neuroendocrine centers, i.e., of the brain and of the prothoracic glands. The inactivation these centers, as has been vividly illustrated in the case of the pupal diapause in *P. brassicae*, is not spontaneous. It is the result of the inhibition processes. It has been also shown that the maintenance of the larval diapause is connected with the active functioning of the corpora allata.

Thus, the hormonal deficiency, characteristic for diapause, is more a consequence than a reason of its onset. The hormonal control of larval and pupal diapause might be carried out under the antagonistic interaction of the factors which stimulate and inhibit the morphogenetic processes.

V. HORMONAL CONTROL OF IMAGINAL DIAPAUSE

Unlike other types, imaginal diapause is primarily associated with reproductive dormancy. The number of investigations of the imaginal diapausing insects is considerably greater as compared with the insects with other diapause types. These studies include insects of Coleoptera, Diptera, Heteroptera, Lepidoptera, and Orthoptera.

The physiology of imaginal diapause is fully studied in the Colorado beetle *Leptinotarsa decemlineata*. Facultative diapause in Colorado beetle is easily controlled by photoperiodic conditions during preimaginal stages (Goryshin, 1956, 1958; Ushatinskaya, 1959; Wilde de, 1954, 1958; Wilde de *et al.*, 1969). This species develops without diapause under photoperiods of over 16 hours of light per day. The diapause occurs if the photoperiod is shorter. Adult beetles predetermined to diapause come out, feed for some time, and then again return to the soil to hibernate. The oocyte growth in such beetles is interrupted and wing muscles degenerate. The eggs are formed in increasing number, although full maturation is not observed until feeding activity is resumed after hibernation.

The role of neuroendocrine elements in the diapause control in *Leptinotarsa decemlineata* was studied by means of microsurgery combined with female rearing under varying photoperiodic conditions.

The brain was shown to be the primary organ controlling the Colorado beetle diapause (Grison, 1949; Wilde de and Boer de, 1961; 1969). The dormant beetles may start to feed and lay eggs, if they receive the brain from nondiapausing donors. The young beetles induced to diapause by rearing in short-day length also may be stimulated to oviposition by the implantation of six corpora allata from active beetles. But this effect disappears if this operation is not supported by simultaneous long photoperiod action (Wilde de, 1958; Wilde de and Boer de, 1961, 1969).

According to Wilde de and Boer de (1969), the short-day effect on Colorado beetle is due to brain inactivation and subsequent corpora allata inactivation. Constant AH influence is necessary to stimulate corpora allata. This influence may be carried out both through corpora cardiaca or hemolymph. The implanted corpora allata may be activated and inactivated by photoperiod.

The suppression of reproductive function, which is characteristic of imaginal diapause, may also be inducted by allatectomy. The growth of ovaries in allatectomized females resumes only after corpora allata implantation and under long-day photoperiod. In the contrast to the allatectomized diapause, the natural diapause does not break by implantation of more than three corpora allata pairs or by retrocerebral complex (Wilde de and Boer de, 1969).

Schooneveld (1970) showed that there were seven types of neurosecretory cells in the protocerebrum in *Leptinotarsa decemlineata*. The neurosecretory material from them accumulates primarily in corpora cardiaca. Axons of A' cells and axons of external neurosecretory cells from corpora cardiaca reach to corpora allata. Comparative study of neurosecretion of several neurosecretory cells at long and short day lengths has shown that neurosecrete production of A' cells in ovipositing females was approximately 7 times more intensive than in diapausing females. Neurosecrete of A' cells in ovipositing females passes during an hour to corpora cardiaca and is liberated from them. It is supposed that the neurosecrete of A' cells is responsible for the synthesis of "female protein," when JH is present; and external neurosecretory cells of corpora cardiaca are responsible for the inhibition of corpora allata activity. Cutting of nerve pathways between these cells and corpora cardiaca considerably interrupts the inactivation of corpora allata and the diapause onset in Colorado beetle.

The alteration corpora allata activity during the diapause leads to considerable changes in JH titer in *Leptinotarsa decemlineata* hemolymph. JH content in hemolymph in diapausing beetles does not exceed 200 *Galleria* units. This low level is maintained until diapause termination. After reactivation JH titer reaches 8000–10,000 *Galleria* units. In nondiapausing beetles reared under long photoperiod the JH titer may increase over 5000 *Galleria* units during the first days of adult life and it stays at this level during all oviposition period.

The diapausing beetles, as well as the castrated ones, have a higher protein level in hemolymph, as compared to active beetles, as the egg maturation during which protein is consumed, does not take place (Emmerich, 1970; Engleman, 1968; Loof de and Wilde de, 1970). The protein content reaches 10% in hemolymph of diapausing Colorado beetle. The protein level falls down slowly during diapause. Allatectomy in active beetles results in sharp increase of the yolk-forming protein content and the appearance of specific "short-day protein," which is absent in the ovipositing beetles. This protein disappears in 2–3 days after JH injection into the diapausing beetles (Loof de and Wilde de, 1970).

The participation of brain, corpora allata, and corpora cardiaca in the imaginal diapause control was also demonstrated in the species, which were not studied in as much detail as the Colorado beetle.

The complete development in *Dytiscus marginatus* and *D. semisul-*

catus depends on the corpora allata activity. Maturation may be stimulated by the implantation of 10 pairs of corpora allata at any period
and the ovaries may regress even during active oviposition immediately
after allatectomy. To maintain the gonadotropic activity, the corpora
allata need brain stimuli, which pass through corpora cardiaca (Joly,
1945).

The general control of maturation is carried out by medial neurosecretory cells of brain in *Calliphora* (Thomsen, 1952; Thomsen and Möller,
1959) and in *Schistocerca gregaria* (Highnam, 1961).

Summer diapause and egg maturation are regulated by different levels
of brain, corpora allata and corpora cardiaca activity in *Galeruca tanaceti*
(Siew, 1966).

The reproductive diapause in locust *Oedipoda miniata* breaks after
the implantation of active corpora allata from *Locusta migratoria migratorioides* males (Broza and Pener, 1969). In females of the latter species
reared under 16-hour photoperiod, corpora allata become inactive or
slightly active. The implantation of additional corpora allata from mature
females reared under short photoperiod may compensate the deficiency
in the activity of the recipient corpora allata (Darjo, 1969).

The above data testify that the imaginal diapause is connected with
the inactivation of brain neurosecretory cells, corpora allata and corpora
cardiaca. But as the insects differ in their biology and reproductive
physiology, the mentioned endocrine organs control their imaginal diapause in somewhat different manner. Corpora cardiaca, but not corpora
allata determine the control of imaginal diapause and seasonal demor-
phism in *Polygonia c-aureum* (Fukuda and Endo, 1966).

The brain neurosecretory cells undoubtedly play the leading role in
the general control of egg maturation. However, this control may manifest itself in two ways: (1) by means of corpora allata activation leading
to the gonadotropic hormone liberation; and (2) by means of the stimulation of ovary development which is independent of corpora allata,
but is dependent only on the activation of protein metabolism (Thomsen,
1952, 1965; Thomsen and Moller, 1960; Highnam, 1961).

It is shown in *Anacridium aegyptium* (Geldiay, 1967) that corpora
allata influence mainly yolk-forming process, whereas the brain neurosecretory cells control the earlier stages of the oocyte growth. According
to Sláma (1964a) brain regulates the trophic function through AH secretion, whereas JH is responsible for the reproductive function.

The majority of the investigaters consider the imaginal diapause as
the syndrome of the hormonal deficiency, arising as a result of the brain
and corpora allata/corpora cardiaca inactivation. But there are facts
which do not agree with this point of view. For instance, during diapause

considerable alterations occur not only in the neuroendocrine system, but they also take place in protein and lipid metabolism and respiration. Protein content in the *Pyrrhocoris apterus* hemolymph increases 2-fold at diapause onset (Emmerich, 1970). Specific "short-day protein," which is absent in active beetles, appears in diapausing *Leptinotarsa decemlineata* (Loof de and Wilde de, 1970). These processes occur gradually during the prediapause period, but the visual changes in neurosecretory cells are observed only when the diapause has begun. It leads to the conclusion that the inactivation of the neuroendocrines controlling the reproduction is sooner a consequence than a cause of the imaginal diapause onset, as it was in the case with the larval and pupal diapause.

As has been noted above, allatectomy causes hormonal deficiency, i.e., artificial diapause. This allatectomized diapause is easily terminated by the active corpora allata implantation, whereas the natural diapause can not be interrupted by the implantation of even several pair of corpora allata (Wilde de and Boer de, 1961, 1969). According to Grison (1949) the activation of diapausing Colorado beetle can be achieved only when the recipient has undergone some chilling. It shows that artificial and natural diapause are not identical and it contradicts the concept explaining diapause as the result of hormonal deficiency.

Based on the experiments with *Hypera postica*, Bowers and Blickenstaff (1966) concluded that a specific inhibitor which takes part in the suppression of corpora allata inactivation is responsible for the diapause onset and maintenance. The recent data (Loof de and Wilde de, 1970) on suppression of corpora allata activity by external neurosecretory cells of corpora cardiaca and a long-established fact of endocrine suppression of neuroendocrine centers by developing ovaries (Engelman, 1962; Nayer, 1957; Scharer, 1959) prove that the inhibiting processes may also play an important part in the imaginal diapause.

VI. HORMONAL CONTROL OF EMBRYONIC DIAPAUSE

The hormonal mechanism of embryonic diapause has been most extensively studied in the silkworm *Bombyx mori*. The strains of this species are divided into uni-, bi- and polyvoltine* ones according to the number of generations per year. The inheritent nature of voltinism is beyond doubt (Morohoshi, 1969a–e). Nevertheless, it has already been shown that diapause induction in bivoltine strain of this species depends on the

* The term "bivoltine" used in the literature on *B. mori* corresponds to our term "polyvoltine species (or strains) with facultative diapause."

temperature and photoperiodic conditions during the incubation period of maternal generation (Fukuda, 1951a, b, c, 1953a, b, c, 1963; Hasegawa, 1952; Kogure, 1933; Morohoshi, 1959, 1969a, b, c, d; Tanaka, 1924, 1953).

If the development of bivoltine strain occurs at low temperatures (15°–20°C) and short photoperiod, the resulting female moths lay non-diapausing eggs. Even though the processes controlling the voltinism begin early, the developmental character of eggs remains nondetermi-nated finally at least in mature larvae and pupae of maternal generation. Umeya (1926) was the first to demonstrate it through ovarial transplanta-tion from larvae, determined to nondiapause development into larvae, determined to diapause and also through reciprocal transplantation. His data were the first evidence that the developmental character of eggs is due to the hormone circulating in hemolymph of silkworm larvae, pupae, and imago. Later it was shown that the genetically and ecologically predetermined voltinism may be changed by decapitation of the pupae immediately after the pupal molt. Determinated to active development decapitated pupae sometimes produce the moths laying the diapausing eggs. The same operation on pupae determined to diapause results in the emergence of moths, which produce mixed egglayings. The *B. mori* moths lay the diapausing eggs in case the brain is absent, but the ganglion of ventral nerve cord remained intact. These facts show that the second neuroendocrine center plays an important role in voltinism determination. The subesophageal ganglion turned to be this center. The neuroendo-crine activity of this organ was confirmed through its implantation in Vth instar larvae determined to nondiapause development (Fukuda, 1951b, c; Hasegawa, 1952).

A detailed study of the subesophageal ganglion has revealed that there are two specific neurosecretory cells which are a source of the diapause-inducing hormone. These cells are located ventrally in the last third of the ganglion near the medial line (Fukuda and Takeuchi, 1967a,b; Park and Yoshitake, 1970a, 1971). In the pupae, determined to nondia-pausing egg production, these cells filled with neurosecrete materials, whereas in diapause pupae they have no neurosecrete products. The liberation of diapause hormone from subesophageal ganglion in diapause pupae is stimulated by the brain through circumesophagel connectives. The neurosecrete product is released from neurosecretory cells as soon as it is formed. On the other hand, the release of these products in nondiapausing pupae is inhibited by the brain and, correspondingly, the neurosecrete product is accumulated in cells.

The interaction between the brain and subesophageal ganglion was clearly demonstrated by Fukuda (1953c). If the brain together with the subesophageal ganglion were removed from a nondiapause donor pupa and were implanted into a nondiapause recipient, the brain under the new conditions continued to inhibit the subesophageal ganglion activity. As a result, the recipient laid only nondiapausing eggs. If the circumesophageal connectives were cut or the isolated brain was implanted, all eggs were diapause or mixed.

According to Morohoshi (1959, 1969a, b, c, d, e; Morohoshi *et al.*, 1969) the seasonal development in *B. mori* is controlled also by corpora allata. The subesophageal ganglion implantation into Vth instar larvae strengthened the diapause tendency and the resulting moths laid diapausing eggs, whereas the corpora allata implantation into the same recipients resulted in appearance of nondiapausing eggs. Thus, the seasonal development in *B. mori* is controlled by two hormones: subesophageal ganglion hormone and corpora allata hormone. The neurosecretion activity of both these organs is controlled by brain through circumesophageal connectives. The character of voltinism in *B. mori* different strains is regulated by the interaction of the above two factors.

It is considered that the level of subesophageal ganglion activity is connected with voltinism characteristics, but that of corpora allata with moltinism characteristics. The strain differences in *B. mori* depend on the antagonistic functions of two gene groups, i.e., on moltinism genes (M) and on voltinism genes (V), which are controlled by maturation genes (L_m). The latter are located in the Z chromosome and control the corpora allata and subesophageal ganglion functions (Morohoshi, 1971). The most active functioning of subesophageal ganglion is observed in univoltine strain; the intermediate one, in bivoltine strain; and the weak one, in polyvoltine strain.

Interacting in such manner, brain, subesophageal ganglion, and corpora allata regulate the general metabolism in bivoltine *B. mori* according to the influence of environmental factors (temperature, photoperiodic condition, food) on diapause or nondiapause development. The influence of external factors on voltinism is especially clear at the earlier stages of larval development. The diapausing eggs are most easily obtained if the young larvae are reared under high temperature and long photoperiod, whereas low temperature and short photoperiod are more favorable for older larvae. The young and older larvae react differently to temperature and light, which is explained by differences in their hormonal balance during larval development. Copora allata release hormone which stimulates metabolism and subesophageal ganglion release hormone which inhibits metabolism. The titers of both hormones gradually decrease to aging instars. But as the two hormones act antagonistically, the young and older larvae respond differently to the same external influences.

It has been shown that subesophageal ganglion production is not specific for sex and species. The ganglia of *B. mori* males and of females in *Lymantria dispar, Antheraea yamamai,* and *Dictiyoploca japonica* (all these Lepidoptera possess embryonic diapause) were capable to induce diapausing or mixed egg-laying, if they were implanted into nondiapause *B. mori* larvae. It was interesting to note that subesophageal ganglion from *Antheraea pernyi* (species with pupal diapause) have same capacity. However, subesophageal ganglia from *Phylosamia cynthia* (with pupal diapause) and from *Dendrolimus pini* (with larval diapause) were inactive (Fukuda, 1951c; Hasegawa, 1952).

An extract containing diapause hormone which had a high activity for the diapause induction was obtained in 1957 (Hasegawa, 1957). Its injection into the nondiapause pupae caused the laying of diapausing eggs by females which emerged from those pupae. Fifteen thousand pupal subesophageal ganglia together with brains from *B. mori* were used to obtain this active substance. An extract having same activity was also obtained from the heads of adult silkworm females.

The hormone of subesophageal ganglion is considered to be an inhibitor of metabolism, e.g., it prolongs the larval development (Morohoshi, 1959). Injection of this hormone suppresses the heart activity in *B. mori* (Morohoshi, 1971) and in *Periplaneta americana* (Ralph, 1962). Corpora allata hormone, on the other hand, produces an opposite influence on these processes.

It has been established (Yamashita and Hasegawa, 1970) that the main target for action of the diapause hormone in *B. mori* pupae are not the gonads as a whole, but separate oocites. They are the most sensitive to the diapause hormone only at a certain stage, i.e., during the second half of the pupal period. Both the premature or delayed hormonal injection is not effective (Yamashita and Hasegawa, 1966). The produced egg will diapause if the quantity of diapause hormone was sufficient during this critical period. A nondiapausing egg is formed if the hormone quantity is insufficient. An interesting phenomenon of disagreement in the normal processes of embryogenesis is observed if the hormone is within its threshold concentration (Fukuda, 1963). It is known that diapausing *B. mori* eggs are dark colored, and nondiapausing eggs are light colored. It is possible to obtain intermediately colored eggs in bivoltine *B. mori* under certain conditions. The examination of such eggs has shown that some embryonic organs have been inhibited in their development, whereas other organs have already passed the stage at which the diapause usually occurs. These data have led Fukuda (1963) to conclude that diapause and nondiapause development in *B. mori* are not to opposite states, but they are conditioned only by quantitative differences in the concentration of the hormone which influences the ovary development during the critical period.

A biochemical study of the diapausing and nondiapausing *B. mori* eggs have revealed their considerable distinctions (Morohoshi *et al.*, 1969). Total nitrogen content in the diapausing eggs is lower than in the nondiapausing eggs. Glycogen content increases in the diapausing eggs in case the corpora allata are removed from Vth instar larvae. The subesophageal ganglion extirpation causes opposite results. Thus, the content of nitrogen and glycogen in the eggs depends on the antagonistic actions of subesophageal ganglion and corpora allata.

According to Chino (1958), at the diapause onset all glycogen in the eggs breaks up as a result of a complex enzymatic reaction, which leads to the appearance of two end products, i.e., glycerin and sorbit. The glycogen resynthesis starts at diapause termination. The presence of glycerin and sorbit in the diapausing eggs facilitates the egg cold-resistance in the same degree as in other diapausing stages.

Bombyx mori eggs developing without diapause hormone have a low content of 3-hydroxykynurenine. The hormone injection into the nondiapause pupae results in an increase of this substance content up to the level in the diapausing eggs, this increase being proportional to the quantity of the injected hormone. The diapausing eggs differ considerably from the nondiapausing eggs also in the content of triglycerides and fatty acid methyl esters (Suzuki *et al.*, 1970).

As it becomes clear from the above the future of the eggs has been already considerably predeterminated at the moment of the egg-laying, but the laid eggs are still bipotential as to their further ways of developing during the first 48 hours (Park and Yoshitake, 1970b; Takami, 1960). According to Takami (1960) metabolism in the *B. mori* laid diapausing egg continues to decrease for 30 days. Maximal metabolism suppression is observed at the thirtieth day after the egg-laying and it remains at this level for 100 days, even under temperature of 25°C.

The resumption of the postdiapause development in embryons of *Locusta migratoria migratorioides* and *Locustana pardalina* (Jones, 1956a, b) starts at the stage when there are no embryonic lateral glands (homologous of prothoracic glands), which are responsible for the hormonal control of develpment during later embryogenesis. The early embryonic diapause seems to be a result of the blocking of the precoded program of embryogenesis. Active embryogenesis in the postdiapause eggs renews according to the predetermined program, which was temporarily suppressed during the diapause. Embryonic own hormonal system controlling growth and differentiation is involved somewhat later.

The hormonal mechanism was studied in detail only in a comparatively small number of species with the embryonic diapause. However, it may be supposed that they are similar to B. mori in this respect.

The embryonic diapause regulation in Orgyia antiqua was found to have much in common with that in B. mori, although these species are ecologically considerably different (Kind, 1965, 1968a, 1969). Orgyia antiqua possesses a photoperiodic reaction of long-day type, whereas B. mori is characterized by a short-day reaction. The O. antiqua larvae reared under 18-hour long day at 25°C produce the moths laying non-diapausing eggs. The larvae reared under short 10- to 12-hour photoperiod produce moths laying only the diapausing eggs (Doskočil, 1961). The regulating role of subesophageal ganglion of O. antiqua in the diapause determination was demonstrated in detail by Kind (1965, 1969, 1971). Two B-type neurosecretory cells responsible for the production of the diapause hormone were found in the O. antiqua subesophageal ganglion. The cauterization of the subesophageal ganglion part, where these cells are located, leads to the same effect as the extirpation of the entire ganglion. The differences in the levels of neurosecretion activity in this ganglion in various geographic populations of O. antiqua are of considerable interest (Kind, 1969, 1971). These differences seem to be identical to the strain differences in the hormonal activity of subesophageal ganglion in B. mori.

The hormonal mechanisms in the other species with the embryonic diapause, as it may be judged by the data on locusts (Jones, 1956a, b), mosquitoes (Vinogradova, 1969), and bugs (Ewen, 1966), seem also to have much in common with that in B. mori.

All the above makes it possible to conclude that the facultative embryonic diapause is controlled at the prediapause stages by the changing balance between the subesophageal ganglion and corpora allata hormones. The subesophageal ganglion hormone is considered to be a metabolism inhibitor, but the corpora allata hormone produces an opposite action on these processes. Thus, the embryonic type of diapause described above is a vivid example of the fine hormonal regulation of seasonal development under the interaction of two antagonistic factors.

VII. BIFACTORIAL PRINCIPLE OF DIAPAUSE HORMONAL CONTROL

In summarizing the brief review of investigations on the hormonal control of insect diapause, the following should be noted. The majority of works on this interesting problem deal with the study of the role of different neuroendocrine organs connected in some way with the diapause control. These works made it possible to clearly define those neuroendocrine elements which are involved in the control of embryonic, larval, pupal, and imaginal diapause. The brain/prothoracic gland system play the leading role in the larval and pupal diapause control. The imaginal diapause is controlled by the brain and corpora allata/corpora cardiaca system. The embryonic diapause, which was studied in more detail in *Bombyx mori* is regulated by neurohormonal stimuli of subesophageal ganglion and corpora allata whose secretion activity is controlled in its turn by the brain.

Some theoretical summations, based on the above experimental data and widely accepted by many physiologists, are of the greatest interest in the present discussion. Many physiological experiments prove that the state of larval, pupal, and imaginal diapause is characterized by the absence of the hormones, which are responsible for the growth or the morphogenesis during the nondiapause development (Beck and Alexander, 1964a, b; Church, 1955; Wilde de and Boer de, 1961, 1969; Williams, 1946, 1947, 1948, 1952). These facts led to the concept of hormonal deficiency, which is often considered to be a direct physiological cause of diapause onset. In fact, the extirpation of the neuroendocrines, which control growth and development, results in the interruption of these processes, i.e., in permanent or artificial diapause, which is easily broken by the reimplantation of the previously extirpated organs. But natural diapause is not terminated so readily because it is not physiologically identical to the artificial one. The natural diapause may be interrupted by the implantation of some neuroendocrines or by the injection of necessary hormone only when the organism becomes physiologically responsive after some period of chilling (Church, 1955; Grison, 1949; Lees, 1955). It is impossible to activate the truly diapausing insect during the period of deep dormancy through the above method because the tissues of such specimens are inhibited and therefore they are not responsive even to specific hormonal triggers (Beck and Shane, 1969; Bowers and Blickenstaff, 1966; Maslennikova, 1970; McDaniel and Barry, 1967). Consequently, the natural diapause cannot be considered as a syndrome of hormonal deficiency only and this is not the cause of a diapause onset.

As is well known, diapause in the species with facultative diapause is induced by influence of certain ecological factors during prediapause period of ontogenesis. It is also known that the diapause may be of various depth and stability, depending on the influencing conditions (Geispitz and Orlovskaya, 1971; Geispitz *et al.*, 1971; Fukuda, 1963; Maslennikova, 1968, 1970). The nature of these differences is quite clear. The quantitative differences in the concentration of hormones controlling the development at the diapause induction depend on external conditions. Taking into consideration these quantitatively different diapause states in *B. mori*, Fukuda has expressed the idea that the diapause and its absence are not qualitatively different states but they are due to the quantity of the diapause hormone which influences the maternal ovaries during diapause induction (Fukuda, 1963).

It might be logically assumed that seasonal development in the species with a facultative diapause is determined only by quantative changes in titer of a hormone, i.e., by the hormone-inhibitor titer in case of the embryonic diapause and by the hormone-activator titer in case of other diapause types. This might lead to a conclusion that the idea of monofactorial hormonal regulation of all diapause types is true. However, the following facts strongly disagree with this attractively simple concept.

1. It has been convincingly shown that during the embryonic diapause in *B. mori* the specific inhibitor hormone controls the diapause of the future eggs in antagonistic interaction with the activator hormone.

2. It has been also shown that some inhibiting system takes part in the regulation of the pupal diapause in *P. brassicae*. This system might be involved in the inactivation of the brain, of the prothoracic glands, and of some other tissues. The participation of some inhibiting systems in the regulation in imaginal and larval diapause seems quite possible according to a few authors (Beck and Alexander, 1964a, b; Bowers and Blickenstaff, 1966; Loof de and Wilde de, 1970; Mitsuhashi and Fukaya, 1960).

3. The potential hormonal activity of the diapause neuroendocrine organs found in a number of species with pupal and larval diapause (Cloutier *et al.*, 1962; McDaniel and Barry, 1967) and, also, the histological pictures of neurosecretion in the diapause brain prove banning or suppression of neurosecretion functioning by some factors, rather than its spontaneous diminishing (Lees, 1955).

Thus, the concept of the hormonal regulation of diapause by two antagonistically acting factors seems to have more reliable factual support presently. At least it gives the explanations for many facts which cannot be explained by monofactorial concept of hormonal deficiency. The bifactorial concept of the hormonal control of diapause fully agrees with the bioscillate hypothesis of photoperiodic reactions in insects (Tyshchenko, 1966) and with many ecological and physiological peculiarities of diapause (Geispitz *et al.*, 1971; Geispitz and Orlovskaya, 1971). It seems that bifactorial principle of hormonal control carried

ERRATUM

LoBue & Gordon: HUMORAL CONTROL OF GROWTH AND DIFFERENTIATION

The subtitle of Volume II should read:

Nonvertebrates, Neuroendocrinology, and Aging

out through fine hormonal balance between hormones activating and inhibiting morphogenesis is true for all known diapause types.

REFERENCES

Andrewartha, H. G. (1952). *Biol. Rev.* 27, 50.

Andrewartha, H. G., and Birch, L. C. (1954). "The Distribution and Abundance of Animals." Chicago Univ. Press, Chicago, Illinois.

Beck, S. D. (1962). *Biol. Bull.* 122, 1.

Beck, S. D. (1968). "Insect Photoperiodism." Academic Press, New York.

Beck, S. D., and Alexander, N. (1964a). *Biol. Bull.* 126, 185.

Beck, S. D., and Alexander, N. (1964b). *Science* 143, 478.

Beck, S. D., and Shane, I. L. (1969). *J. Insect Physiol.* 15, 721.

Beck, S. D., Shane, I. L., and Colvin, I. B. (1965). *J. Insect Physiol.* 11, 297.

Belozerov, V. N. (1962). *Vestnik. Leningrad St. Univ.* 9, 54.

Bowers, W. S., and Blickenstaff, C. C. (1966). *Science* 154, 1673.

Broza, M., and Pener, M. P. (1969). *Experientia* 25, 414.

Bünning, E., and Joerrens, G. (1959). *Naturwissensonaften.* 46, 518.

Burdette, W., and Bullock, M. (1963). *Science* 140, 1311.

Butenandt, A., and Karlson, P. (1954). *Naturforsch.* 96, 389.

Chefurka, W., and Williams, C. M. (1952). *Anat. Rec.* 113, 562.

Chino, H. (1958). *J. Insect Physiol.* 2, 1.

Church, N. S. (1955). *Can. J. Zool.* 33, 339.

Claret, J. (1966). *Ann. Endocrinol.* 27, 311.

Clever, U. (1964). *Naturwissenschaften* 51, 449.

Cloutier, E. J., Beck, S. D., McLeod, D. G. R., and Silhacek, D. L. (1962). *Nature (London)* 195, 1222.

Danilevsky, A. S. (1949). *Entomol. Obozr.* 30, 194.

Danilevsky, A. S. (1950). *Tr. Leningrad. Obshchestva Estestvoisp.* 70, 90.

Danilevsky, A. S. (1961). "Photoperiodism and Seasonal Development in Insects." Leningrad State Univ., Leningrad. English transl.: Oliver and Boyd. London (1965).

Danilevsky, A. S., Goryshin, N. I., and Tyshchenko, V. P. (1970). *Ann. Rev. Entomol.* 15, 201.

Darjo, A. (1969). *C.R. Acad. Sci. Paris D* 268, 337.

Doskočil, J. (1961). *Vestnik. Czesk. Zool. Spol.* 25, 105.

Emme, A. M. (1967). *Bull. Moskow Obshch. Estestvoisp.* 72, 117.

Emmerich, H. (1970). *J. Insect Physiol.* 16, 725.

Engelman, F. (1962). *Gen. Comp. Endocrinol.* 2, 183.

Engelman, F. (1968). *Ann. Rev. Entomol.* 13, 1.

Ewen, A. B. (1966). *Experimentia* 22, 470.

Fukaya, M., and Mitsuhashi, J. (1957). *Jap. J. Appl. Entom. Zool.* 1, 145.

Fukaya, M., and Mitsuhashi, J. (1958). *Jap. J. Appl. Zool.* 2, 183.

Fukuda, S. (1951a). *Zool. Mag. Japan.* 60, 119.

Fukuda, S. (1951b). *Proc. Imp. Acad. Japan.* 27, 582.

Fukuda, S. (1951c). *Proc. Imp. Acad. Japan.* 27, 672.

Fukuda, S. (1953a). *Proc. Imp. Acad. Japan.* 29, 381.

Fukuda, S. (1953b). *Proc. Imp. Acad. Japan.* **29**, 385.
Fukuda, S. (1953c). *Proc. Imp. Acad. Japan.* **29**, 389.
Fukuda, S. (1963). *Bull. Soc. Zool. France* **LXXXVIII**, 151.
Fukuda, S., and Endo, K., (1966). *Proc. Imp. Acad. Japan.* **42**, 1082.
Fukuda, S., and Takeuchi, S. (1967). *Proc. Imp. Acad. Japan.* **43**, 51.
Fukuda, S., and Takeuchi, S. (1967). *Embriologia* **9**, 333.
Geispitz, K. F. (1953). *Entomol. Obozr.* **33**, 17.
Geispitz, K. F., and Orlovskaya, E. I. (1971). *In* "Cold-resistance in Insects and Mites," pp. 16–20. Tartu, Inst. Zool. and Botany.
Geispitz, K. F., Sapognikova, F. D., and Taranetz, M. N. (1971). *Entomol. Obozr.* **50**, 275.
Geldiay, S. (1967). *J. Endrocrinol.* **37**, 63.
Gersch, M. (1961). *Amer. Zoologist* **1**, 53.
Gersch, M. (1964). "Vergleichende Endocrinologie der Wirbellosentiere," p. 535. Academische Verlagsgesellschft, Geest and Portig, Leipzig.
Gersch, M. (1968). "Neuroendokrinologie der Insecten." Academie-Verlag, Berlin.
Gersch, M., and Berger, H. (1962). *Naturwissenschaften* **49**, 292.
Gersch, M., and Richter, K. (1963). *Zool. Jahrb. Abstr.* **70**, 301.
Gersch, M., Under, H., Fisher, F., and Kabitza, W. (1963). *Z. Naturforsch.* **18**, 587.
Gersch, M., Under, H., Fischer, F., and Kabitza, W. (1964). *Zool. Jahrb. Abstr.* **70**, 455.
Gilbert, L. (1964). *In* "The Physiology of Insecta." Academic Press, New York.
Gilbert, L., and Schneiderman, H. (1960). *Trans. Amer. Microsc. Soc.* **79**, 38.
Gilbert, L., and Schneiderman, H. (1961). *Gen. Comp. Endocrin.* **1**, 453.
Goryshin, N. I. (1956). *Dokl. Akad. Nauk S.S.S.R.* **109**, 205.
Goryshin, N. I. (1958). *In* "Colorado Beetle" p. 2. Acad. Nauk. Moscow.
Goryshin, N. I., and Tyshchenko, V. P. (1968). *In* "Photoperiodic Adaptation in Insects and Acari," p. 192. Leningrad State Univ., Leningrad.
Grison, P. (1949). *C.R. Acad. Sci. Paris* **228**, 428.
Harvey, W. R. (1962). *Ann. Rev. Entom.* **7**, 57.
Hasegawa, K. (1952). *J. Fac. Agr. Tottori Univ.* **1**, 83.
Hasegawa, K. (1957). *Nature (London)* **179**, 1300.
Herman, S. W., and Gilbert, L. I. (1965). *Nature (London)* **205**, 9267.
Highnam, K. C. (1961). *Quart. J. Microsc. Sci.* **102**, 27.
Highnam, K. C. (1968). *New Sci.* **38**, 353.
Hinks, C. F. (1971). *J. Entom. A* **46**, 13.
Hinton, H. E. (1953). *Entomologist* **86**, 279.
Hsiao, C., and Hsiao, T. H. (1969). *Life Sci.* **8**, 767.
Joly, P. (1945). *Arch. Zool. Exp. Gen.* **84**, 49.
Jones, B. M. (1956a). *J. Exp. Biol.* **33**, 174.
Jones, B. M. (1956b). *J. Exp. Biol.* **33**, 685.
Karlson, P. (1957). *Zool. Anz.* **20**, 203.
Karlson, P. (1967). *Pure Appl. Chem.* **1**, 75.
Kind, T. V. (1965). *Entomol. Obozr.* **44**, 534.
Kind, T. V. (1968a). *In* "Photoperiodic Adaptation in Insects and Acari." p. 153. Leningrad State Univ., Leningrad.
Kind, T. V. (1968b). *Zool. J.* **47**, 1489.
Kind, T. V. (1969). *Dokl. Acad. Nauk S.S.S.R.* **187**, 226.
Kind, T. V. (1971). *Dokl. Acad. Nauk S.S.S.R.* **198**, 254.

Kogure, M. (1933). *J. Dept. Arg. Kyushu Univ.* **4**, 1.

Kozanchikov, I. V. (1938). *Bull. Entom. Res.* **29**, 253.

Kozanchikov, I. V. (1948). *Izv. Acad. Nauk S.S.S.R.* **6**, 653.

Kroeger, H. (1963). *Int. Congr., 16th* **4**, 251–255.

Kurland, G., and Schneiderman, H. (1959). *Biol. Bull.* **116**, 136.

Lees, A. D. (1955). "The Physiology of Diapause in Arthropods." Cambridge Univ. Press. London and New York.

Lees, A. D. (1956). *Ann. Rev. Entomol.* **1**, 1.

Lees, A. D. (1960). *J. Insect Physiol.* **4**, 154.

Lees, A. D. (1961). *Proc. Symp. Cryptobiotic Stages Biol. Systems, 5th Biol. Conf. Oholo, 1960* **14**, 132. Elsevier Monogr. Biol. Subseries.

Lees, A. D. (1963). *J. Insect Physiol.* **9**, 153.

Lees, A. D. (1968). Photoperiodism in Insects. *In* "Photophysiology," Vol. 4, p. 47. Academic Press, New York.

Loof, A. de, and Wilde, J. de (1970). *J. Insect Physiol.* **16**, 1455.

Lozina-Lozinsky, L. K. (1963). *In* "The Cell Reactions on Extremal Conditions," pp. 34–54. Leningrad. Acad. Nauk S.S.S.R.

Maslennikova, V. A. (1961). *Dokl. Acad. Nauk S.S.S.R.* **139**, 249.

Maslennikova, V. A. (1968). *Entomol. Obozr.* **47**, 429.

Maslennikova, V. A. (1970). *Dokl. Acad. Nauk S.S.S.R.* **192**, 942.

Maslennikova, V. A. (1971). *Tr. Peterg. Biol. Inst.* **21**, 46.

McDaniel, C. N., and Beery, S. J. (1967). *Nature (London)* **214**, 1032.

Meyer, A. (1965). *Nature (London)* **206**, 4981, 272.

Mitsuhashi, J., and Fukaya, M. (1960). *Jap. J. Appl. Entom. Zool.* **4**, 127.

Morohoshi, S. (1959). *J. Insect Physiol.* **3**, 28.

Morohoshi, S. (1969a). *Proc. Jap. Acad.* **45**, 621.

Morohoshi, S. (1969b). *Proc. Jap. Acad.* **45**, 733.

Morohoshi, S. (1969c). *Proc. Jap. Acad.* **45**, 739.

Morohoshi, S. (1969d). *Proc. Jap. Acad.* **45**, 797.

Morohoshi, S. (1969e). *Procid. Jap. Acad.* **45**, 937.

Morohoshi, S. (1971). *In* "Insect Endocrines," pp. 131. Academia, Praha.

Morohoshi, S., Iijima, T., Kikuchi, S., and Ikeda, S. (1969). *Proc. Japan. Acad.* **45**, 328.

Müller, H. J. (1958). *Zool. Anz.* **160**, 294.

Müller, H. J. (1965). *Verhand. Deutsch. Zool. Ges. Jena* **29**, 192.

Müller, H. J. (1970). *Nova Acta Leopold.* **35**, 27.

Nayer, K. K. (1957). *Proc. Nat. Inst. Sci. India.* **22**, 171.

Novák, V. J. A. (1957). *Acta Soc. Entomol. Csl.* **54**, 269.

Novák, V. J. A. (1966). "Insect Hormones." Methuen, London.

Novák, V. J. A. (1969). *Gen. Comp. Endocrinol. Suppl.* **2**, 439.

Novák, V. J. A. (1970). *J. Obshch. Biol.* **31**, 14.

Panov, A. A. (1962a). *Dokl. Acad. Nauk SSSR* **143**, 471.

Panov, A. A. (1962b). *Dokl. Acad. Nauk SSSR* **145**, 1409.

Panov, A. A. (1964). *Entom. Obozren.* **43**, 789.

Panov, A. A. (1969). *J. Obshch. Biol.* **30**, 87.

Panov, A. A. (1971). *In* "Insect Endokrines," p. 91. Academia Praha.

Panov, A. A., and Kind, T. V. (1963). *Dokl. Acad. Nauk SSSR* **153**, 1186.

Park, K. E., and Yoshitake, N. (1970a). *J. Insect Physiol.* **16**, 1655.

Park, K. E., and Yoshitake, N. (1970b). *J. Insect Physiol.* **16**, 2223.

Park, K. E., and Yoshitake, N. (1971). *J. Insect Physiol.* **17**, 1305.

Pflügfelder, O. (1958). "Entwicklungsphysiologie der Insecten." Leipzig.
Pridantzeva, E. A., Drabkina, A. A., and Zysin, U. S. (1971). *Usp. Sovrem. Biol.* **71**, 292.
Raabe, M. (1964). *Ann. Amer. Endocrinol.* **25**, 107.
Raabe, M. (1965). *C. R. Acad. Sci.* **260**, 6710.
Raabe, M. (1971). In "Insect Endocrines," p. 105. Academia Praha.
Ralph, C. L. (1962). *J. Insect Physiol.* **8**, 431.
Rehm, M. (1952). *Rev. Suisse Zool.* **59**, 173.
Röller, H., and Dahm, K. H. (1968). *Rec. Progr. Hormone Res.* **24**, 651.
Röller, H., Bjerke, J., and McShan, W. H. (1965). *J. Insect Physiol.* **11**, 1185.
Röller, H., Dahm, K. H., Sweely, C. C., and Trost, B. H. (1967). *Angew. Chem.* **79**, 190.
Röller, H., Bjerke, J., Holthans, L. M., Norgard, D. W., and McShan, W. H. (1969). *J. Insect Physiol.* **15**, 379.
Scharer, B. (1959). *Symp. Comp. Endocrinol.* p. 134. New York.
Schneiderman, H. A. (1957). In "Physiological Triggers and Discontinuous Rate Processes," p. 46. Washington, D. C.
Schneiderman, H. A., and Williams, C. M. (1952). *Anat. Rec.* **113**, 79.
Schneiderman, H. A., and Williams, C. M. (1953). *Biol. Bull.* **105**, 320.
Schneiderman, H. A., and Williams, C. M. (1954a). *Biol. Bull.* **106**, 210.
Schneiderman, H. A., and Williams, C. M. (1954b). *Biol. Bull.* **106**, 238.
Schneiderman, H. A., and Williams, C. M. (1955). *Biol. Bull.* **109**, 123.
Schneiderman, H. A., Ketchel, M., and Williams, C. M. (1953). *Biol. Bull.* **105**, 188.
Schooneveld, H. (1970). *Neth. J. Zool.* **20**, 151.
Siew, Y. C. (1966). *Trans. Roy. Entomol. Soc. London* **118**, 359.
Sláma, K. (1959a). *Acta Soc. Entomol. Csl.* **56**, 113.
Sláma, K. (1959b). *Acta Symp. Evol. Ins., Prague* p. 195.
Sláma, K. (1959c). *Acta Symp. Evol. Ins., Prague* p. 222.
Sláma, K. (1960). *J. Insect Physiol.* **5**, 341.
Sláma, K. (1964a). *J. Insect Physiol.* **10**, 283.
Sláma, K. (1964b). *J. Insect Physiol.* **10**, 773.
Suzuki, M., Kobayashi, M., and Kekawa, N. (1970). *Lipids* **5**, 539.
Takami, T. (1960). *Rev. Ver. Soie.* **12**, 315.
Tamarina, I. A. (1966). *Usp. Sovrem. Biol.* **62**, 413.
Tanaka, Y. (1924). *Genetics* **9**, 479.
Tanaka, Y. (1953). *Advan. Genet.* **5**, 239.
Thomsen, E. (1952). *J. Exp. Biol.* **29**, 137.
Thomsen, E., and Möller, J. (1959). *Nature (London)* **183**, 1401.
Thomsen, M. (1965). *Z. Zellforsch.* **67**, 693.
Truman, J. W., and Riddiford, L. M. (1970). *Science* **167**, 1624.
Tyshchenko, V. P. (1966). *J. Obshch. Biol.* **27**, 209.
Umeya, Y. (1926). *Bull. Seric. Exp. St. Chosen* **1**, 1.
Ushatinskaya, R. S. (1957). "The Cold Resistance in Insects." Acad. Nauk SSSR, Moscow.
Ushatinskaya, R. S. (1959). *Int. Symp. Colorado Beetle. Acad. Nauk SSSR* **57**.
Vinogradova, E. B. (1969). "The Diapause in Mosquitos and its Control." Acad. Nauk SSSR, Leningrad.
Waku, J. (1960). *Sci. Rept. Tohoku Univ.* **26**, 327.
Wigglesworth, V. B. (1954). "The Physiology of Insect Metamorphosis." Cambridge Univ. London and New York.

Wigglesworth, V. B. (1966). *In* "Cell Differentiation and Morphogenesis." p. 180. North-Holland Publ., Amsterdam.
Wigglesworth, V. B. (1967). "Endocrine Genetics," p. 77. Cambridge Univ. Press.
Wigglesworth, V. B. (1970). "Insect Hormones." Oliver and Boyd, Edinburgh.
Wilde, J. de (1954). *Say. Arch. Neer. Zool.* **10,** 375.
Wilde, J. de (1958). *Proc. Int. Congr. Entoth.* **2,** 213.
Wilde, J. de (1961). *Bull. Res. Counc. Israel.* **10,** B, 36.
Wilde, J. de (1962). *Ann. Rev. Entomol.* **7,** 1.
Wilde, J. de, and Boer, T. A. de (1961). *J. Insect Physiol.* **6,** 152.
Wilde, J. de, and Boer, T. A. de (1969). *J. Insect Physiol.* **15,** 661.
Wilde, J. de, Staal, G. B., Kort, C. A. de, Loof, A. de, and Baard, G. (1968). *Proc. Koninkl. Nederl. Acad. Wet.* **71,** 321.
Wilde, J. de, Duinter, C. S., and Mook, L. (1969). *J. Insect Physiol.* **3,** 75.
Williams, C. M. (1946). *Biol. Bull.* **90,** 234.
Williams, C. M. (1947). *Biol. Bull.* **93,** 89.
Williams, C. M. (1948). *Biol. Bull.* **94,** 60.
Williams, C. M. (1952). *Biol. Bul.* **103,** 120.
Williams, C. M. (1956a). *Nature (London)* **178,** 212.
Williams, C. M. (1956b). *Biol. Bull.* **110,** 201.
Williams, C. M. (1958). *Sci. Amer.* **198,** 67.
Yamashita, O., and Hasegawa, K. (1964). *J. Seric. Sci. Japan* **33,** 115.
Yamashita, O., and Hasegawa, K. (1966). *J. Insect Physiol.* **12,** 325.
Yamashita, O., and Hasegawa, K. (1970). *J. Insect Physiol.* **16,** 2377.
Yamazaki, M., and Kobayashi, M. (1969). *J. Insect Physiol.* **15,** 1981.

HUMORAL INFLUENCES IN HYDRA DEVELOPMENT

Hans Bode

I. Introduction 35
II. Growth 36
    A. Growth Patterns of the Tissue Layers 36
    B. Maintenance of the Ratios of Cell Populations 38
III. Head Regeneration 40
    A. Promotors of Hypostome Formation 41
    B. Inhibitors of Hypostome Formation 44
IV. Cell Differentiation 46
    A. Interstitial Cell Differentiation 46
    B. Position-Dependent Cell Differentiation 50
V. Conclusion 53
    References 55

I. INTRODUCTION

The processes of growth, cell differentiation and morphogenesis are undoubtedly coordinated in some way. Hydra is a potentially interesting model system for studying the nature of such interrelated controls, or control mechanisms common to more than one process. Because the animal is relatively simple, the complexity of control is probably relatively simple too. Hydra also has an advantage in that the adult conveniently displays many of the characteristics of the embryonic state. It is constantly growing, cell differentiation takes place continuously, and almost any part of the body (even aggregates of cells, Gierer *et al.*, 1972) is capable of regeneration. That some of these processes are interrelated is vividly illustrated during head regeneration when a region of the body column is rearranged to form a new hypostome and tentacles accompanied by a marked change in the pattern of cell differentiation.

At present a great deal is known about the phenomenology of growth, cell differentiation, and head regeneration in hydra. Less is known about the nature of the control, although several indications of hormonal control exist. The object of this chapter is to describe each of these processes and show where hormonal influences are possible (growth) and probable (cell differentiation) and to discuss those reported (head regeneration).

II. GROWTH

Growth in hydra can be described in terms of the overall patterns of the growth of the two tissue layers, or the sizes of the individual cell populations within each layer. Both will be examined for clues of hormonal control. Growth, as used here, will mean an increase in tissue mass as a result of an increase in cell numbers, since the size of hydra cells is not affected by developmental or environmental changes. Thus, the concern is with control of cell division.

A. Growth Patterns of the Tissue Layers

A hydra is shaped like a hollow tube and made up of two epithelial layers that are separated by the acellular mesoglea. The layers resemble other animal epithelia in that they are in continual growth. In regularly fed animals the tissue mass increases exponentially (see, for example, Brien and Reniers-Decoen, 1949). Yet, the adult animals maintain a constant size by controlling morphogenesis, not growth.

As the tissue layers expand they move into new buds (85%, Campbell, 1967b) and to a lesser extent out toward the extremities, the ends of the tentacles, the hypostome, and basal disk (see Fig. 1), where the cells are eventually sloughed. Since the tissue moves in opposite directions down the column toward the basal disk and up the column into the head (Tripp, 1928; Brien and Reniers-Decoen, 1949; Burnett, 1961; Campbell, 1967b), there must be a point in between where there is no movement. This apparently stationary spot is the subhypostomal region (Fig. 1) in budding animals (e.g., Brien and Reniers-Decoen, 1949).

This pattern of growth led to the idea that cell division occurs primarily in the subhypostomal region, called the "growth zone" by Burnett (1961), and that the newly produced tissue was forced either up or

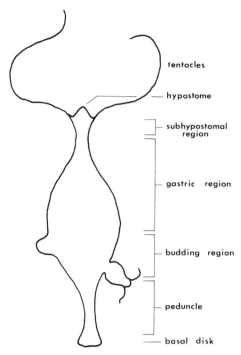

tentacles

hypostome

subhypostomal region

gastric region

budding region

peduncle

basal disk

Fig. 1. Diagram of hydra showing the several body regions. The head region consists of hypostome and tentacles and the foot region is the basal disk.

down the body column (Tripp, 1928; Brien and Reniers-Decoen, 1949; Burnett, 1961). In recent years a great deal of evidence (reviewed by Webster, 1971) indicates this view to be erroneous. Cell division occurs at a constant rate and occurs uniformly throughout the body though declining in the lower peduncle and ceasing in the basal disk and tentacles (e.g., Campbell, 1965, 1967a). At least 90% of the population of the epithelial cell type of each of the two layers is in the proliferative phase (David and Campbell, 1972). This same uniform pattern of cell division occurs throughout the bud (Clarkson and Wolpert, 1967; Webster and Hamilton, 1970). Moreover, the mitotic rate is the same as in the adult (Webster and Hamilton, 1970), which is not very surprising since the bud is formed by evagination of the ever-growing parent tissue.

Perhaps more surprising is that regeneration has no significant effect on the mitotic rate. Removal of the upper part of the body will cause the remainder to regenerate the head and form a normally proportioned, though smaller, animal. This process involves changes in the arrangement of the tissue and in the pattern of cell differentiations (e.g., Bode *et al.,* 1973). However, the rate of cell division remains unaltered both locally in the wounded region and throughout the animal (Clarkson and Wolpert, 1967; Park *et al.,* 1967; David, unpublished results). This is in strong contrast to other regenerating systems such as amphibian limbs (Goss, 1969a) or rat liver (Goss, 1964), where the mitotic rate increases and remains higher than normal until the limb or organ has regained the appropriate size.

These observations on adult, budding and regenerating animals lead to the conclusion that the rate of cell division is unaffected by developmental events. As a consequence there is no indication of a control necessary to either stimulate or inhibit growth. Hence, the question of hormonal control remains unresolved except for one point. The hydra epithelia behave much like the self-renewing epithelia of mammals. However, the resemblance is only partial, for if a portion of mammalian epidermis is removed, the mitotic rate (the percentage of cells in the proliferative phase) initially increases sharply, and as the surface is recovered, slowly returns to normal (Bullough and Laurence, 1960). Chalones (see Bullough, Vol. I) are thought to control the mitotic activity of such cells. These tissue-specific molecules are produced by the cells themselves and act by blocking mitosis when in high concentration. When a tissue is wounded, the cell density drops, the chalone concentration is lowered, and consequently the mitotic rate increases. Because removal of tissue does not affect the mitotic rate, chalones are apparently absent in hydra.

As might be expected, the growth rate of the animal is affected by environmental factors such as feeding (e.g., Loomis, 1954; Stiven, 1962; Webster and Hamilton, 1970), symbiotic algae (Muscatine and Lenhoff, 1965), ion composition of the medium (Lenhoff and Bovaird, 1960; Lenhoff, 1966), temperature (e.g., Loomis, 1954; Stiven, 1962) and density of animals in culture (Lenhoff and Loomis, 1957). Whether these influences act on cell division rates by affecting general metabolism only or are mediated by growth principles as well is unknown.

B. Maintenance of the Ratios of Cell Populations

Examination of the cell composition of the two tissue layers reveals phenomena that must be under some sort of control. The ratio of the population sizes for any two cell types remains constant over many generations for a given feeding condition (Bode *et al.,* 1973). For example, the epithelial-muscular cell and the digestive cell are roughly in a 1:1 ratio and the interstitial cell to epithelial-muscular cell ratio is 2:1. The problem that arises is how these cell type ratios are maintained in an ever-growing tissue. It is clearly illustrated by the relationships among the four self-renewing cell populations: epithelial-muscular, digestive, gland, and interstitial cells.

As the whole animal mass grows the two tissue layers, gastrodermal and ectodermal, must grow at the same rate. If they did not, after a few generations hydra would no longer consist of roughly equal amounts of each layer. Since the epithelial-muscular cell (ectoderm) and the digestive cell (gastroderm) are the principal tissue building cells of each layer, this question is reduced to one of the growth rates of these cell types. As David and Campbell (1972) have shown, both have an average generation time of 80 hours in *H. attenuata*. Thus, the control here is to keep the two cell types dividing at the same rate.

A more striking problem exists in the ectoderm. The interstitial cells are twice as frequent throughout the body column as the epithelial-muscular cells and have a generation time of 40 hours (David and Campbell, in preparation). This is half of the 80-hour generation time of the epithelial-muscular cell. Without some sort of control the interstitial cell: epithelial-muscular cell ratio would continuously increase and not remain constant at the observed ratio of 2:1. A part of the solution lies in the function of the interstitial cell which is to produce nerves and nematocytes by differentiation (see Section III,A). Enough interstitial cells may be drained off into these differentiation pathways to maintain the balance. Unless one assumes that the percentage of interstitial cells differentiating is by chance sufficient to maintain the interstitial cell: epithelial-muscular cell ratio when these two cell types have widely differing cell cycle times, the suggestion is strong that some regulation must exist to keep the balance. The control could be on the numbers of interstitial cells differentiating, or on the cell division rate of one or the other or both cell types. That the G-2 phase of the epithelial-muscular cell cycle is unusually variable in length, as much as 5 times (David and Campbell, 1972), hints of a mechanism for modulating the division rate of this cell type, and supports the argument for control of cell division.

A similar problem exists in the gastroderm with respect to the digestive and gland cells. The ratio of gland:digestive cell decreases from 0.4 in the gastric region to 0.2 in the peduncle, and finally to zero in the basal disk (e.g., Bode *et al.*, 1973). How this gradient of ratios is maintained in a continuously growing tissue as it moves down the column through the various body regions is not clear. The gland cells are not just selectively siphoned off into developing buds since the gland:digestive cell ratio in the bud is no different than that of the tissue from which it is derived. Besides these ratios are roughly the same in non-budding animals (Bode *et al.*, 1973). Simply loss of the cell, due to being "used up", similar to the loss of expended nematocytes is also an inadequate explanation, for such a process would eventually over several generations lead to the complete loss of the gland cell type from the tissue. A more likely explanation is based on Campbell's (1967a) observation that the mitotic index of the gland cells declines faster from the budding region to the basal disk than that of the digestive cell. Such a slowdown of the cell division rate of the gland cell with respect to the digestive cell would give rise to this gradient of ratios, and coupled with loss, would lead to no gland cells in the foot. In some manner this change of division rate must be controlled.

Growth in the extremities presents another problem. The mitotic index of cells in the tentacles and the basal disk is zero (e.g., Campbell, 1967a). Thus, epithelial-muscular cells and digestive cells proliferating in the body column close to the tentacles abruptly cease cell division upon moving into the tentacle. The change is more gradual for these two cell types as they move toward the basal disk. Nothing is known as to how this change takes place, but for the epithelial-muscular cells it is accompanied by differentiation into battery cells in the tentacles and into secretory cells in the basal disk. The digestive cells flatten in shape as they move into either extremity, and lose their ability to absorb nutrients when moving through the peduncle toward the basal disk. Since differentiation of a cell is often accompanied by loss of its proliferative capacity, this may be a question of control of differentiation and/or growth.

In summary, though nothing at the tissue level indicates hormonal influence there is a need for some control to maintain the ratios of the self-renewing cell types within each of the two tissue layers. Controls may operate here by regulating cell division rates, triggering differentiation, or inducing selective loss of a cell type. Hormones are readily imagined as the controls, but remain only a possibility at present.

II. HEAD REGENERATION

After removal of both head and foot from a hydra, the remaining piece always regenerates a new head at the upper end and a new foot at the lower end. Regeneration is said to be polarized. For a long time (see Child, 1941) this polarity of regeneration has been explained in terms of gradients of one kind or another. More recently with the use of some elegant grafting experiments, Webster and Wolpert have described some of the characteristics of two gradients involved in hypostome (or head) formation. One is a gradient of hypostome formation potential and the other of inhibition of hypostome formation, both with a high point in the hypostome and low point in the basal disk. Though gradients of diffusible substances have been postulated as the underlying mechanisms for the development of tissues or limbs in many systems, only in hydra has the work progressed to the point where reasonable evidence for such gradients exists. The evidence for these gradients and for diffusible substances as the basis of the gradient will not be presented here because it requires a lengthy discussion, and has been brilliantly reviewed by Webster (1971) and by Wolpert *et al.* (1971). A number of people have attempted to isolate substances responsible for the gradients. This is the only aspect of hydra development for which direct isolation of molecules with hormonal qualities has been attempted.

A. Promotors of Hypostome Formation

In considering an assay for a molecule that is supposedly involved in a gradient of hypostome formation, it is necessary to define the effects such a substance should evoke. Since regions closer to the head are capable of regenerating a head faster than those further away (e.g., Rulon and Child, 1937), above a certain concentration a promotor substance would be expected to speed up regeneration. Second, if the head formation potential is reduced, a typical result is a partially formed head or a reduced number of tentacles (Webster and Wolpert, 1966). An excess of a substance might then be expected to enhance development of extra tentacles or even extra hypostomes. However, this assay must be used with care, for unless large pieces of nonbudding animals are used, a variety of abnormal regenerations occur anyway (Müller and Spindler, 1971; Schaller, unpublished results).

A number of people (Lentz, 1965; Lesh and Burnett, 1964, 1966; Lenicque and Lundblad, 1966b; Müller and Spindler, 1971; Schaller, 1973) have reported activities obtained from extracts of hydra which promote formation of extra tentacles or extra hypostomes and/or accelerate head regeneration. Similar activities have also been reported for other hydroids, *Hydractinia* (Müller, 1969a, b) and *Clava* (Lenicque and Lundblad, 1966a). However, in every case but one the activity was found in an impure fraction. Lentz (1965) describes an activity that sediments at 35,000 g and stimulates regeneration of heads at both ends of the piece. No one else has reported such a dramatic effect. The pieces of tissue Lentz (1965) used were rather short, and as Schaller (personal communication) has shown that short pieces taken from the gastral region often give a high incidence of bipolar animals in untreated controls. Lesh and Burnett (1964, 1966) boiled the animals to remove nematocyst toxins before homogenization and found the activity in a 25,000 g supernatant. Lenicque and Lundblad (1966a) separated material obtained by NH_4SO_4 precipitation of hydra extract by gel filtration and reported the excluded fraction contained an activity with MW $>$ 40,000 Daltons. Müller and Spindler (1971) also fractionated a 105,000 g supernatant of hydra extract by gel filtration and found several active fractions. In a separate experiment they discharged the nematocytes by exposing animals to dilute acetic acid. Material isolated from the solution surrounding the treated hydra stimulated formation of extra tentacles leading Müller and Spindler (1971) to claim that the activity was due to nematocyst toxin. However, acetic acid will cause not only nematocytes to discharge, but will release a variety of material from the animal.

Until the material is further purified, these results must be considered preliminary and treated with caution, since similar effects can be obtained with a variety of substances. EDTA, guanidine HCl (Schaller, 1973), LiCl (Ham *et al.*, 1956) in the concentration range 10^{-3}–10^{-6} M, and degradative enzymes such as trypsin or peptidases (Schaller, 1973) all stimulate formation of extra tentacles during regeneration, probably through some indirect nonspecific action of the given molecule. In all four reports the impure material was used in high concentration so that the activity may have been caused by nonspecific substances.

In only one case has an extensive purification been carried out. Schaller (1973) concentrated a material $\times$100,000 that accelerates the rates of both head regeneration and budding, and increases the number of tentacles. The purified material is active at a concentration of 10^{-11} M which is 5–8 orders of magnitude below the concentration of the nonspecific substances mentioned above. This suggests the activity is quite specific, acting directly on hypostome formation. Chemical characterization of the activity indicates it is resistant to boiling, sensitive to trypsin and pronase, but insensitive to deoxyribonuclease and ribonuclease. The molecular weight as estimated by Diaflow ultrafiltration, gel filtration, and sedimentation equilibrium centrifugation is around 900 Daltons. The stimulating activity was apparently due to a low molecular weight peptide.

Though the material is very impure, it is of interest that the hypostome stimulating activity reported by others has characteristics similar to Schaller's (1973) "activator." Lesh and Burnett (1964, 1966) found their "heteropolarity factor" to be dialyzable, resistant to boiling, and sensitive to trypsin. Also, Müller's (1969a, b) activity in *Hydractinia* extracts bears an even closer resemblance in that besides being heat-stabile and trypsin-sensitive, it elutes after salt on a Sephadex G-25 column as does Schaller's (1973) material. The fact that Lesh and Burnett (1964, 1966) and Müller (1969b) were able to find activity only at high concentrations may be attributed to the impurity of the material. Schaller (1973) found both the material in crude extracts and the purified to be active in low concentrations. Despite this last difference, these results suggest that several workers may be dealing with a similar substance, and that it occurs in several different organisms.

Schaller and Gierer (1973) extended the investigation and examined the distribution of the "activator" in the animal. Isolation of the activity from animals cut in quarters indicates a gradient that is highest in the hypostome and lowest in the foot. Lesh and Burnett (1966) also reported more activity in the upper half than in the lower half of the body. The concentration of activity found in the tentacles was less than half that found in the hypostome. Therefore, nematocyst toxin is probably not responsible for the effects because virtually all nematocysts of the head region are in the tentacles and not in the hypostome (Schaller and Gierer, 1973). Since the absolute numbers of nerve cells are also distributed in a gradient down the body column (Bode *et al.*, 1973), it was an obvious assumption that the activity actually did originate in the nerves. To demonstrate this, cells obtained by macerating hydra were separated on a glycerol gradient (Schaller and Gierer, 1973). They found a fraction that contained only 10% of all the macerated cells, but was enriched ×10 for nerve cells, and contained more than 80% of the activity. This represented a ×10 increase in specific activity compared to that of the mixture of macerated cells. No other cell type was enriched in that fraction. Further, vesicles with the same dimensions as those of neurosecretory granules in hydra (see Lentz, 1966) were purified from homogenates by sedimentation through sucrose gradients and shown to contain most of the activity of the homogenate (Schaller and Gierer, 1973). Lentz (1965) had also found most of his hypostome-forming activity in a fraction that contained such vesicles. This evidence along with the distribution indicates the activity originates in nerve cells found not only in the head region, but throughout the animal, with the exception of the foot.

Though the activity is distributed in a gradient throughout the body column, the amounts found in the body are at least ×1000 higher (Schaller, 1973) than the amount needed to evoke an acceleration of regeneration or formation of extra tentacles. This is evidence that most of the material exists in a bound form, probably stored in the vesicles of neurosecretory cells. The active gradient in the animal could be produced by short range diffusion of the molecule released from the vesicles (Schaller and Gierer, 1973). Finally, since the distal end of a regenerating piece is destined to be the new hypostome, the hypostome formation potential of the regenerating tip will rise to the level found in the hypostome (Webster, 1966). In terms of diffusible sustances the amount of activity or specific activity of this tip should rise. As Schaller and Gierer (1973) have shown the specific activity (activity/mg protein) rises $\times 4$ in 24 hours. Moreover, this effect is specific to head regeneration and is not just due to the stimulation of a cut surface, because no such increase can be found at the proximal end of a piece regenerating a foot.

That the peptide is consistent with what is known about the gradient of hypostome formation potential is very encouraging and supports the concept of diffusible substances being involved in the gradient. Obviously, more will have to be learned about the properties of gradients, and the effects of the gradient at a cellular level to know whether the peptide is directly involved.

B. Inhibitors of Hypostome Formation

In a similar fashion attempts have been made to find materials that will inhibit hypostome formation. The three reports dealing with inhibition of growth and regeneration involve crudely fractionated extracts of hydra (Lenicque and Lundblad, 1966a; Davis, 1967) or material isolated from cultures of hydra grown under crowded conditions (Davis, 1966). Davis (1967) found that inhibitors of regeneration of the hydroids he studied were concentrated in the tentacles and suggests nematocyst toxins are responsible for the effects. Quarternary ammonium compounds found in hydra tentacles (Davis, 1967) inhibit regeneration as does a protein isolated directly from nematocysts (Kline, 1961). It is quite possible that the trypsin-sensitive material Davis (1966) obtained by growing animals at a density of roughly 100/ml may well be this protein, because discharge of nematocysts under these highly crowded conditions is very likely. Because regeneration is blocked by many different inhibitors of metabolism (e.g., Ham and Eakin, 1958; Clarkson 1969a, b) as well as nematocyst toxins, knowledge of the purity and specificity of the inhibiting material is important. No evidence is presented to suggest specificity.

In contrast, the literature on inhibitors of head (hydranth) formation in *Tubularia,* a marine hydroid, is more extensive. Though no inhibitors have been purified, some of the evidence is interesting and has a bearing on the gradient of hypostome inhibition in hydra. A good deal of the early work is inconclusive because of problems with bacterial contamination (Steinberg, 1954, 1955; Tardent, 1955; Tardent and Eymann, 1958; Tweedall, 1958; Fulton, 1959). Tardent (1956) and Tardent and Eymann (1959) demonstrated that inhibitors of hydranth formation that are of a diffusible nature do exist. By grafting a hydranth to the proximal end of a regenerating stalk they were able to prevent hydranth regeneration at the distal end where it normally occurs. Therefore, as in hydra, the experiment suggests an inhibitory substance produced in the hydranth that diffuses through the tissue.

Beyond this, Rose (1957, 1963) has shown that transmission of some types of inhibition is polarized. Distal hydranth structures of tubularia can inhibit more proximal tissue from forming distal structures. The proximal tissue forms proximal hydranth structures, which presumably in turn inhibit the stem from forming either. In a series of experiments, Rose (1966, 1967) and Rose and Powers (1966) attempted to obtain different inhibitor activities that would explain these initial results. They tested the effects of electrophoretic fractions obtained from homogenates of either different parts of adult hydranths or different parts of regenerating stems. Fractions of distal homogenates inhibited formation of distal structures of the hydranth, but not proximal structures. Fractions of proximal homogenates inhibited formation of both. In both cases the activity is positively charged and trypsin-sensitive. The results are interesting, but as with a great deal of coelenterate work, still very preliminary. Not excluded is the possibility that the effects, some or all, are due to nematocyst toxins. These results are not consistent with a gradient of inhibition in any obvious way. Indeed, Rose interpreted them in terms of a theory of specific inhibition (1952). Recently, Webster (1971) has reinterpreted the results in terms of gradients assuming polarized transport of a substance that is also capable of free diffusion.

Because of the unhappy history of searching for specific inducers in development, looking for molecules that may be the diffusible substances of gradients governing polarity in hydra is a treacherous undertaking. Nevertheless, grafting experiments (Wilby and Webster, 1970a, b; Wolpert et al., 1972) indicate that diffusible promotors and inhibitors of hypostome formation should be present. The inhibition phenomena in *Tubularia* clearly support the idea of diffusible substances. Therefore, it is reasonable to seek such molecules, and, indeed, an apparent promoter of hypostome formation has been isolated (Schaller, 1973). There is good evidence that more than two gradients are involved in hydra pattern formation because an analogous pair of gradients appears to control basal disk formation. A basal disk implanted in the gastric region will organize a second axis (Wolpert et al., 1971) with the disk at the apex in the same way a piece of hypostome induces a second axis with a hypostome at the apex (Browne, 1909; Webster, 1966). Also, a basal disk can inhibit the formation of a second disk in its vicinity (MacWilliams and Kafatos, 1968; MacWilliams *et al.*, 1970). Thus, the polarity of regeneration may well be under the control of a series of diffusible substances or hormones.

III. CELL DIFFERENTIATION

The relative simplicity of hydra is reflected in its cell composition. It has only seven classes of cells and altogether 15–20 cell types. Consequently the variety of cell differentiations that take place are relatively few, and fall roughly into two groups. First, there is a multipotent stem cell, the interstitial cell, which differentiates into three other cell classes. The second group of differentiations are positional. As cells are moved in the ever-growing tissue to new positions in the body, they change their shape, function, or proliferative ability accordingly. Some of the interstitial cell differentiations can also be included in this category. At present a good deal of interesting phenomenology exists and there is some evidence that suggests hormonal control of differentiation.

A. Interstitial Cell Differentiation

The interstitial cells are distributed throughout the ectoderm of most of the body column. They serve as stem cells for four kinds of nematocytes (Slautterback and Fawcett, 1959), several types (2–4?) of nerves (Lentz, 1966), and the two types of gametes (Brien, 1953). The interstitial cells must continuously differentiate into nematocytes and nerves to maintain these nondividing populations in the continuously growing tissue. This is similar to the erythropoietic system in which a stem cell continuously gives rise to the red blood cells (Goss, 1969b, Erslev, 1971), except that the interstitial cell is the stem cell for not one, but several cell types. In addition, when hydra switches from asexual to sexual reproduction, the interstitial cell also gives rise to gametes.

Before examining the differentiation phenomena, a primary question must be answered: is the interstitial cell multipotent, or are there several morphologically indistinguishable interstitial cells, each serving as a stem cell for a particular differentiated product? The general acceptance of the view that the interstitial is multipotent is based on a mass of histological observation (e.g., Brien, 1953). Though these observations do not exclude the possibility of several interstitial cells, there is some experimental evidence that does. For example, in some species of hydra (Brien, 1961; Burnett and Diehl, 1964), budding ceases as the animal enters the sexual phase. Examination of these animals shows few nematocytes and nerves, no intersitial cells, and large numbers of gametes. Presumably, most of the interstitial cells not only ceased dividing, but also switched their differentiation from somatic cell types to gametes.

Another example involves nerve formation during regeneration. In the gastric region about 20–40% of the interstitial cell differentiations are to nerve cells, while in regenerating animals initially severed in this region the interstitial cell-nerve transitions have risen sharply to 50–80% (David, unpublished results). Possibly, the stimulus of regeneration diverted interstitial cells away from nematocytes to nerve differentiation. Though this evidence is indirect, it certainly supports the view that the interstitial cell is multipotent. The direct experiment of following the fate of a single interstitial cell through several generations and examining the various daughter cells for indications of at least two different product cells has yet to be undertaken.

Accepting its multipotency, a number of interesting questions arise concerning the control of interstitial cell differentiation. During asexual growth, how does an individual interstitial cell know whether to divide or differentiate into one of the several cell types? What is the nature of the system of control that insures that the numbers of nematocytes and nerves are produced in the proper ratios to maintain the cell composition in the dynamic situation of constant growth? Are the controls external to the cell or is the interstitial cell "internally preprogrammed"? During the sexual phase, how does the interstitial cell switch from somatic cell to gamete production? Though very little is known about these questions, it is easy to imagine hormonal control of the various interstitial differentiations. For those differentiations investigated, the evidence does suggest that they are controlled by diffusible substances.

Two examples involve the formation of nematocytes. One case has been studied both by Brien (1961) and Burnett and Diehl (1964). When *H. pirardi* is grown at 8°C, asexual reproduction ceases and the animals enter the sexual phase completing the development in about 30 days. By that time interstitial cells usually distributed evenly throughout the ectoderm are only found close to ovaries or testes. At the same time nerve cells and nematocytes have been markedly reduced while nematoblasts have completely disappeared. Apparently in entering the sexual phase, a control is exercised to stop interstitial cell differentiation into somatic cell types and instead channel them into formation of gametes. This effect can be reversed by grafting tissue from animals grown at 20°C to the sexual animal and returning the graft to 8°C. The interstitial cell near the gonads begins nematocyte differentiation. The implication is that tissue grown at 20°C contains a diffusible factor which signals the interstitial cell to begin nematocyte differentiation.

The second example concerns the production of the four nematocyte types in the proper ratios to maintain a constant cell composition. Since all of the nematocytes arise by interstitial cell differentiation, it is a possibility that the control of these processes is of a feedback nature much as erythropoiesis is controlled by feedback inhibition and feedback stimulation (e.g., Goss, 1969b, Erslev, 1971). Zumstein and Tardent (1971) have reported evidence supporting this view. They were able to selectively remove almost all of one type of nematocyte, the stenotele, and examine the reaction of the animal at the cellular level. If nematocyte production were under feedback control, a response to replace the depleted stenoteles to restore the normal ratios of nematocytes would be expected. The population was partially restored very rapidly as though a reserve of stenoteles existed. A similar immediate response had been found earlier (Lenhoff and Bovaird, 1961) after discharge of 10–20% of the nematocytes due to excessive feeding. A second response, detectable only several days later, was an increased number of stenotele precursor cells. The more obvious conclusion that an increased number of interstitial cells were committed to stenotele production is probably incorrect. The first step in nematocyte differentiation is for the interstitial cell to undergo two- to four-cell divisions, with the resulting cells remaining together in a "nest." Zumstein and Tardent (1971) found that the average "nest" size for stenotele precursors had increased, thus implying that the cells in the 'nest" had undergone an extra division to increase the cell number before nematocyst formation began. Though there is no obvious effect on the interstitial cell, both responses to the perturbation demonstrate the feedback nature of the control of maintaining nematocyte populations. Since diffusible substances are usually involved in feedback control systems, hormonal control in some parts of the nematocyte differentiation is strongly suggested.

Another type of phenomenon implying hormonal control has to do with sex determination. Some species of hydra are clearly hermaphroditic, some clearly dioecious and some unstably dioecious. An animal of a species in the last category, such as *H. attenuata,* usually maintains its sex stably, but occasionally undergoes spontaneous sex inversion in either direction. At a cellular level this means the interstitial cell which previously has differentiated into one gamete now produces only the other gamete. Tardent (1968) investigated this problem of sex determination by making reciprocal grafts between halves of male and female *H. attenuata.* The resulting animals were all male. Grafts in which only the head or the stalk (peduncle and basal disk) (see Fig. 1) of male

e.g., feedback. Nevertheless, the thrust of the model that diffusible substances underlie the gradients and play a role in controlling cell differentiations is a tempting hypothesis since it provides a mechanism for coordinating many diverse differentiation events.

Attempts have been made to directly demonstrate that the substances which affect regeneration also affect interstitial cell differentiation. Lesh (1970) reported that different concentrations of the "heteropolarity factor" will stimulate interstitial cells in pieces of tissue to undergo differentiation into various product cells in accordance with Burnett's model (1966). Burnett *et al.* (1968) reported similar effects with the same factor on cells of *Tubularia* explants. These results are interesting but must be taken cautiously due to the impurity of the material. It has been shown that a variety of ions and metabolites can affect cell differentiation in hydra (Lenhoff and Bovaird, 1960; Loomis, 1964; Macklin and Burnett, 1966) as well as in other animals (e.g., Barth, 1965).

Schaller and Gierer (unpublished results) have noted that their peptide speeds up the rate of appearance of nerve and mucous cells in the distal tips of regenerating animals. Because the material is highly purified and active in such small concentrations it is very probable that the peptide is involved in the effect. But, regeneration is undoubtedly complex and whether the peptide affects the rate of cell differentiation directly or indirectly is not yet known.

In summary, position-dependent differentiation clearly exists in hydra. Whether these differentiations are related to the postulated gradients and are thereby controlled by diffusible substances is uncertain. Attempts to answer these questions have been suggestive though inconclusive. In contrast, interstitial cell differentiations not obviously related to position may well be under hormonal control. In some cases, namely gamete formation and maintenance of nematocyte populations the evidence indicates control by diffusible factors.

V. CONCLUSION

Having examined these three developmental processes separately it is worthwhile viewing them as a coordinated whole, and asking how the controls might be interrelated. Since the tissue layers are continuously moving toward the extremities, every piece of tissue undergoes changes in shape, cell division rates, and cell composition. A piece of tissue in the body column upon moving into the head will be reshaped into tentacle or perhaps hypostome. Here morphogenesis provides a check on indefinite elongation. Cells within that piece change from a dividing state to a nondividing differentiated one. For example, the epithelial-muscular cells cease division and become the battery cells of the tentacle, and the interstitial cells, which were dividing as well as differentiating in the body column, move into the hypostome and only differentiate into nerve cells. These changes are tightly coordinated, suggesting either tightly coordinated or common controls.

Our knowledge of the nature of the control of these developmental processes in hydra is at best sketchy, but enough is known to attempt to show how some controls may be common. The most prominent feature is a pair of gradients that are involved in the control of the polarity of regeneration. The evidence is very suggestive that diffusible substances are the basis for these gradients. Further, the only molecule with hormonal qualities isolated from hydra may be involved in the gradient of hypostome formation potential.

These gradients are always present, not only during regeneration (Webster, 1966, 1971). Their function may be to continually instruct the cells of the tissue as to their position in the column as the layer moves (say) up the body (Wolpert, 1969). One response to the received "positional information" (Wolpert, 1969) is to differentiate. Thus, an epithelial-muscular cell upon reaching the head would "read" the gradient and differentiate into a battery cell. An interstitial cell moving into the hypostome would "read" the gradient and form only nerve cells.

However, the gradients are probably not the only influence on cell differentiation. Nematocyte production, which takes place throughout the body column from the subhypostomal region to well into the peduncle, is only partly influenced by position. For example, the size of a stenotele in *H. attenuata* is large or small depending on its proximity to the head (Tardent *et al.*, 1971). But, stenotele production is also subject to a feedback control as the replacement kinetics of stenoteles in depleted tentacles illustrated (Zumstein and Tardent, 1971). The two influences could be affecting different aspects of stenotele production. The feedback system would control the quantity of stenoteles produced, while their position dictated the size. Whether the feedback control affects the size of the stenotele produced remains to be determined.

Similarly gamete formation is probably under more than one control. Ovaries and testes appear only in specific locations along the body column reflecting the possibly influence of the gradients. However, the particular gamete that an interstitial cell will form is controlled by something of a diffusible nature which is clearly unrelated to position (Tardent, 1968). Also, it is unlikely that the induction of gamete formation is dependent on the gradient since the sexual state occurs infrequently and is affected by environmental conditions.

It would appear that the gradient affects the various cell differentiations to varying degrees. Those cell differentiations vital to the structural integrity of the animal which must occur at very specific positions in the body column may be strongly influenced by the gradient. These cells are bound into the tissue either before or at the latest after differentiation. Cell types that arise by interstitial cell differentiation whose numbers can vary somewhat without impairing the structure (nematocytes), or those occurring infrequently (gametes) may be affected by the gradients, but are undoubtedly influenced by other hormones in addition.

How the gradients control growth, except at the extremities where a cell may "read" its position and cease dividing, is not apparent. In fact, it is hard to speculate as to what kinds of controls, much less interrelated controls, govern growth since the overall growth rates of the tissue remain constant regardless of developmental changes. The only discernible need for control of growth is a more subtle one. Somehow the growth rates of the individual cell populations must be orchestrated to maintain the constant cell distribution which exists generation after generation.

This brief sketch indicates that there is a plausible argument to be made that diffusible substances controlling head regeneration also affect cell differentiation. Whether this or other examples of common control correspond to reality awaits further investigation. Until now most of the knowledge gained about gradients, and indications of control by diffusible substances have come from experiments carried out at the tissue level. Questions of hormonal control and common controls must now be asked at a cellular level, and indeed, attention has shifted in this direction.

ACKNOWLEDGMENT

I would like to thank Richard Campbell, Howard Lenhoff, Chica Schaller, and Patricia Bode for many valuable discussions and for reviewing the manuscript.

REFERENCES

Barth, L. (1965). *Biol. Bull.* **129**, 471.
Bode, H., Berking, S., David, C. N., Gierer, A., Schaller, H., and Trenkner, E. (1973). *Wilhelm Roux' Arch.* **171**, 269.

Brien, P. (1953). *Biol. Rev.* **28**, 308.
Brien, P. (1961). *Bull. Biol. Fr.-Belg.* **95**, 301.
Brien, P., and Reniers-Decoen, M. (1949). *Bull. Biol. Fr.-Belg.* **83**, 293.
Browne, E. (1909). *J. Exp. Zool.* **7**, 1.
Bullough, W. S., and Laurence, E. B. (1960). *Exp. Cell Res.* **21**, 394.
Burnett, A. L. (1961). *J. Exp. Zool.* **146**, 21.
Burnett, A. L. (1966). *Amer. Natur.* **100**, 165.
Burnett, A. L. (1968). *In* "The Stability of the Differentiated State" (H. Ursprung, ed.), pp. 109–127. Springer-Verlag, New York.
Burnett, A. L., and Diehl, N. A. (1964). *J. Exp. Zool.* **157**, 237.
Burnett, A. L., Ruffing, F. E., Zongker, J., and Necco, A. (1968). *J. Embryol. Exp. Morphol.* **20**, 73.
Campbell, R. D. (1965). *Science* **148**, 1231.
Campbell, R. D. (1967a). *Develop. Biol.* **15**, 487.
Campbell, R. D. (1967b). *J. Morphol.* **121**, 19.
Child, C. M. (1941). "Patterns and Problems of Development." Univ. of Chicago Press, Chicago, Illinois.
Clarkson, S. G. (1969a). *J. Embryol. Exp. Morphol.* **21**, 33.
Clarkson, S. G. (1969b). *J. Embryol. Exp. Morphol.* **21**, 55.
Clarkson, S. G., and Wolpert, L. (1967). *Nature (London)* **214**, 780.
David, C. N., and Campbell, R. D. (1972). *J. Cell Sci.* **11**, 557.
Davis, L. V. (1966). *Nature (London)* **212**, 1215.
Davis, L. V. (1967). *J. Exp. Zool.* **164**, 187.
Erslev, A. J. (1971). *Amer J. Pathol.* **65**, 629.
Fulton, C. (1959). *Biol. Bull.* **116**, 232.
Gierer, A., Berking, S., Bode, H., David, C. N., Flick, K., Hansmann, G., Schaller, H., and Trenkner, E. (1972). *Nature Biol.* **239**, 98.
Goss, R. (1964). "Adaptive Growth." Academic Press, New York.
Goss, R. (1969a). "Principles of Regeneration." Academic Press, New York.
Goss, R. (1969b). *Advan. Cell Biol.* **1**, 233.
Gross, J. (1925). *Naturwissenschaften* **26**, 73.
Ham, R. G., and Eakin, R. E. (1958). *J. Exp. Zool.* **139**, 55.
Ham, R. G., Fitzgerald, D. C., and Eakin, R. E. (1956). *J. Exp. Zool.* **133**, 559.
Kanaev, I. I. (1952). "Essays on the Biology of Freshwater Polyps." Sov. Acad. of Sci., Moscow.
Kline, E. S. (1961). *In* "The Biology of Hydra and of Some Other Coelenterates" (H. M. Lenhoff and W. F. Loomis, eds.), pp. 153–168. Univ. of Miami Press, Coral Gables, Florida.
Lehn, H. (1951). *Z. Naturforsch. B* **6**, 388.
Lenhoff, H. M. (1966). *J. Exp. Zool.* **163**, 151.
Lenhoff, H. M., and Bovaird, J. (1960). *Exp. Cell Res.* **20**, 384.
Lenhoff, H. M., and Bovaird, J. (1961). *Develop. Biol.* **3**, 227.
Lenhoff, H. M., and Loomis, W. F. (1957). *Anat. Rec.* **127**, 429 (Abstr.).
Lenicque, P. M., and Lundblad, M. (1966a). *Acta Zool.* **47**, 185.
Lenicque, P. M., and Lundblad, M. (1966b). *Acta Zool.* **47**, 227.
Lentz, T. L. (1965). *Science* **150**, 633.
Lentz, T. L. (1966). "The Cell Biology of Hydra." North-Holland Publ., Amsterdam.
Lesh, G. (1970). *J. Exp. Zool.* **173**, 371.
Lesh, G., and Burnett, A. L. (1964). *Nature (London)* **204**, 492.
Lesh, G., and Burnett, A. L. (1966). *J. Exp. Zool.* **163**, 55.

Loomis, W. F. (1954). *J. Exp. Zool.* **126**, 223.
Loomis, W. F. (1964). *J. Exp. Zool.* **156**, 289.
Macklin, M., and Burnett, A. L. (1966). *Exp. Cell Res.* **44**, 665.
MacWilliams, H. K., and Kafatos, F. C. (1968). *Science* **159**, 1246.
MacWilliams, H. K., Kafatos, F. C., and Bossert, W. H. (1970). *Develop. Biol.* **23**, 380.
Müller, W. A. (1969a). *Wilhelm Roux' Arch.* **163**, 334.
Müller, W. A. (1969b). *Wilhelm Roux' Arch.* **163**, 357.
Müller, W. A., and Spindler, K. D. (1971). *Wilhelm Roux' Arch.* **167**, 325.
Muscatine, L., and Lenhoff, H. M. (1965). *Biol. Bull.* **129**, 316.
Park, H. D., Sharpless, N. E., and Ortmeyer, A. B. (1965). *J. Exp. Zool.* **160**, 247.
Park, H. D., Ortmeyer, A., and Blankenbaker, D. (1967). *Amer. Zoologist* **7**, 750 (Abstr.).
Rich, F., and Tardent, P. (1969). *Rev. Suisse Zool.* **76**, 779.
Rose, S. M. (1952). *Amer. Natur.* **86**, 337.
Rose, S. M. (1957). *J. Morphol.* **100**, 187.
Rose, S. M. (1963). *Develop. Biol.* **7**, 488.
Rose, S. M. (1966). *Growth* **30**, 429.
Rose, S. M. (1967). *Growth* **31**, 149.
Rose, S. M., and Powers, J. A. (1966). *Growth* **30**, 419.
Rulon, O., and Child, C. M. (1937). *Physiol. Zool.* **10**, 1.
Schaller, H. (1973). *J. Embryol. Exp. Morphol.* **29**, 27.
Schaller, H., and Gierer, A. (1973). *J. Embryol. Exp. Morphol.* **29**, 39.
Slautterback, D. B., and Fawcett, D. W. (1959). *J. Biophys. Biochem. Cytol.* **5**, 441.
Steinberg, M. S. (1954). *J. Exp. Zool.* **127**, 1.
Steinberg, M. S. (1955). *Biol. Bull.* **108**, 219.
Stiven, A. E. (1962). *Ecology* **43**, 325.
Tardent, P. (1955). *Rev. Suisse Zool.* **62**, 289.
Tardent, P. (1956). *Rev. Suisse Zool.* **63**, 229.
Tardent, P. (1968). *Develop. Biol.* **17**, 483.
Tardent, P., and Eymann, H. (1958). *Acta Embryol. Morphol. Exp.* **1**, 280.
Tardent, P., and Eymann, H. (1959). *Wilhelm Roux' Arch.* **151**, 1.
Tardent, P., Rich, F., and Schneider, V. (1971). *Develop. Biol.* **24**, 596.
Tripp, K. (1928). *Z. Wiss. Zool.* **132**, 476.
Tweedall, K. S. (1958). *Biol. Bull.* **114**, 255.
Webster, G. (1966). *J. Embryol. Exp. Morphol.* **16**, 123.
Webster, G. (1971). *Biol. Rev.* **46**, 1.
Webster, G., and Hamilton, S. (1970). *J. Embryol. Exp. Morphol.* **27**, 301.
Webster, G., and Wolpert, L. (1966). *J. Embryol. Exp. Morphol.* **16**, 91.
Whalen, R. E. (1968). *In* "Perspectives in Reproduction and Sexual Behavior" (M. Diamond, ed.), pp. 303–340. Indiana Univ. Press, Bloomington, Indiana.
Wilby, O. K., and Webster, G. (1970a). *J. Embryol. Exp. Morphol.* **24**, 583.
Wilby, O. K., and Webster, G. (1970b). *J. Embryol. Exp. Morphol.* **24**, 595.
Wolpert, L. (1969). *J. Theoret. Biol.* **25**, 1.
Wolpert, L., Hicklin, J., and Hornbruch, A. (1971). *Symp. Soc. Exp. Biol.* **25**, 391.
Wolpert, L., Clarke, M. R. B., and Hornbruch, A. (1972). *Nature Biol.* **239**, 101.
Zumstein, A., and Tardent, P. (1971). *Rev. Suisse Zool.* **78**, 705.

3

AGGREGATION FACTORS OF MARINE SPONGES

*William J. Kuhns, George Weinbaum, Robert Turner,
and Max M. Burger**

I. Introduction 59
II. Sponge Differentiation and Development 60
III. Sponge Cell Aggregation and Aggregation Factor (AF) 62
IV. Characterization and Properties of Sponge Aggregating Materials .. 66
V. Studies on *Microciona* Sponge Receptor Sites Reactive with AF.. 69
VI. Immunological Studies of Cell–Cell Interactions 71
VII. Sponge Aggregation: Summary 75
VIII. Conclusion 76
References 77

I. INTRODUCTION

Sponges are considered to be the most primitive of the multicellular animals. Their origin is uncertain, possibly from free swimming colonial flagellates, and they diverged early from the main line of metazoan evolution. They have given rise to no other members of the animal kingdom, and in this respect may be considered a dead-end phylum. They are devoid of mouth and digestive cavity and the epidermal layer

* This work was supported in part by NCl Grant CA10151 to M. M. B. and NSF Grant 25170 and USPHS Grant NS-07268 to G. W. G. W. is a Career Development Awardee of the National Institute of General Medical Sciences Grant, 5 K4-07259-04. R. S. T. is a postdoctoral Fellow of the Damon Runyon Fund for Cancer Research, Inc. Work carried out by these authors was accomplished at Marine Biological Laboratories, Woods Hole, Massachussetts.

is poorly developed: the whole body is built about a special water-canal system. Indeed, the discovery of internal water currents within this system first established the animal nature of this species, since early zoologists had expressed doubts in view of the sessile nature of sponges. Most sponge species live in marine environments, but a small number are inhabitants of fresh water (Barnes, 1968).

Why the recent upsurge of interest in sponges? Primarily because as a loose federation of primitive but specialized cells, it is an extremely useful model system for studying specific sorting of cell types into aggregates, and eventually the build up of the latter into differentiated tissues. In the presence of divalent cations, mainly calcium, sponge aggregation is promoted by a soluble extracellular material which can be considered as a primordial type of humoral agent, possibly important in growth and development and amenable to chemical and physical characterization.

The experimental studies on sponge cell interactions are also highly relevant to cellular interactions in higher animals, since there is now considerable evidence for tissue specific aggregation among vertebrate cells, particularly embryonic cells, in which aggregation can be mediated by supernatant fluid from cultured monolayer cells (Lilien, 1968; Garber and Moscona, 1972). In addition, aggregation in certain mammalian cell lines is promoted by soluble materials produced by these lines resembling hyaluronic acid, which reacts with membrane sites (Pessac and Defendi, 1972).

This chapter summarizes present knowledge of materials which promote sponge aggregation, particularly in light of their interaction with receptors on the cell surface, since the latter may be instrumental in developmental organization of cells into tissues and organs. For orientation purposes, a brief statement on sponge development will be presented. Comprehensive treatment of the sponge and its cellular interactions is available from the works of Moscona (1963, 1968), Humphreys (1967, 1970), Lilien (1969), Fry (1970), Webb (1935), Hymen (1940), Brien (1943), Jones (1956), Levi (1957), and Turner and Burger (1973).

II. SPONGE DIFFERENTIATION AND DEVELOPMENT

A schematic diagram of an organized sponge and its cellular arrangement is depicted in Fig. 1. When dissociated sponge cells begin to reaggregate, the initial stage is mainly reassociation of the different cell types. Selective reaggregation according to cell type is suggested by recent work of Leith and Steinberg (1972). The cell suspension was stained with Nile Blue Sulfate, which preferentially stained cells with amoeboid properties. Cells were allowed to reaggregate. The stained amoebocytic cells formed a compact internal mass surrounded by unstained cells in a less compact aggregate.

This process of cellular reassembly proceeds for a time without any appreciable cell mutiplication (Galtsoff, 1925b). Further development consists in the formation of a skeleton and mesenchyme, in the growth of canals, and in the formation of dermal membrane and flagellated chambers (Wilson, 1910, Galtsoff, 1925b).

Archeocytes play a very important part, since they are thought to be embryonic or germinal cells which may differentiate into most other cell types with the possible exception of the flagellated choanocytes or collar cells (Rasmont, 1962; Galtsoff, 1925b).

Sponge tissue can form gemmules, masses of embryonic cells surrounded by a thick protective covering which is later capable of dissolution at a more favorable time, permitting the formation of new sponge tissue from progenitor cells (Rasmont, 1962).

Archeocytes as well as choanocytes possess the power of locomotion and an adhesive cell surface. Both properties may be important in deter-

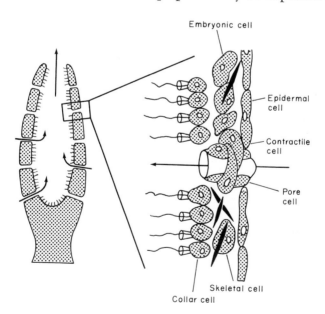

Fig. 1. Organization of a simple sponge. Cross-sectional view showing the flow direction of water (left). Detail of a portion of the body wall (right). (From Paul B. Weisz, "The Science of Biology." Copyright 1959, McGraw-Hill Book Company. Used with permission of McGraw-Hill Book Company.)

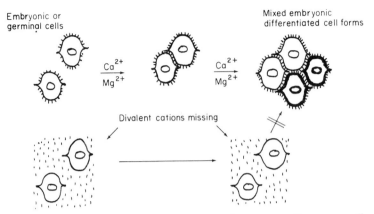

Fig. 2. Simplified scheme depicting importance of AF in specific sponge cell adhesion and differentiation. Small black lines surrounding cells indicate AF; heavily lined cells depict differentiated cell forms. In reality, aggregated cells need to achieve a certain critical mass at optimal temperature before development into mature sponge occurs.

mining their activities which probably include aggregation and differentiation into specialized cell forms (Fig. 2).

Techniques for the laboratory study of sponge tissue explants and sponge cells in culture have been described (Simpson, 1963, Spiegel and Metcalf, 1957, Rasmont, 1961). Under laboratory conditions, outgrowths some 10 mm in size have been observed from explants as early as 4 days (Simpson, 1963). Sponge cells dissociated in seawater quickly reaggregate to form small masses; the latter can form complete sponges within 6–7 days. When cultures are suspended in coarse wire containers from a dock, development is more rapid (Fell, 1967). Conditions which appear to favor sponge reconstitution have been discussed by Fell.

III. SPONGE CELL AGGREGATION AND AGGREGATING FACTOR (AF)

Andrews (1897a,b) observations on the pseudopodial activities of various forms of metazoan cells raised the possibility of a viscous substance extending outward from the cell which by contact with the medium, could encourage contact with like pseudopodia on adjacent cells. Wilson (1907, 1910) was influenced by this concept in explaining his own observations on the coalescence of aggregated sponge cells and their regeneration into mature sponge tissue. The cell types which formed aggregates were (a) large, granular ameboid cells (archeocytes), (b) flagellated cells (collar cells) with ameboid activity, (c) a heterogeneous mixture consisting of cells of different sizes and internal composition. Of these disaggregated cells which had been placed in suspension by mechanical means, (a) appeared to exhibit the most prompt tendency to reaggregate and appeared to play the most important role in regeneration since the aggregates they formed collected other cell types and they also appeared to differentiate into other cell forms following aggregation. When suspensions of cells from *Microciona* and *Lissodendoryx* were mixed, the aggregates which subsequently formed contained only cells of a single species. These findings were confirmed later by Galtsoff (1923, 1925a,b).

Humphreys (1963) and Moscona (1962) working with the sponges *Microciona prolifera* and *Haliclona occulata,* confirmed earlier observations indicating that sponge reaggregation was species specific. They observed that this specificity might be mediated by a soluble material with a high carbohydrate content. Their conclusions were based on experiments in which the cells were dissociated by treatment with Ca^{2+}- and Mg^{2+}-free seawater (CMF-SW). The resultant cells were unable to reaggregate at low temperatures. However, the supernatant from such CMF-SW treated cells contained a labile component, termed aggregation factor (AF), which restored this property in the presence of divalent cations. Increases in temperature to 25°C were also capable of restoring the aggregating property. AF caused adhesion of CMF-SW treated cells fixed in formalin, which indicated that AF effects were mediated through the cell periphery and were not dependent on metabolic activities of the cells (Moscona, 1963, 1968). The activity was felt to be specific in that AF from *Microciona prolifera* or *Haliclona occulata* cells would cause aggregation of only the homologous cells. The capacity of aggregates to reconstitute new sponges, commented upon earlier by Galtsoff, depended upon the ability to form cell groups of a certain size generally about 1.0 mm in diameter and possessing about 2000 cells.

Results of a different nature were reported by MacLennan and Dodds (1967) who prepared aggregation promoting factors from dissociated cells of the following three species: *Hymeniacedon, Halichondria,* and *Ficulina.* They found that aggregation of the cells from each species alone was promoted by *Hymeniacedon* AF, and that in the case of *Halichondria* and *Ficulina* promotion by homologous factors was lesser in degree than the promotion achieved by the cross-reacting factor from *Hymeniacedon.*

In accounting for this apparent discrepancy, MacLennan and Dobbs felt that the closeness of taxonomic relationships between their species might have been responsible for the heterologous reactions, since *Hy-*

meniacedon, Halichondria, and *Ficulina* represent families within an order. On the other hand, *Microciona* and *Haliclona* represent two distinct orders. Therefore, they proposed that cell surface structure differs more markedly between the two species from different orders than among the three species in the same order.

McClay (1971) has recently carried out an autoradiographic analysis to support the idea that sponge aggregation in mixtures will occur in specific fashion if the species employed are sufficiently diverse. Five taxonomically diverse species of the class Dempospongiae were utilized, three within separate orders of the subclass Monaxonida, and two within the subclass Keratosa. Chemically disaggregated cells were tested for their ability to form species-specific contacts using the assay of Roth and Weston (1967). Radioisotope labeled single cells were incubated with preformed aggregates of cells from the same or a different species. The separation of the labeled and unlabeled cells upon reaggregation could then be determined by autoradiography. Labeled cells were observed to initially attach nonspecifically to the aggregates, but within a few hours species specificity was established as indicated by a sharp boundary between labeled and unlabeled heterologous cells and, therefore, aggregation according to species could be demonstrated.

The question of specificity of AF preparations from different species may assume additional dimensions in view of studies of Turner and Burger (1973a,b) on AF preparations from *Microciona* and *Haliclona.* Their experiments were organized such that various concentrations of homotypic and heterotypic AF preparations were tested for ability to aggregate *Microciona* cells. These experiments yielded a graded form of response in which homotypic cell-AF reactions were more efficient than those in which heterotypic reactions were involved, but cross reactions did occur (Fig. 3). The findings of Leith and Steinberg (1972), referred to earlier, suggest specificity according to cell type as well as species as has been demonstrated by Moscona and his co-workers for vertebrate cells (Lilien and Moscona, 1967; Garber and Moscona, 1972; Goldschneider and Moscona, 1972). It seems clear that our thoughts on the exact nature of specificity will require revision as we come closer to understanding the molecular basis for these reactions.

Immunological studies of rabbit antisera prepared against the cell surface of the three species described by MacLennan (1967) supported the closeness of the taxonomic relationship in his experiments, since the sera formed cross-reacting precipitins. On the other hand, antisera prepared by us against *Microciona* and *Haliclona,* cell surface material formed precipitins only against the homologous AF preparation, and no evidence of cross reactivity was observed (Figs. 4 and 5). A paral-

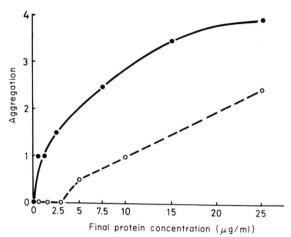

Fig. 3. Aggregation of *Microciona prolifera* cells by *Microciona prolifera* aggregation factor (●——●) and by *Haliclona occulata* aggregation factor (○---○). The original protein concentration of the *Microciona* factor preparation was 750 μg/ml and the *Haliclona* factor preparation contained 150 μg/ml protein, which means that adding 0.10 ml of each factor brings the final protein concentration to 5 μg/ml in the case of *Haliclona* factor and 25 μg/ml in the case of *Microciona* factor. The degree of aggregation indicated was attained in 20 minutes (at 22°C) and there was no significant increase in aggregation during the subsequent 60 minutes. Fifty μg/ml of *Haliclona* factor did cause complete aggregation (4) in 20 minutes. (From Turner *et al.*, 1972).

lelism between specific precipitins against AF preparations on the one hand, and their relative specificities in mediating sponge cell aggregation deserve comment. Thus, sponge specific antisera were able to inhibit re-aggregation of dissociated homotypic cells but not of heterotypic cells (Spiegel, 1954). This could be accomplished in the presence of crude AF-cell mixtures, but not when enriched AF preparations were employed, presumably because titers of antibody directed against AF were insufficient (Kuhns and Burger, 1972). Immunization experiments are currently in progress to verify the effectiveness of high titered preparations in this assay. If confirmed, they would suggest the use of immunological markers for the localization of AF activity on sponge cells. If more than one AF exists within a species, differences between multiple forms could be detected using sera which had first been absorbed with separated Nile blue sulphate positive or Nile blue sulphate negative sponge cells (Leith and Steinberg, 1972). The results would not necessarily indicate that the antigenic determinants were identical to those involved in sponge adhesion, since they could be spatially related but chemically

Figs. 4 and 5. Gel diffusion experiments. Figure 4 indicates in center well, immune rabbit serum #833 prepared against *Microciona prolifera* crude aggregation factor. Peripheral wells contain, starting at top and proceeding clockwise, *Haliclona* AF, *Microciona* AF #1, *Halichondria* AF, *Microciona* AF #2 and *Microciona* AF #3. Clear precipitation bands were formed against all *Microciona* factor preparations, but not against *Haliclona* or *Halichondria* factors. Figure 5 indicates in center well, immune rabbit serum #837 prepared against *Haliclona occulata* crude aggregation factor. Peripheral wells contain, starting at top and proceeding clockwise. *Microciona* AF, empty well, *Halichondria* AF, *Haliclona* AF #1, *Haliclona* AF #2. Precipitating bands formed against *Haliclona* factor preparations, but not against *Microciona* or *Halichondria* factors.

discrete. In any event, the ability of specific antisera to inhibit sponge aggregation suggests the existence of complementarity of the antigen–antibody type between extracellular material and cell receptor sites (Tyler, 1946; Weiss, 1947).

IV. CHARACTERIZATION AND PROPERTIES OF SPONGE AGGREGATING MATERIALS

Chemically dissociated cells are prepared by cutting sponge tissue into small pieces and soaking these fragments in cold (0°) calcium- and magnesium-free seawater (CMF-SW) (Fig. 6). The tissue is placed in fine bolting silk and cells are squeezed out of the matrix into CMF-SW at 0°C; the cell suspension is kept in motion on a gyratory shaker at 4°C. Aggregation factor is prepared by spinning down the cells at low speed and then recentrifuging the supernatant at 12,000 rpm. The clear supernatant solution is termed crude AF and immediately recalcified. It is important to recalcify without delay, since otherwise AF preparations appear to lose their effects rapidly as judged by the poor aggregation properties of preparations to which Ca^{2+} has been added belatedly.

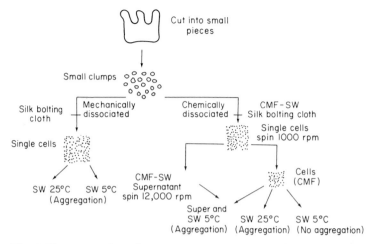

Fig. 6. Extraction of AF from sponge tissue. (See text for explanation.)

An enriched material can be prepared by pelleting the calcified factor for 90 minutes at 110,000 g (0°C) (Humphreys, 1963) and resuspending the pellet in seawater.

The test for AF is carried out as follows: chemically dissociated cells are placed in mixture with putative AF and Ca^{2+} and mixed on a gyratory shaker at 4°. In the presence of AF, visible aggregation is generally observed within 30 minutes. It can be shown that the time required for aggregation and the degree to which this phenomenon proceeds is dose related. Thus 20 μg to 75 μg (as protein) derived from a resuspended 110,000 g mucoid pellet causes aggregation within 15 minutes. Partial and less prompt aggregation follows from the use of crude AF derived from material spun at 12,000 rpm.

Crude or enriched AF loses activity rapidly when stored at room temperature and when incubated for short periods of time at 56°C. Chemical analyses of *Microciona* pellets indicate the presence of hexose, uronic acid, and protein in an approximate ratio of 45:5:50 (Turner et al., 1972). The primary sugars in AF preparations have been reported as galactose, fucose, glucose, hexosamine, and hexuronic acid with a considerable degree of variability between species (MacLennan, 1970). The content of ribose and dexoxyribose is small (i.e., below 5%) (Margoliasch et al., 1965). In preliminary tests, the resuspended 110,000 g pellet which contained AF activity moved as a simple band on sucrose density gradient with a sedimentation constant of 90–100 S. This finding suggested a molecular weight of greater than 5×10^6 (Humphreys, 1965). Examination of *Microciona* and *Haliclona* factors by electron microscopy by Margoliasch *et al.* (1965) has shown roughly spherical particles 20–25 Å in diameter; Humphreys' (1971) examination of *Microciona* factor indicated an arrangements of fibers about 45 Å in diameter arranged in a novel sunburst configuration. Upon prolonged treatment with EDTA, this macromolecular complex dissociated into smaller pieces (Humphreys, 1971). Humphreys feels that removal of calcium produces subunits which can reassociate upon readdition of calcium by bonds involving calcium bridges to form a large structure, the latter providing a functional supramolecular structure for cell to cell adhesion.

The role of glucuronic acid in the biological activity of *Microciona* AF may be especially critical as judged from the destructive effects upon AF by *Helix pomatia* glusulase. These effects are abrogated when the enzyme mixture is mixed with glucuronic acid prior to digestion of AF, suggesting that the effects produced by *Helix* preparations are indeed produced by glucuronidase (Turner *et al.*, 1972). Polymers which contain glucuronic acid, such as hyaluronic acid and heparin, as well as several other monosaccharides, were unable to inhibit the ability of the glusulase to destroy AF activity. Hale stain has been employed to indicate the presence of acid polysaccharides on the surface of mechanically dissociated cells; however such materials were found to be greatly reduced or absent on chemically dissociated cells. Similar findings were observed when stains such as toluidine and Alcian blue were employed. The material which stained as acid polysaccharide was resistant to the action of ribonuclease. The role of sugars in the activity of *Microciona* AF is also suggested by the sensitivity of AF to periodate oxidation (Turner and Burger, 1973a,b).

Studies by Gasic and Galanti (1966) on an AF derived from *Haliclona variabilis* indicated a lower carbohydrate:protein ratio than other workers have found for *Microciona*. Margoliasch *et al.* (1965) compared preparations from *Microciona prolifera* and *Haliclona occulata* and found a lower proportion of carbohydrate:protein in *Haliclona* AF than in *Microciona* AF. However, the amino acid composition of both preparations was essentially similar, particularly high values being found for aspartic and glutamic acids. Proteins and intact disulfide groups appear to possess some importance in aggregation, at least in one species examined (Gasic and Galanti, 1966), since AF activity was sensitive to chymotrypsin, K-zyme and keratin.

The specific role of the polysaccharides in reaggregation has been questioned by MacLennan (1970) who conducted exhaustive study of extracts derived from 17 British species of frozen and thawed sponges or macerated tissue derived from these sponges. Dry residues (DR) exhibited the following ranges of protein and carbohydrate contents:

protein 21–80%, neutral carbohydrate 23–55%, hexuronic acid 2–10%, and hexosamine 4.3–12.9%. MacLennan (1970) indicated that sponge surface antigens contained specificities shared with DR carbohydrates from the same species, and thus the latter material might compete with cell receptor sites for AF, thereby inhibiting its aggregating promoting activity. However, DR carbohydrates proved ineffective as inhibitors of aggregation factor. In considering causes for the negative inhibition tests noted by MacLennan, it is possible that certain spatial arrangements between sugars may be required for biological activity, and that these are disrupted by processing the sponge cells for chemical analyses.

V. STUDIES ON *MICROCIONA* SPONGE RECEPTOR SITES REACTIVE WITH AF

In order to establish how the factor functions one must determine how factor interacts with the cell surface. To this end, Weinbaum and Burger (1971) have isolated a binding site or baseplate with which *Microciona* aggregation factor interacts. This cell receptor site was isolated from chemically dissociated cells by hypotonic shock with 0.08 M NaCl. The shock supernatant accumulated a component which inhibited factor induced aggregation of chemically dissociated cells when preincubated with AF. This baseplate preparation also stimulated aggregation of hypotonically shocked chemically dissociated cells when preincubated with AF. Both responses are probably mediated through a baseplate-factor complex which cannot bind to chemically dissociated cells since baseplate is already present but can be reincorporated into the surface of shocked cells. It has been shown that baseplate can reassociate with hypotonically shocked cells in the absence of AF or Ca^{2+} (Weinbaum and Burger, 1973). Additional studies by these authors have shown that the baseplate–AF interaction is inhibited by glucuronic acid, suggesting that the recognition of AF by receptor site is through the glucuronic acid moiety of AF. Further studies on the baseplate showed that it is destroyed by periodate, it is not sedimented at 110,000 g, it is not removed from whole *Microciona* cells by hypertonic shock, nor is it removed from fragmented cells by hypotonic shock.

Once the baseplate was isolated, the problem of the study of the early molecular events occurring during cell recognition and aggregation was more amenable to experimental probing. For example (Weinbaum and Burger, 1973), AF was chemically bound to Sepharose 4B beads using the cyanogen bromide technique of Cuatrecasas and Anfinsen (1971). Also, baseplate was bound to a separate set of beads by the same technique, thus establishing an acellular system for the investigation of the early reactions in aggregation. The specificity of the system was established by the fact that AF-coated beads only bound chemically dissociated cells, while baseplate-coated beads preferentially bound hypotonically shocked chemically dissociated cells. With such AF-coated beads a number of interesting questions can be asked: (1) Is there preferential association of one cell type with the bead, or do all cell types interact with aggregation factor? (2) Is there an early ultrastructural change in the surface membrane or in the microtubules or microfilaments during cell–bead interaction? (3) If there is an ultrastructural change, is it different when chemically dissociated cells interact with AF beads than when shocked cells bind to baseplate coated beads?

Finally, factor or baseplate coated beads can be examined for their ability to specifically purify the complementary component from a heterogeneous mixture of glycoproteins which may be present in the CMF-SW supernate or in the 0.08 M NaCl shock fluid. Preliminary experiments (Weinbaum and Burger, personal communication) have been done using AF-coated beads in an attempt to purify baseplate from the 0.08 M NaCl shock fluid. When the crude baseplate preparation was passed over a column of AF coated beads, an average of 75% of

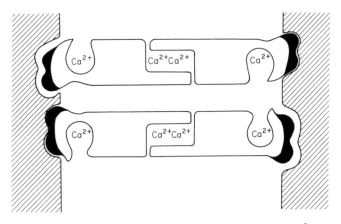

Fig. 7. Working model of the association of sponge aggregation factor with the cell surface. The filled portion of the factor symbol indicates the location of the factor carbohydrates which seem to be involved in the factor-cell interaction. A tight and a loose factor-cell interaction are depicted to account for the fact that mechanically dissociated cells retain factor on their surfaces. Washing the cells in calcium-free seawater is thought to cause a conformational change in the factor which destabilizes the factor-cell interaction by loss of the calcium nearest to the cell surface. The calcium involved in factor–factor interactions is necessary for maintaining the biological activity of factor and for the agglutination of factor beads.

the 280 μm absorbing material was bound to the beads, while an average of 18% bound nonspecifically to untreated Sepharose 4B beads. Biological assay of the column eluates showed that all baseplate activity was bound to the factor column, while none was bound to the control column. The problem presently under study is the development of a specific elution technique to remove baseplate without altering aggregation factor.

In any case a number of new techniques are presently available which may shed light on the problem of cell recognition and early reactions occurring during cellular aggregation. Based upon evidence outlined above, we have proposed the working model described in Fig. 7. This is a tentative model which will undergo revision as new information about sponge reaggregation is obtained.

VI. IMMUNOLOGICAL STUDIES OF SPONGE CELL–CELL INTERACTIONS

Immunological studies of AF using rabbit antisera have been described in Section III. The studies did not relate immunological activity to the chemical content of AF. However, the high carbohydrate content of AF suggested that one or more of its component sugars might possess haptenic or antigenic activity. Lectins or plant agglutinins have been useful in studies on surface membrane carbohydrates, because many lectins contain specific saccharide binding sites (Sharon and Lis, 1972). In view of this, we used lectins to carry out studies to determine whether sponge AF contained sugar groups complementary to sponge cell receptor sites. It was found that the nature of these combinations appear to be similar to those which characterize erythrocyte-lectin combinations resulting in agglutination (Boyd and Shapleigh, 1954).

The lectins tested were not able to agglutinate sponge cells, but in some instances agglutination of indicator red blood cells with lectins could be inhibited with AF. In such cases, the sugars known to react with this particular lectin could be tested for their effects on sponge aggregation by AF.

An examination of lectin-erythrocyte reactions indicated common sugar complementarity in the case of lectins from (a) lentils, green peas, fava beans, and (b) different varieties of field beans, i.e., kidney bean, white navy bean, Great Northern bean, pinto bean, and lima bean (Table I). *Microciona* AF, but not *Haliclona* or *Cliona* AF was capable of inhibiting the reactions between group (a) and jack bean (Con A) lectins and their indicator erythrocytes.

TABLE I
LECTIN PANEL

Origin of reagent	Indicator erythrocytes	Most sensitive erythrocyte	Principal inhibitory sugars
1. Lentils, green peas, fava beans	Human O or Baboon	Human Cord O trypsin or ficin treated[a]	α-Methyl-D-mannose 3-o-Methyl-D-glucose N-Acetyl-D-glucosamine D-Glucose, D-mannose
2. Jack bean meal	Horse Baboon or chicken	Chicken trypsin or ficin treated	α-Methyl-D-mannose N-Acetyl-D-glucosamine D-Glucose, D-mannose
3. Wheat germ lipase	Human Chicken	Human trypsin or ficin treated	N-Acetyl-D-glucosamine N-Acetyl-D-galactosamine (chicken)
4. Castor bean (Ricin)	Human Chicken	Human	D-Galactose, D-fucose 2-Deoxy-D-galactose
5. Kidney bean, white navy bean, Great Northern bean	Human or horse	Horse	N-Acetyl-D-galactosamine
6. Pinto bean	Horse or chicken	Horse	N-Acetyl-D-galactosamine
7. Lima bean	Human A	Human A	N-Acetyl-D-galactosamine
8. *Amaranthus* sp.	Baboon, horse, or human	Horse	N-Acetyl-D-galactosamine
9. Osage Orange	Human, sheep	Human or sheep	D-Galactose, N-Acetyl-D-galactosamine, 2-Deoxy D-galactose
10. *Ulex europaeus*	Human O	Human O	L-Fucose, lactose

[a] Chuba, J. V. (1972). unpublished findings.

TABLE II

INHIBITION BY LECTINS OF REACTION BETWEEN SPONGE CELLS
AND SPONGE CELL AGGREGATION FACTORS

Name of lectin	Name of aggregation factor	Aggregation with sponge cell		
		Microciona	*Haliclona*	*Cliona*
Lentil	*Microciona*	—		
O	*Microciona*	+		
Lentil	*Haliclona*		+	
Lentil	*Cliona*			+

Reactions between the other lectins and their corresponding red cells were not affected by any of the three aggregation factors. Table II indicates that the reaction between *Microciona* cells and *Microciona* AF was inhibited by lentil agglutinin preparations to a titer of 1:16. Extracts from green peas, fava beans, and jack beans (Con A) also proved inhibitory. No other lectins examind were inhibitory for *Microciona* reactions, nor were other sponge species we tested inhibited by these extracts. Since the principal inhibitory sugars for each of these lectins were known, it would be assumed that one or more of these sugars (α-methyl-D-mannose, 3-o-methyl-D-glucose, N-acetyl-glucosamine, D-glucose, and D-mannose) was complementary to AF sugars involved in aggregation.

Figure 8 summarizes the effects of various sugars upon the biological reactivity of Microciona cell AF, and indicates that α-methyl-D-mannose and 3-o-methyl-D-glucose inhibited sponge cell aggregation. The result is in accordance with the inhibition reactions shown by lectin extracts which possessed complementarity for these two sugars. However, it should be noted that a more marked inhibition of AF activity was obtained with cellobiuronic acid and glucuronic acid. AF sugars complementary to these two sugars would not be detected by the use of lectins because there is no lectin presently available which binds cellobiuronic or glucuronic acid.

The limited number of factor specimens examined (two for each species) emphasizes the need for caution in interpretation. However, the results using *Microciona* AF suggest that four lectins, because of their complementarity to terminal or inner sugars, i.e., α-methyl-D-mannose and 3-o-methyl-D-glucose were of use in screening for sugar constitutents necessary for AF activity.

Lectins are ordinarily employed as agglutinators of erythrocytes, and

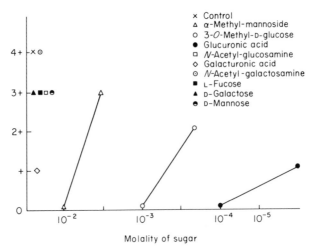

Molality of sugar

Fig. 8. Reagent sugars for inhibition tests were made up as 10% (w/v) stock solutions in phosphate buffered solution at pH 7.3. Appropriate dilutions of the latter were prepared in mixture with sponge cells to achieve a volume of 1.5 ml. and the mixture was stirred on a gyratory shaker at 0°C. To this mixture was added the appropriate quantity of calcified AF. The results of such tests were compared with controls which lacked sugars. Each point is derived from the results of duplicate experiments.

in some cases are able to define discrete forms of blood group specificity depending upon the composition of their heterosaccharide determinants. Thus, extracts of *Phaseolus limensis* possess a specificity for blood groups A_1 and A_1B due to a complementarity for N-acetyl-D-galactosamine, the primary immunodeterminant sugar for blood group A. Similarly extracts derived from *Ulex europaeus* are complementary for L-fucose, the immunodeterminant sugar for group H(O). The ability of *Microciona* AF to inhibit reactions between four lectins and their indicator red cells suggest the presence of AF-like sites upon the latter. In addition, the ability of these lectins to inhibit *Microciona* AF from aggregating *Microciona* cells implies a complementarity between lectin reactive sites and *Microciona* AF. Among the determinant sugars α-methyl-D-mannose and 3-o-methyl-D-glucose were specified by the appropriate lectin inhibition experiments. In addition, glucuronic acid possessed strong inhibitory properties. Therefore, these three sugars, of many examined, appear to be involved in the biological activity of AF. Others as yet untested, may prove to react similarly. This evidence, in addition to that provided by serum inhibition of aggregation (cited earlier, p. 65) support the concept of antigen-antibody complementarity to account for adhesions of AF to sponge cell receptor sites. The possibilities to account for such reactivity might comprise spatial and chemical features, among the latter the various forms of nonconvalent linkages, i.e., hydrogen bonds, charge transfer, etc.

VII. SPONGE AGGREGATION: SUMMARY

The evidence previously cited from homotypic reaggregation, from sorting of cells and from immunological studies all support the argument that specific adhesion of sponge cells is necessary to achieve a degree and arrangement of multicellularity which is conducive to further differentiation and growth. Experiments with chemically dissociated cells indicate that this adhesion may be a cell surface phenomenon mediated by soluble materials capable of promoting aggregation. Thus, a supernatant from chemically dissociated sponge cells will in the presence of divalent cations cause aggregation of the cell suspension from which it was derived. *Microciona* AF will preferentially aggregate *Microciona* cells. *Haliclona* AF will behave similarly in regard to *Haliclona* cells. The extent to which this form of selectivity can be determined may be related to taxonomic differences and possibly to differences between cell types as well. The isotope labeling studies of McClay suggest that the initial stages involve random collisions or contacts with adjacent cells, during which time it may be that cells of like type and species can recognize each other. AF molecules may function based upon multiple combining groups which are complementary to those on cell surface receptor sites and thus may act to bring cells together in a fashion analogous to antigen-antibody interactions. It is not clear yet whether AF plays the role of antigen or of antibody, but the former possibility is suggested on the basis of the lectin studies described above. Weinbaum and Burger have evidence (cited above) that the cell surface contains a receptor molecule for AF which they could release from chemically dissociated cells by hypotonic treatment (1971). Humphreys has found a similar receptor which he isolated after proteolytic treatment of chemically (EDTA) dissociated cells (Humphreys, personal communication, 1972).

Further studies such as those initiated by Leith and Steinberg (1972) may prove that cell sorting occurs during sponge reaggregation. This sorting out process may be due to both random mobility and to quantitative differences in the general adhesiveness of the cells (Steinberg, 1963a,b). The latter could be a function of the number and/or arrangement of cell receptor sites, or of the existence of more than one AF to correspond to the various cell types. The exact chemical nature, both of the receptor molecule and the AF ligand is of key importance to our further understanding of this phenomenon. Preliminary observations indicate that the receptor is also a glycoprotein, as AF, but of much smaller size; while AF has a molecular weight in Ca^{2+} of several millions, the receptor is probably less than 100,000 Daltons (Weinbaum and Burger, personal communication).

Calcium may function as a stabilizing agent for both the AF molecule itself as well as its interaction with the host membrane. If Ca^{2+} is removed from an isolated AF preparation it breaks down into biologically inactive subunits. In this regard, the need to preserve structural integrity of functional combining groups, whatever they may be, will require a careful selection of analytic methods.

VIII. CONCLUSION

Cell adhesion is a selective process, proposed to possess a specific relationship to morphogenesis and tissue differentiation, and assumed to be a fundamental importance in the metamorphosis from genotype to phenotype. Therefore, an understanding of the chemical and physical nature of the binding mechanism may be expected to provide some insight into the molecular basis for tissue and organ formation. Cell binding is influenced by the structure and function of cell membranes and the intercellular material which serves as a cement for adjacent cells. The extent to which these noncellular or humoral substances influence development in cells and tissues is only beginning to be understood.

The important advantages of sponges in these studies deserve emphasis. (1) Multicellularity within a relatively loose organization of embryonic and differentiated cells which are easily disaggregated; (2) sponges are mostly marine organisms; thus, manipulation on the chemical composition of the supporting medium (sea water) is made simpler than similar alterations in the complex media required in the study of vertebrate cells; (3) the effect of such a medium deficient in divalent cations upon sponge cells is to remove a soluble material, aggregation factor (AF), which is then capable of promoting specific cell aggregation. The fact that release of AF takes place when Ca^{2+} and Mg^{2+} are lacking raises the possibility that these ions provide cross links for AF so that AF can maintain structural continuity between cells. Indeed this system may prove to be of great use in elucidating a biological role for Ca^{2+} reactive polyelectrolytes such as exemplified by the mucopolysaccharides (Manery, 1966, 1969).

It is possible that AF may perform membrane functions in intact sponge tissue which in higher vertebrates is attributed to hormones. Thus,

Rasmussen (1970) has suggested a relationship between hormone, hormone receptor sites, calcium ion, and cyclic adenosine monophosphate (cyclic AMP). The latter substance can modulate the activities of many macromolecules, and is generated from adenosine triphosphate (ATP) following hormone stimulation of membrane-bound adenyl cyclase. If such mechanisms are operative in the sponge, the conditions employed for disaggregation such as removing calcium may adversely influence a form of modulation which is concerned with continued morphogenesis.

The following problems require urgent solution before this concept can be profitably applied to sponge growth. (1) The relationship of AF to cell type and more satisfactory approaches to the question whether more than one factor is casually related to reaggregation in a single species; (2) studies of the relationship of chemical structure to biological activity of AF; (3) studies on the relationship of AF to the binding site on the cell surface in order to evaluate specificity on molecular grounds; (4) studies on the exact location of AF *in vivo,* and its changes in relation to the presence or absence of divalent cations; (5) a satisfactory, more quantitative assay to overcome the semiquantitative aggregation assay; (6) systems for the continuous propagation of sponge tissue to minimize the variability of the aggregation system encountered by several laboratories so far. Success in this last area would permit detailed study of progeny of single cells or cell types and would permit observations on the role of AF on this process.

We hope that progress on the molecular basis of sponge cell aggregation may give us some insight and possibly new concepts concerning specific cell recognition in other biologically important processes like tissue and organ formation, morphogenesis, and perhaps even the loss of growth control in interacting transformed cells.

REFERENCES

Andrews, G. F. (1897a). *J. Morphol.* 12, 367–389.
Andrews, G. F. (1897b). *J. Morphol. Suppl.* 12, 1–176.
Barnes, R. D. (1968). "Invertebrate Zoology." Saunders, Philadelphia, Pennsylvania.
Boyd, W. C., and Shapleigh, E. (1954). *J. Immunol.* 73, 226–231.
Brien, P. (1943). L'embryologie des Eponges. *Bull. Musee Roy. Hist. Nat. Belg.* 19, Article No. 16.
Burger, M. (1972). Unpublished findings.
Burger, M., Lemon, S. M., and Radius, R. (1971). *Biol. Bull.* 141, 380.
Cuatrecasas, P., and Anfinsen, C. B. (1971). *Methods Enzymol.* 22, 345.
Curtis, A. S. G. (1970). *In* "Problems and some solutions in the study of cellular aggregation" Biology of the *Porifera*-XXV *Symp. Zoolog. Soc. London* (W. G. Fry, ed.). Academic Press, New York.

Fell, P. (1967). Sponges. *In* "Methods in Developmental Biology" (F. Wilt and N. Wessells, eds.). Crowell, New York.

Fry, W. G. (ed.) (1970). "The Biology of the *Porifera*." Academic Press, New York.

Galtsoff, P. (1923). *Biol. Bull.* **45**, 153–161.

Galtsoff, P. (1925a). *J. Exp. Zool.* **42**, 183–222.

Galtsoff, P. (1925b). *J. Exp. Zool.* **42**, 223–254.

Garber, B., and Moscona, A. (1972). *Develop. Biol.* **27**, 235–243.

Gasic, G., and Galanti, N. (1966). *Science* **151**, 203–205.

Goldschneider, I., and Moscona, A. (1972). *J. Cell Biol.* **53**, 435–449.

Humphreys, S. (1971). *J. Cell. Biol.* **51**, 133.

Humphreys, T. (1963). *Develop. Biol.* **8**, 27–47.

Humphreys, T. (1965). *Exp. Cell Res.* **40**, 539–551.

Humphreys, T. (1967). *In* "The Specificity of Cell Surfaces" (B. D. Davis and L. Warren, eds.), p. 195. Prentice Hall, Englewood Cliffs, New Jersey.

Humphreys, T. (1970). *In* "Biochemical analysis of sponge cell aggregation" Biology of the *Porifera*-XXV *Symp. Zoolog. Soc. London* (W. G. Fry, ed.). Academic Press, New York.

Humphreys, T., Humphreys, S., and Moscona, A. A. (1960a). *Biol. Bull.* **119**, 294.

Humphreys, T., Humphreys, S., and Moscona, A. A. (1960b). *Biol. Bull.* **119**, 295.

Hymen, L. N. (1940). "The Invertebrates," Vol. 1, pp. 284–312. Protozoa through Ctenophora. McGraw Hill, New York.

Jones, W. C. (1956). *Quart. J. Microsc. Sci.* **97**, 269–285.

Kuhns, W. J., and Burger, M. (1972). Unpublished results.

Leith, A., and Steinberg, M. (1972). *Biol. Bull.* (in press).

Levi, C. (1957). *Syst. Zool.* **6**, 174–183.

Lilien, J. E. (1968). *Develop. Biol.* **17**, 657–678.

Lilien, J. E. (1969). *In* Current Topics in Developmental Biology Vol. 4 (A. Moscona and A. Murray, Eds.) "Toward a molecular explanation for specific cell adhesion" Academic Press, New York.

Lilien, J., and Moscona, A. (1967). *Science* **157**, 70–72.

MacLennan, A. P. (1970). *In* "Polysaccharides from sponges and their possible significance in cellular aggregation" Biology of the *Porifera*-XXV *Symp. Zoolog. Soc. London* (W. G. Fry, ed.). Academic Press, New York.

MacLennan, A. P., and Dodd, R. Y. (1967). *J. Embryol. Exp. Morphol.* **17**, 473–480.

Manery, J. F. (1966). *Fed. Proc.* **25**, 1804–1810.

Manery, J. F. (1969). "Calcium and Membranes in Mineral Metabolism" (C. L. Comar and Felix Bromer, eds.), Vol. 3. Academic Press, New York.

Margoliasch, E., Schenck, H. M., Burokas, S., Richter, W., Barlow, G., and Moscona, A. A. (1965). *Biochem. Biophys. Res. Commum.* **20**, 383–388.

McClay, D. (1971). *Biol. Bull.* **141**, 319–334.

Morgan, W. T. J. (1960). *Proc. Royal Soc. B* **151**, 308–347.

Moscona, A. A. (1962). *J. Cell Comp. Physiol. Suppl. 1* **60**, 65–80.

Moscona, A. A. (1963). *Proc. Nat. Acad. Sci.* **49**, 792–747.

Moscona, A. A. (1968). *Develop. Biol.* **18**, 250–277.

Pessac, B., and Defendi, V. (1972). *Nature New Biol.* **238**, 13–15.

Ramussen, H. (1970). *Science* **170**, 404–412.

Rasmont, R. (1961). *Ann. Soc. Roy. Zool. Belg.* **91**, 147–155.

Rasmont, R. (1962). Physiology of Gemmulation in Fresh Water Sponges. *In* "Regeneration for the Study of Development and Growth" (D. Rudnick, ed.). Ronald, New York.

Roth, S. A., and Weston, V. A. (1967). *Proc. Nat. Acad. Sci.* **58**, 974–980.
Sharon, N., and Lis, H. (1972). *Science.* **177**, 949–959.
Simpson, T. L. (1963). *J. Exp. Zool.* **154**, 135–151.
Spiegel, M. (1954). *Biol, Bull.* **107**, 120–148.
Spiegel, M., and Metcalf (1957). *Biol. Bull.* **113**, 356–361.
Steinberg, M. (1963a). *Exp. Cell Res.* **30**, 257–79.
Steinberg, M. (1963b). *Science* **141**, 401–408.
Turner, R. S. and Burger, M. M. (1973a). *Nature New Biology.* (In press).
Turner, R. S., and Burger, M. M. (1973b). *Rev. Physiol.* **68**, 121–155.
Tyler, A. (1946). *Growth* **10**, 7–15.
Watkins, W. (1959). *Vox. Sang.* **4**, 97–119.
Webb, D. (1935). *Quart. J. Microsc. Sci.* **78**, 51–70.
Weinbaum, G., and Burger, M. M. (1971). *Biol. Bull.* **141**, 406.
Weinbaum, G., and Burger, M. M. (1973). *Nature New Biology.* (In press).
Weiss, P. (1947). *Yale J. Biol. Med.* **19**, 235–278.
Wilson, H. V. (1907). *J. Exper. Zool.* **5**, 245–258.
Wilson, H. V. (1910). *Bull. Bur. of Fisheries* **30**, 3–30.

HORMONES IN SOCIAL AMOEBAE

John Tyler Bonner

I. Hormones which Orient Morphogenetic Movements	84
A. Cell Repulsion	84
B. Cell Attraction	85
C. Repulsion of Cell Masses	86
II. Hormones which Affect Pattern	88
A. Spacing Hormone	88
III. Hormones which Affect Differentiation	90
A. Microcyst Inducing Hormone	90
B. Cyclic AMP and Differentiation	90
IV. Hormones which Affect Spore Germination	92
V. Discussion	92
A. A Summary of Slime Mold Hormones	92
B. Timing Sequences of Hormone Actions	93
C. Hormone Controlled Spacing Mechanisms	94
D. Cyclic AMP and the Evolution of Hormonal Control Mechanisms	95
References	97

The cellular slime molds are a curious group of common soil organisms. They were first discovered just over 100 years ago by Brefeld (1869), but their importance in nature and their significance as experimental organisms was not fully appreciated until the pioneer work of K. B. Raper beginning with his first paper in 1935. In that study he described a new species, *Dictyostelium discoideum* which has become the most commonly used organism in experimental studies of development, beginning with his own work (1937, et seq. see Bonner, 1967 for complete references). Since it is essential to understand the rudiments of the life cycle in order to follow the discussion of hormones in their development, let me begin by briefly describing the life cycle of *D. discoideum.*

The spores of this species are capsule-shaped, from 6 to 14 μm long. Each spore contains a single, uninucleate amoeba which emerges after the splitting of the spore case in suitable moist conditions which favor germination. Immediately upon its emergence the amoeba begins to engulf bacteria by phagocytosis; in fact, the amoebae of the cellular slime molds are one of the principal predators involved in the consumption of soil bacteria. The amoeba will soon divide by mitosis to form two separate amoebae; the generation time is about 3 hours under optimal conditions.

When the food supply is totally depleted in one area, the amoebae enter their social phase. They become attracted to central collection points; they aggregate into cell masses. After the aggregation stage this particular species is characterized by a migration stage. The amoebae of *D. discoideum* form a sluglike cell mass which crawls about the substratum for varying periods of time depending upon the environmental conditions (e.g., high humidity favors prolonged migration). Ultimately the migrating slug will stop, right itself, and rise into the air. The anterior third of the slug forms stalk cells, while the posterior two-thirds forms spores. The early signs of this differentiation are already evident in the slug stage, but as the cell mass rises, the mature stalk cells are formed at the very tip, one on top of another, and by a sort of reverse fountain movement the stalk is slowly extended at the apex, at the same time pulling up with it the spores which mature as they rise. The final fruiting body consists of a delicate stalk made up of a tapering cellulose cylinder and enclosing large, vacuolate, thick, cellulose-walled stalk cells and an apical mass of small, dense spores, each one encased in a cellulose capsule. (Fig. 1. For a more complete description and full references, see Bonner, 1967.)

As experimental organisms for developmental studies the cellular slime molds have many advantages. In the first place they separate in time the processes of growth from the other developmental processes. Growth occurs first, and this is followed by a series of morphogenetic movements as well as the differentiation of stalk and spore cells. Not only is there such a separation in time, but also the spatial relations are relatively open to observation and experimentation. This is especially true of the aggregation stage where the cells are spread flat over the surface, and a three-dimensional organism is only produced after the cell mass is

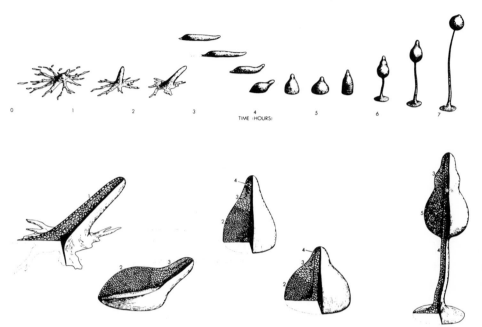

TIME (HOURS)

Fig. 1. Development in *Dictyostelium discoideum*. Above, The aggregation, migration, and culmination stages shown in an approximate time scale. Below, Cutaway diagrams to show the cellular structure of different stages. 1, Undifferentiated cells at the end of aggregation; 2, prespore cells; 3, prestalk cells; 4, mature stalk cells; 5, mature spores. (Drawing by J. L. Howard, courtesy of Scientific American.)

formed. With all these advantages, it is not, as some authors insist, a suitable "model system" for development; it is a developmental process in its own right. If some of the properties of its development have application to other organisms, this is of great interest, and this is equally true of such applications in the reverse directions.

In the discussion that follows of the role of hormones in slime mold development, we will not be concerned with growth itself, but mainly the subsequent stages in the life history. By hormones we mean, in this particular discussion, substances which are given off by the cells themselves, which pass to other cells, and which have a specific effect on the morphogenetic movements or the differentiation of the organism. The word hormone is used synonymously with the very general notion of a chemical signal.

The surprising thing, as I shall show, is that so many of the steps in development of the slime molds are hormone controlled. In some cases we know very much more about the details than others, but I will nevertheless give all of those that we have good reason to suspect at the moment, giving a summary of the details where we have them. We will see that both volatile and nonvolatile hormones play a part, and that the hormones are of key importance for morphogenetic movements, pattern formation, and differentiation. Finally, there is cyclic AMP which acts as a key hormone in the slime molds, both in differentiation and in morphogenesis, and this is of special interest because of its widespread role in the action of mammalian hormones.

I. HORMONES WHICH ORIENT MORPHOGENETIC MOVEMENTS

A. Cell Repulsion

Samuel (1961) was the first to suggest that the amoebae gave off a substance which caused a mutual repulsion, much as Twitty (1949) suggested for amphibian chromatophores. The key experiment of Samuel was to put a few amoebae on agar and place near them a small slip of cellophane (coated with agar) which had a larger concentration of amoebae. The isolated amoebae would soon begin to move away from the cellophane slip. If the slip was then moved to a new location on another side of the isolated amoebae, the latter again turned away from the slip which contained a much larger concentration of cells.

A very strking way to see the phenomenon is to inoculate a streak of bacteria at one end with amoebae. After they consume the bacteria in one region, the amoebae suddenly lose their tendency to stay together in the bacterial streak, and move rapidly, in an oriented fashion, away from the depleted streak. It is assumed that the significance of this phenomenon is to promote the most effective grazing of the bacteria. When a bacterial colony has been devoured it is clearly advantageous to wander far and wide to find more food.

We know that the cells are attracted to bacteria by substances given off by bacteria (Konijn, 1961; Samuel, 1961; Konijn et al., 1969; Bonner et al., 1970), and the question has been asked whether the cells might not be suddenly wandering at random, once the bacteria are gone and can no longer hold the amoebae by their attractants. The answer is relatively simple to establish; the paths of the cells are not what one would expect from random diffusion, but they are oriented away from the other cells.

A good test to discriminate between random diffusion and negative chemotaxis produced by a cell repellent was suggested to me by Dr. Lee Segel. By placing a concentrated drop of cells on an agar surface, one can measure the rate at which the peripheral cells move outward;

if their rate of net outward movement falls off as the square of the distance from the center, then the cells are moving out by straightforward diffusion kinetics. It was easy to show that the rate of outward movement of the cells stays constant, that is, there is a linear relation between outward movement and distance. This rules out random diffusion and decisively supports the presence of a repellent hormone (negative chemotaxis).

One possibility that has been suggested (Bonner et al., 1969) is that the cells give off an attractant (cyclic AMP) and an enzyme that makes the attractant ineffective (a phosphodiesterase which converts the cyclic AMP to 5'-AMP). Since the substrate is a much smaller molecule it will diffuse outward more rapidly than the enzyme and the result will be a ring of high concentration of substrate with an inner disk-shaped area of a lower concentration. This results in an outward gradient of attractant which will orient the cells away from the central concentration of cells. Theoretically this possibility could account for the repulsion phenomenon, but there is a strong reason to suspect that there is also a separate agent given off by the cells of postfeeding D. discoideum that affects the cells directly by negative chemotaxis. At this stage the amoebae are known to produce very little, if any, cyclic AMP and furthermore they are only slightly responsive to it at this stage (Bonner et al., 1969). Therefore, the more likely explanation is that there is a separate repellent for preaggregating amoebae. At the moment nothing is known of its chemical nature.

B. Cell Attraction

The fact that the cells attract each other is the whole basis of the aggregation process (for a review of the early work see Bonner, 1967). It has been shown recently that cyclic AMP acts as an attractant, an "acrasin" (Konijn et al., 1967) and that the amoebae of D. discoideum secrete both cyclic AMP (Barkley, 1969; Konijn et al., 1969) and a specific phosphodiesterase (Chang, 1968).

Many key aspects of the mechanism remain totally obscure, but one can briefly summarize what is known. As was pointed out previously, the vegetative amoebae produce little or no extracellular cyclic AMP in D. discoideum. At the beginning of the aggregation phase there is a great surge of cyclic AMP secretion, and at the same time the amoebae become extraordinarily sensitive to cyclic AMP as an attractant. In some unknown way a center of aggregation is produced and the cells form streams which flow in toward the center. The acrasin is given off both by the center and the streams. Furthermore, it is given off in pulses at the beginning of aggregation. The function of these pulses is not yet clear; we do not know if they are essential to aggregation, or the result of some sort of superfluous oscillation due, for instance, to momentary oscillatory excesses of enzyme and substrate (cyclic AMP), a suggestion made to me by Professor Fritz Lipmann. It has been argued by various authors that the presence of extracellular phosphodiesterase (or acrasinase) produces steeper gradients for attraction by chemotaxis and prevents the total concentration of acrasin from becoming too high. (See Bonner, 1967, 1971, for reviews and detailed references to all these points.)

A number of other species appear to be attracted by cyclic AMP, but there are at least three species (and probably more) that produce it and do not respond chemotactically to it. These species (*Polysphondylium violaceum, P. pallidum,* and *D. minutum*) are also characterized by beginning their aggregation with a founder cell (Shaffer, 1961, Francis, 1965; Gerisch, 1964), a cell that is a keystone of the initial center. We now have some recent evidence which supports Shaffer's (1953, 1957) view that these species have a separate acrasin that is not cyclic AMP (Bonner *et al.,* 1972).

If we look to our list of important, unanswered questions, we want to know the nature of this second acrasin. We also want to understand how cyclic AMP orients a cell. This latter problem is complicated by the fact that there are still many questions which surround the process of amoeboid motion itself. Clearly one wants to observe the effects of cyclic AMP on permeability, and it has been shown by Chi and Francis (1971) that cyclic AMP causes an efflux of calcium in slime mold amoebae. That there is a connection between calcium permeability and cyclic AMP is also supported by an independent study of Mason *et al.* (1971). We do not know anything of the adenyl cyclase in amoebae, nor anything of a kinase which might bind to cyclic AMP. This list of unanswered questions could be greatly extended.

C. Repulsion of Cell Masses

The fact that cell masses repel each other was first observed in *D. discoideum* by Rorke and Rosenthal (1959). If two rising fruiting bodies are close to each other they will diverge as they go up into the air. Even a solitary fruiting body, if placed in an artificial cleft in the agar, will orient so that as it rises, it is equidistant from the agar surfaces. It is presumed that this is a mechanism whereby the fruiting bodies can orient in cavities in the soil or humus in such a way that they will be optimally placed for effective spore dispersal.

The evidence that this phenomenon is the result of a gas, a volatile hormone, was shown most clearly by the fact that if a piece of activated charcoal is placed near a culminating fruiting body, the fruiting body will bend into it. (Fig. 2). (Bonner and Dodd, 1962b). The volatile substance is apparently adsorbed by the charcoal, and therefore there is less of the gas on the charcoal side, and as a result the elongating stalk bends away from the higher concentration on the opposite, or noncharcoal side.

We do not yet know what the volatile substance might be, although Lonski (1973) is presently working on this problem in our laboratory. He has shown that besides CO_2 and NH_3 (the latter known from a previous study of Gregg *et al.*, 1954) ethane, ethanol, acetaldehyde, and ethylene are also given off during the culmination stage; one of these substances might be the active agent in cell mass orientation.

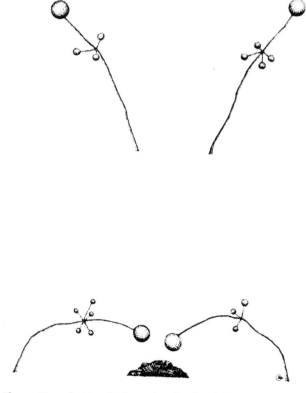

Fig. 2. Above, Two fruiting bodies of *Polysphondylium pallidum* bending away from each other. Below, Two similar fruiting bodies bending towards some activated charcoal. (Drawing by Rogene Gillmor.)

It should be emphasized that the substance is not affecting growth, but in some way speeding up cell movement or cellulose deposition on one side, thereby causing the bending. The growth has occurred before the aggregation stage and culmination is entirely a morphogenetic movement.

Besides spacing in cavities in the soil, this gas-induced orientation also causes spacing between fruiting bodies that arise close to one another, again effecting an optimal distribution for dispersal. Finally, some species are branched, such as members of the genus *Polysphondylium*, and the orientation of each branch at the whorls is presumably governed by this volatile hormone.

II. HORMONES WHICH AFFECT PATTERN

A. Spacing Hormone

In a sense the two repellents described above (cell and cell mass) have an effect on pattern. But here we are dealing with a substance that is even more specific in its pattern effects; it is responsible for the distribution of the aggregation centers with respect to one another.

The initial observation that led to the conclusion that such a hormone existed was the fact that if the amoebae of different species of slime molds were grown on various densities of food, one could get a wide range of corresponding amoeba densities. Nevertheless, the numbers of fruiting bodies per culture dish, or to put it another way, the territory size of each aggregate, remained constant. The only effect of cell density was on the size of the fruiting bodies; the more amoebae per unit area, the larger the fruiting bodies (Bonner and Dodd, 1962a).

Kahn (1968) has pointed out that there are three factors which could affect the spacing pattern: (1) the formation of a center in one area could reduce the number of amoebae in another, thereby producing a nonrandom distribution of centers. This would mean that the timing of center formation would be of key importance, for if they all formed at the same time there would be many, but if they followed one another in a slow sequence, the centers would be few and far between because the first centers would subtract the cells from the outlying areas. (2) The second factor is an inhibitor of some sort produced by the center which diffuses outward and inhibits the formation of closely neighboring centers. (3) There is still another possibility, suggested first by Francis (1965) that a steep acrasin gradient will inhibit a center already formed at the periphery of a well-established center.

Let me say immediately that there is excellent evidence that all three factors are involved; they are not mutually exclusive. As was known from the early work of Harper (1932) and Raper (1940), light produces more numerous, smaller fruiting bodies (smaller territories) and that it causes the centers to appear sooner (and therefore in closer succession in time). Konijn showed conclusively that in *D. discoideum* an older center could cause the disintegration of a younger center close by (an observation first made by Arndt, 1937, on *D. mucoroides*) and that the same process could be exactly duplicated with an artificially applied gradient of cyclic AMP (Konijn *et al.*, 1968).

The second possibility, that there is a special center-inhibiting hormone, is the one that concerns us here. The first clear evidence of such an inhibitor came from the work of Shaffer (1961, 1963) who showed that *P. violaceum* center was initiated by a founder cell. If a founder cell is destroyed, then a number of surrounding cells immediately turn into founder cells. They obviously had been suppressed by the original founder cell. He also showed the inhibition by opposing two layers of agar (so that they touch), one containing established aggregates, and the other with cells about to aggregate. The older centers prevented any centers being founded near them. All the new centers were between the old centers.

We were able to show a similar sort of inhibition using nonfounder cell species (Bonner and Hoffman, 1963). There was evidence that the substance (which we called spacing substance) was volatile. In *D. mucoroides* we could oppose agar surfaces with a 5-mm air space between them, and the new centers never formed near an older center on the opposite side; our results were essentially similar (except for the air space) to those of Shaffer cited above. We speculated that the substance might be CO_2 but subsequently Feit (1969) showed this to be unlikely. He repeated the experiment, but did it two ways: the older centers above (on the ceiling of the agar chamber) and the reverse, with them below (on the floor). The amount of inhibition was far greater when the older centers were below, indicating that the volatile substance was lighter than air and rose. This ruled out CO_2, and for a number of other reasons he suspected NH_3 as the possible spacing substance. There is, of course, no reason to suppose that the spacing substance is the same for all species. It is quite possible, for instance, that those species with founder cells have a different spacing substance.

As before, the *raison d'être* of this kind of spacing hormone seems clear, provided we assume that it is selectively advantageous to disperse spores as effectively as possible. The even spreading of the aggregation centers will produce a regular distribution of the fruiting bodies, which we have already shown will also be spaced, as they rise into the air. The question of whether the spacing substance and the cell-mass repellent are the same volatile substance remains a possibility which hopefully will be determined soon.

III. HORMONES WHICH AFFECT DIFFERENTIATION

A. Microcyst Inducing Hormone

In a number of species, including *P. pallidum,* sometimes aggregation and fruiting body formation are circumvented and the amoebae become directly incorporated individually into small spherical cysts. These microcysts seem to form in some strains more than in others, and they are clearly affected by external physical conditions, such as light.

Recently these have been studied in detail by Lonski (1973) who showed that crowding of the amoebae produced a pronounced increase in the number of microcysts in a particular strain of *P. pallidum*. Since he could produce the same effect in sparse amoeba populations provided they were in the same small enclosed space with a dense population, he concluded that the effect was produced by a volatile substance given off by all the amoebae and which accumulates to a high level in the presence of many cells. One of the volatile substances given off by the amoebae is NH_3 and if physiological concentrations of NH_3 are reintroduced over separate amoebae they can be made to switch entirely into the formation of microcysts.

In this case there are two developmental pathways: the amoebae can go through aggregation and subsequent differentiation into stalk and elliptical spore cells, or the amoebae can go directly to small, spherical, sporelike microcysts. One assumes that depending upon the environmental circumstances, each of these alternatives has its own survival value. The switch that is thrown to choose between these two pathways is perhaps triggered by NH_3 or at least volatile substances given off by the amoebae. Therefore, dense concentrations of the slime mold will automatically favor microcysts, while sparse concentrations will favor true spores held up in the air in masses by delicate stalks.

B. Cyclic AMP and Differentiation

In the normal development of *D. discoideum* there is good evidence that cyclic AMP is involved in differentiation. If the preaggregation amoebae are placed on agar containing a high concentration of cyclic

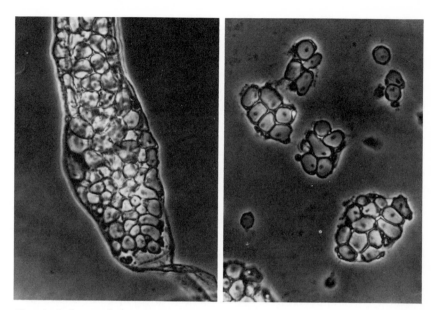

Fig. 3. Left, A phase-contrast microphotograph of the base of a normal stalk on 2% agar. Reading upwards one first sees the slime sheath, then a flat (somewhat one-sided in this case) basal disk, and at the top one can see the stalk partially surrounded by the disc. Right, A group of artificially induced stalk cells on 2% agar containing 10^{-3} M cyclic AMP. Both photographs are the same magnification and both were taken 48 hours after the amoebae were placed on agar. Magnification: the rounded single cell in the upper-right hand corner is approximately 10 μm in diameter.

AMP (10^{-3} M), in 24–48 hours they will form clusters of separate stalk cells directly without ever having passed through aggregation or migration (Fig. 3) (Bonner, 1970). (The effective concentration must be much lower since the cells are constantly secreting phosphodiesterase.) It has been known from some earlier work (Bonner, 1949) that the site of high production of acrasin (and ultimately the only detectable site) is the stalk cell-forming region at the tip. This would suggest that perhaps cyclic AMP is involved in stalk cell formation during normal development. It would be especially interesting to know, in this instance, if the presence of cyclic AMP promotes the activities of any enzymes which are specifically associated with stalk cell differentiation.

There is another less direct link between differentiation and cyclic AMP. As was mentioned previously, it has been shown that certain species, especially species of the genus *Polysphondylium* do not respond chemotactically to cyclic AMP, yet they secrete it and also synthesize a specific phosphodiesterase. It is hard to believe this is a substrate and an enzyme without a function; they must be doing something else in the cell mass, and that something else could be an involvement in differentiation.

There is a related point of interest. *Polysphondylium* is characterized by branches in whorls, and they differ in their predifferentiation stages from *Dictyostelium*. In the latter genus there are early signs of differentiation in the slug stage: there is a clear, histochemically distinct prestalk zone at the anterior end, and a prespore zone at the posterior end (Bonner *et al.*, 1955). This is totally absent in *Polysphondylium:* all the differentiation occurs rapidly at the last moment. Mechanically it is understandable that there must be total reorganization at the last minute since the groups of cells that are left behind on the stalk subsequently separate into the whorls of branches, and each branch is a discrete, minute fruiting body. Therefore, those species which use cyclic AMP for both aggregation and differentiation have slow differentiation which begins early, the reverse is true of those species that do not use cyclic AMP as an aggregation attractant. The timing sequence of developmental events appears to differ in these two groups of cellular slime molds.

IV. HORMONES WHICH AFFECT SPORE GERMINATION

In order to make our list complete, it should be added that there is evidence for a substance given off by the spores which inhibits the germination of the spores (Russell and Bonner, 1960; Cohen and Ceccarini, 1967; Cotter and Raper, 1968). This is a common phenomenon found in bacteria, fungi, and even higher plants, apparently serving the purpose of preventing germination in the spore mass (or seed pod) and allowing it to occur after spore dispersal. Cohen and Ceccarini (1967) have evidence that there are at least two such inhibiting substances (both water soluble) among the different species they tested.

V. DISCUSSION

A. A Summary of Slime Mold Hormones

Perhaps the most striking feature of slime mold development is the variety of extracellular chemical messengers that are known (and we can safely assume there are others which are as yet undiscovered). Slime molds differ from mammals in that maturity is equivalent to total quiescence; the spores are dormant and the vacuolated stalk cells are rigid and dead. Therefore, all its active periods are developmental, and these periods are characterized by a succession of chemical signals.

The chemicals involved are both volatile and nonvolatile. Of the former, NH_3 is one suspect, but there are very likely others as well for the cells give off a variety of volatile substances. Volatile hormones are characteristic of other sessile or slow-moving organisms, the role of ethylene in higher plants being an excellent example.

Of the nonvolatile substances, the only one for which we have direct evidence is cyclic AMP. There is no reason to doubt that there might be large macromolecules involved as well. All we can say is that thus far none have been demonstrated.

The functions of the hormones are numerous. As we have seen, they involve germination inhibition, orientation of individual cells, and cell masses; they are responsible for blocking out the territories of individual aggregations, and they appear to control the switching on (and perhaps off) of different kinds of differentiation. One example of the latter was where a volatile substance, possibly NH_3 will, depending on its concentration, push the development toward microcyst formation, or normal development with stalk cells and elliptical spores. Another example was the direct stimulus of stalk cell differentiation by cyclic AMP.

B. Timing Sequences of Hormone Action

I have already made the point that we want to know what other hormone controlled processes might exist for the development of these organisms, and for each of the ones we have or will discover, we want to learn the mode of synthesis and secretion of the hormone, and more particularly, understand the chemical details of its mode of action. But besides this, there is an even more important goal. It is a goal that involves the whole of the developmental process, the entire fabric upon which it is woven. We must find out how the succession of chemical signals follows in time and space so that a perfect development is achieved.

Translated into more specific terms, this means we need to know what determines *when* a particular chemical signal is emitted, and once given off, we want to know its spatial relations; what causes it to be produced in one place and not another, and how does it produce a specific pattern in the surroundings.

As far as the timing is concerned, we have a few fragmentary bits of information. We know, for instance, that starvation results ultimately in the production of acrasin and the increased reactivity of cells to acrasin. In the case of those species that presumably use cyclic AMP as an acrasin, we want to know further how the chemical changes induced by starvation lead to a shift in the cyclic AMP activities.

The importance of timing is also shown in the case we cited where the role of differentiation follows a different course in time in those species that use cyclic AMP as an acrasin and those that do not. We have a whole series of different chemical steps, each one of which, we assume, can be altered in its rate, and these alterations will give the different characteristics of the various species.

Timing and spacing are not necessarily two separate phenomena, a point made in a most sophisticated fashion by Goodwin and Cohen (1969). They show that theoretically one can produce all sorts of patterns by having oscillating events that might be occurring within cells in a developing system. Cohen and Robertson (1971a, b) have used a variation of this model to suggest that the pulsations of acrasin might be responsible for some of the subsequent patterns of differentiation, an hypothesis that will certainly have to be considered seriously along with others as we proceed with our experiments.

Finally, there is another aspect of timing that should be mentioned. Many of the timed events are geared to environmental changes, especially light. A most obvious example which we discussed is the effect of light on precipitating the onset of aggregation. The fact that the timing of slime mold hormones are geared to environmental changes brings them in line with all other organisms. Consider the effect of day length on sexual activity in most mammals—a situation that is basically very similar.

C. Hormone Controlled Spacing Mechanisms

Turning to spacing, the most obvious cases are between centers of aggregation among masses of slime molds, and in the subsequent orientation of the rising fruiting bodies. But there is an even more important example in the internal division between stalk and spore cells in the slug. We have shown that the ratio is constant (allometric) for a particular species regardless of the size of the slug (review: Bonner, 1967). The question of how this is determined is of great importance. There are many possible hypotheses, including the one of Cohen and Robertson mentioned above, but we have few facts. That the tip should be producing more cyclic AMP does allow for the possibility that again these hormones are in some way responsible for the process, but we have a long way to go to find out if this is so.

D. Cyclic AMP and the Evolution of Hormonal Control Mechanisms

This brings us to the last point. The evidence we have so far from the slime molds seems to indicate that cyclic AMP is itself the hormone, the chemical messenger, while in mammals all those cases that are well understood seem to imply that the chemical messenger exerts its effect through cyclic AMP; in the terms of Sutherland and his colleagues, cyclic AMP becomes the "second messenger", not the primary hormone signal. How are we to reconcile this rather basic difference?

It may be helpful to add here the fact that in bacteria it is known that cyclic AMP acts directly on the genome by combining with a specific protein, which in turn affects the promoter site of the chromosome so that catabolite repression can be removed (review: Pastan and Perlman, 1970). In *Escherichia coli* the cells secrete cyclic AMP, and therefore this removal of catabolite repression can spread from cell to cell, and in this sense cyclic AMP is a bacterial hormone.

In mammals there is excellent evidence that the cyclic AMP also reacts with a protein. Whether or not this protein acts directly on the genome in some cases is not known, but in others it acts to stimulate a cascade of reactions which ultimately will lead to a specific result, such as the conversion of glycogen to glucose-1-phosphate (review: Pastan and Perlman, 1971). Furthermore, in mammals, there is, at least in all known cases, a primary messenger which activates the production of the cyclic AMP.

Consider all these facts from an evolutionary point of view. At all levels we presume that the cyclic AMP combines with a specific protein which, as a result, stimulates some reaction. In the case of bacteria, the cyclic AMP serves as a hormone and directly affects the genome. Starvation seems to be the prime mode of stimulating cyclic AMP synthesis.

The production of cyclic AMP is perhaps achieved by the same mechanism in slime molds, but there it has a variety of effects, some of which are clearly not directly on the genome (such as chemotaxis). Therefore, as evolution has progressed, the specific proteins that can combine with cyclic AMP have increased in number, and each presumably produces a separate effect.

In mammals the process has become even more complex; first because there are a large number of specific proteins which combine with cyclic AMP and carry out special tasks. Second, because the cyclic AMP does not generally travel from cell to cell, but remains inside the cells, and another substance, the *primary* hormone does the external traveling and stimulates a particular group of target cells to synthesize cyclic AMP.

In this case we have a whole new opportunity to produce specific reactions. Apparently adenyl cyclase can differ in various ways, and as a result one cyclase can only respond to one specific primary hormone and no other. There is a cyclase which responds to epinephrine, and another which responds to TSH, and so forth; in each instance cyclic AMP is produced. But since a particular cell type will possess a particular cyclase, then when its cyclic AMP is produced by the primary hormone, the protein which combines with the cyclic AMP can only produce one reaction, and one ultimate end product.

This does raise the question of how the proposed different actions of cyclic AMP act in the slime molds, for cell differentiation is not so far advanced that one has the equivalent of liver or thyroid cells, each with its own cyclase which can react in a specific way. One might imagine that as the cells change with time, passing from the preaggregation state to the aggregating condition, and then finally into the slug condition, that there is a progressive change in the proteins that can react with cyclic AMP; new ones arise, and old ones disappear. In this way cyclic AMP will have different effects at different times in the development of the organism. This hypothesis is consistent with the fact that the amoebae of *D. discoideum* at the end of the vegetative stage show little chemotactic response to cyclic AMP, but at a later stage they respond vigorously.

In the most general terms, we are suggesting that cyclic AMP and its parent enzyme, adenyl cyclase have provided perfect on–off switches for a great variety of different reactions. First they achieved this by having specific proteins localized in specific cells in the body which, when combined with cyclic AMP, would initiate a special reaction. Secondly, the adenyl cyclase itself developed variant forms so that it could be triggered by a certain substance only, and no other. In this way one might account for the astonishing fact that cyclic AMP seems to be involved in such a large diversity of different life processes.

ACKNOWLEDGMENTS

I am very grateful to Dr. Joseph Lonski for his helpful comments and criticisms. All the work done in our laboratory was supported by grants from the National Science Foundation and the National Institutes of Health.

REFERENCES

Arndt, A. (1937). *Roux' Arch. Entwicklungsmech. Organ.* **136**, 681–747.
Barkley, D. S. (1969). *Science* **165**, 1133–1134.
Bonner, J. T. (1949). *J. Exp. Zool.* **110**, 259–271.
Bonner, J. T. (1967). "The Cellular Slime Molds," 2nd ed. Princeton Univ. Press, Princeton, New Jersey.
Bonner, J. T. (1970). *Proc. Nat. Acad. Sci. U.S.* **65**, 110–113.
Bonner, J. T. (1971). *Ann. Rev. Microbiol.* **25**, 75–92.
Bonner, J. T., and Dodd, M. R. (1962a). *Biol. Bull.* **122**, 13–24.
Bonner, J. T., and Dodd, M. R. (1962b). *Develop. Biol.* **5**, 344–361.
Bonner, J. T., and Hoffman, M. E. (1963). *J. Embryol. Exp. Morphol.* **11**, 521–589.
Bonner, J. T., Chiquoine, A. D., and Kolderie, M. Q. (1955). *J. Exp. Zool.* **130**, 133–158.
Bonner, J. T., Barkley, D. S., Hall, E. M., Konijn, T. M., Mason, J. W., O'Keefe, G., III, amd Wolfe, P. B. (1969). *Develop. Biol.* **20**, 72–87.
Bonner, J. T., Hall, E. M., Sachsenmaier, W., and Walker, P. B. (1970). *J. Bacteriol.* **102**, 682–687.
Bonner, J. T., Hall, E. M., Noller, S., Oleson, F. B., Jr., and Roberts, A. B. (1972). *Develop. Biol.* **29**, 402–409.
Brefeld, O. (1869). *Abhandl. Seuckenberg, Naturforsch. Ges.* **7**, 85–107.
Chang, Y. Y. (1968). *Science* **160**, 57–59.
Chi, Y. Y., and Francis, D. W. (1971). *J. Cell Physiol.* **77**, 169–174.
Cohen, A., and Ceccarini, C. (1967). *Ann. Bot.* **31**, 479–487.
Cohen, M. H., and Robertson, A. D. J. (1971a). *J. Theoret. Biol.* **31**, 101–118.
Cohen, M. H., and Robertson, A. D. J. (1971b). *J. Theoret. Biol.* **31**, 119–130.
Cotter, D. A., and Raper, K. B. (1968). *J. Bact.* **96**, 1680–1689.
Feit, I. (1961). Evidence for the Regulation of Aggregate Density by the Production of Ammonia in the Cellular Slime Molds. Ph.D. thesis, Princeton Univ.
Francis, D. W. (1965). *Develop. Biol.* **12**, 329–346.
Gerisch, G. (1964). *Roux' Arch. Entwicklungsmech. Organ.* **155**, 342–357.
Goodwin, B. C., and Cohen, M. H. (1969). *J. Theoret. Biol.* **25**, 49–107.
Gregg, J. H., Hackney, A. L., and Krivanek, J. O. (1954). *Biol. Bull.* **107**, 226–235.
Harper, R. A. (1932). *Bull. Torrey Botan. Club.* **59**, 49–84.
Kahn, A. (1968). *Develop. Biol.* **18**, 149–162.
Konijn, T. M. (1961). Cell Aggregation in *Dictyostelium discordeum*. Ph.D. thesis, Univ. of Wisconsin.
Konijn, T. M., van de Meene, J. G. C., Bonner, J. T., and Barkley, D. S. (1967). *Proc. Nat. Acad. Sci. U.S.* **58**, 1152–1154.
Konijn, T. M., Barkley, D. S., Chang, Y. Y., and Bonner, J. T. (1968). *Amer. Natur.* **102**, 225–233.
Konijn, T. M., van de Meene, J. G. C., Chang, Y. Y., Barkley, D. S., and Bonner, J. T. (1969). *J. Bacteriol.* **99**, 510–512.
Lonski, J. (1973). Evidence for the regulation of development by ammonia, for induction of encystment by methionine starvation and for the presence of a new chemotactic substance in the cellular slime molds. Ph.D. thesis, Princeton Univ.
Mason, J. W., Rasmussen, H., and DiBella, F. (1971). *Exp. Cell. Res.* **67**, 156–160.

Pastan, I., and Perlman, R. (1970). *Science* **169**, 339–344.

Pastan, I., and Perlman, R. (1971). *Nature (London)* **229**, 5–7.

Raper, K. B. (1935). *J. Agr. Res.* **50**, 135–147.

Raper, K. B. (1937). *J. Agr. Res.* **55**, 289–316.

Raper, K. B. (1940). *J. Elisha Mitchell Sci. Soc.* **56**, 241–282.

Rorke, J., and Rosenthal, G. (1959). Influences on the Spatial Arrangements of *Dictyostelium discoideum*. Senior thesis, Princeton Univ.

Russell, G. K., and Bonner, J. T. (1960). *Bull. Torrey Botan. Club.* **87**, 187–191.

Samuel, E. W. (1961). *Develop. Biol.* **3**, 317–335.

Shaffer, B. M. (1953). *Nature (London)* **171**, 975.

Shaffer, B. M. (1957). *Amer. Natur.* **91**, 19–35.

Shaffer, B. M. (1961). *J. Exp. Biol.* **38**, 833–849.

Shaffer, B. M. (1963). *Exp. Cell Res.* **31**, 432–435.

Twitty, V. C. (1949). *Growth, Symp.* **9**, 133–161.

II

PLANT GROWTH AND DIFFERENTIATION

5

HORMONAL CONTROL OF SEXUALITY IN ALGAE

William H. Darden, Jr.

I. Sexual Differentiation in Algae 102
  A. Chemotactic Attraction of Gametes 102
  B. Mating-Type Agglutinins 103
  C. Induction of Sexual Differentiation 103
II. Control of Differentiation in *Volvox* 104
  A. *Volvox* as an Experimental Organism 104
  B. General Description of *Volvox* 105
  C. Biochemistry of Development 107
  D. Variations in Developmental Patterns 107
  E. Susceptibility to Induction 111
  F. Summary, Hypothesis, and Speculation 112
  References 117

The hormonal control of development in lower plants is manifested in a variety of ways. Space does not permit a comprehensive review of even a majority of these systems. With this in mind, the present discussion is limited to hormonal systems which are involved in the control of sexual differentiation in algae. Control systems found in the fungi have been recently reviewed (Barksdale, 1969; Bonner, 1971), and Green (1969; Green *et al.*, 1970) has discussed the physical basis of cell morphogenesis.

The first section of the chapter includes a brief summary of the various types of mechanisms which are known to control sexual differentiation in the algae. The summary is not intended to be a comprehensive review of the literature, but rather is presented to emphasize the variety and levels at which the control of sexual development may operate. For a more comprehensive treatment the reader is directed to the excellent review of Wiese (1969).

The second, and largest, section of the chapter deals in considerable depth with the control of development in the genus *Volvox*. This system was chosen for critical analysis because it is under active investigation in a number of laboratories; it shows exceptional promise as a model system for the elucidation of developmental problems; and it has not been as extensively reviewed as the other major developmental systems.

I. SEXUAL DIFFERENTIATION IN ALGAE

A. Chemotactic Attraction of Gametes

Chemotactic systems where gametes of one sex or mating type secrete chemical substances which attract gametes of the opposite type are found widely distributed among the algae. Such systems are known in *Chlamydomonas* (Tsubo, 1957, 1961; Tschermak-Woess, 1959, 1962), *Oedogonium* (Hoffman, 1960; Rawitscher-Kunkel and Machlis, 1962), *Ectocarpus* (Berthold, 1881), *Fucus* (Cook and Elvidge, 1951; Cook *et al.*, 1948), *Cosmarium* (Brandham, 1967), *Cutleria* (Hartmann, 1950), *Hydrodictyon* (Mainx, 1931), *Colpomenia* (Kunieda and Suto, 1938), and *Sphaeroplea* (Pascher, 1931). There are numerous other algae in which such substances probably exist. These include *Derbesia* (Ziegler and Kingsbury, 1964) and *Caulerpa* (Goldstein and Morrall, 1970).

The chemical nature of most of the chemotactic substances is incompletely known. Tsubo (1961) showed that a variety of substances including carbon monoxide, carbon dioxide, ethylene, acetylene, hydrogen sulfide, and buffers with a pH above 9 or below 5 are attractive to gametes of some species of *Chlamydomonas*. The natural attractant is volatile but is stable when heated at 130°C for 30 minutes in a sealed ampule. The attractant from *Oedogonium* is dialyzable, resistant to drying, freezing and treatment with 4 N HCl. Some activity was retained after boiling for 15 minutes (Hoffman, 1960). Cook and Elvidge (1951) and Cook *et al.* (1948) found the attractant produced by female gametes of *Fucus* was a volatile substance which could be extracted in chloroform or ether. A number of hydrocarbons as well as some ethers and esters showed chemotactic activity. Cook and Elvidge (1951) found that *n*-hexane was the best substitute for the natural attractant. It has recently been demonstrated (Hlubucek *et al.*, 1970) that *n*-hexane is found in the female tips of *Fucus vesiculosus*.

The sperm attractant from female gametes of *Ectocarpus siliculosus* has been isolated (Müller, 1967, 1968), identified (Müller *et al.*, 1971), and synthesized (Jaenicke *et al.*, 1971). It is a hydrocarbon ($C_{11}H_{16}$) with the structure of an *allo-cis*-1-(cyclohepta-2′,5′-dienyl)but-1-ene. The compound is similar to dictyopterenes isolated from *Dictyopteris* (Pettus and Moore, 1970).

B. Mating-Type Agglutinins

When compatible gametes of many algae are mixed, the cells clump together to form aggregates. The aggregation is followed by pairing and subsequent fusion of gametes. The initial clumping is caused by an agglutination of the flagella of gametes of opposite mating type. Mating-type substances which are responsible for the flagella agglutination are known in a number of algae including *Tetraspora* (Geitler, 1931), *Dunaliella* (Lerche, 1937), *Pandorina* (Coleman, 1959), and *Chlamydomonas* (Förster and Wiese, 1954).

The most completely studied algal agglutinins are those of *Chlamydomonas*. The mating-type substances are localized in the flagella (Wiese, 1969). The agglutinins can be isolated from the flagella and when added to gametes of opposite mating type cause isoagglutination (clumping of gametes of the same mating type) (Förster and Wiese, 1954, 1955). The isoagglutinins were purified and characterized as glycoproteins of approximately 10^8 particle weight (Förster *et al.*, 1956). The male (+) and female (−) substances differ somewhat as to protein content and sensitivity to temperature and pH (Wiese, 1969). They also differ with respect to sensitivity to trypsin (Wiese and Metz, 1969) and concanavalin (Wiese and Shoemaker, 1970). The male substance also contains lipid and a volatile component while the female substance contains 3.9% sulfur (Förster and Wiese, 1954; Förster *et al.*, 1956; Wiese, 1961). The sugars are similar in both substances and include rhamnose, galactose, xylose, arabinose, and mannose (Förster *et al.*, 1956; Wiese and Shoemaker, 1970). The minus mating-type substance of *C. moewusi* is sensitive to trypsin, pronase, and sulfatases. It contains most of the amino acids with serine, threonine, and aspartic acid found in highest amounts (Wiese and Wiese, 1971).

C. Induction of Sexual Differentiation

Hormones which initiate gametogenesis or the *de novo* formation of reproductive structures are known in several algae. One interesting complex of hormones is found in the green alga, *Oedogonium*. In addition to the chemotactic substances already mentioned, other hormones are present which are responsible for the directional growth of male filaments and for the delimitation of oogonia (Rawitscher-Kunkel and Machlis, 1962).

Gametogenesis in *Glenodinium lubiniensiforme* (Diwald, 1938) and *Pandorina morum* (Coleman, 1959) can be induced in one strain by cell-free filtrates of the culture medium of the opposite mating type. Gametogenesis is also initiated in a number of other algae only after two different mating types are mixed (Wiese, 1969). Although direct evidence is lacking in most of these cases, it is likely that substances are produced by one strain which induce gametogenesis in the other.

The inducing systems of *Volvox* have been the most thoroughly investigated; they are discussed in detail in Section II.

Little is known regarding the chemical nature of the inducing substances. The substances from *Glenodinium* are inactivated by heat and are pH-sensitive (Diwald, 1938) while those of *Volvox* are believed to be proteinaceous (Darden, 1970; Darden and Sayers, 1971).

II. CONTROL OF DIFFERENTIATION IN *Volvox*

There are more than 15 recognized species of *Volvox* (Smith, 1944; Starr, 1968). The present discussion will be primarily limited to the species which have been most thoroughly investigated relative to the control of their developmental processes. These include *V. aureus,* *V. carteri, V. rousseletii,* and *V. gigas.*

A. *Volvox* as an Experimental Organism

Volvox is a haploid colonial flagellate which is especially suited for developmental studies. It can be grown axenically in a completely defined volvox medium (Provasoli and Pintner, 1959) with a short generation time (3–5 days in most species). *Volvox* exhibits true cellular differentiation but contains only two types of cells, reproductive and somatic. The small number of cell types minimizes cellular interactions and renders the organism more amenable to experimental study and interpretation. Chemical systems which control the differentiation and/or development of *Volvox* are also known in several species. The developmental patterns can therefore be specifically and predictably altered at will. The genus offers a variety of forms and species which makes comparative studies possible. Finally, several mutants have been isolated by R. C.

Starr at Indiana University and E. R. Sayers at the University of Alabama which can be utilized in genetic studies.

B. General Description of *Volvox*

1. Organization of Asexual Colonies

The colonies (coenobia) of *Volvox* consist of a number of biflagellate cells embedded in a gelatinous matrix and arranged as a single layer on the periphery of a spherical or ovoidal colony. The individual cells are chlamydomonad in structure and may or may not be interconnected by cytoplasmic strands. Most of the cells in a colony are somatic and do not function in reproduction. Depending upon the species, from 8–60 enlarged reproductive cells (gonidia) are found in the posterior portion of a colony.

2. Asexual Reproduction

Asexual reproduction occurs inside the parental colony by enlargement and repeated division of gonidia to form a multicellular mass of the cells (the embryo). The embryo is still enclosed within the parent colony at the completion of cell division and consists of a hollow sphere of cells with an opening (phialopore) at one end. At this time the anterior, flagellated ends of the cells are directed toward the interior of the sphere. The normal orientation of the cells is achieved by an inversion process in which the embryo turns itself inside out through the phialopore. This process, first observed by Powers (1908), has been described in detail for a number of species (Darden, 1966; Kochert, 1968; Starr, 1968; McCracken and Starr, 1970; Vande Berg and Starr, 1971). The young inverted colony continues to enlarge, primarily by expansion of the gelatinous matrix in which the cells are embedded. Mature young colonies are eventually liberated by a breakdown of the parent coenobia. Asexual reproduction occurs in both male and female strains in heterothallic species and only under special conditions are gametes formed.

3. Sexual Reproduction

a. Formation of Gametes. Sexual reproduction in *Volvox* is oogamous. Sperm packets are formed by mitotic divisions of single-celled androgonidia (Deason *et al.*, 1969, 1971), while eggs differentiate from undivided reproductive cells. The same type of reproductive cell may function either as a gonidium or gamete initial in some species, while in others special androgonidia and eggs are formed (Starr, 1970).

b. Fertilization and Zygote Germination. Sperm packets are released from males and attach to egg-bearing female colonies. The colonial matrix of the female breaks down in the region of contact (presumably due to some enzymatic secretion from the sperm packet). This produces an opening in the female matrix through which the individual sperm enter the female colony. The actual fusion of egg and sperm has been observed only in *V. carteri* (Starr, 1970); however, genetic data indicate that true fertilization does occur in a number of other species. Thick-walled zygotes are formed and in many species germinate readily.

Meiosis presumably occurs during zygote germination. Three of the meiotic products degenerate, and the single surviving product divides to form a new haploid colony (Zimmermann, 1921).

4. Induction of Gametes

All of the five predominant species discussed here have chemical systems which are known to induce the differentiation of male and/or female gametes. The inducers are prepared by Millipore filtration of medium from sexual cultures (Darden, 1966).

TABLE I
Gamete Induction in *Volvox*

Source of inducer	Induced by filtrate	Reference
V. aureus M5 (homothallic)	Sperm in M5	Darden, 1966
V. carteri f. *nagariensis*		
Male strain	Eggs in female strain Sperm in male strain	Starr, 1969
Female strain	No evidence for production of an inducer	
V. carteri f. *weismannia*		
Male strain	Eggs in female strain Sperm in male strain	Kochert, 1968 and personal communication
Female strain	No evidence for production of an inducer	
V. rousseletii		
Male strain	Eggs in female strain Sperm in male strain	McCracken and Starr, 1970
Female strain	No evidence for production of an inducer	
V. gigas		
Male strain	Eggs in female strain Sperm in male strain	Vande Berg and Starr, 1971
Female strain	No evidence for production of an inducer	

Bioassays used to detect the various inducers are basically similar. Young vegetative colonies are placed into inducer-containing or control medium. After an appropriate period of development, the assays are scored for males and/or females, and the results are recorded as a percentage of the total number of colonies. The assay may be made quantitative by assaying serial dilutions of the inducer-containing medium and observing the dilution at which activity is lost or reduced (Darden, 1970).

Asexual colonies which are exposed to the inducer differentiate gametes, while only asexual structures are found in uninduced controls. A summary of these induction systems is given in Table I.

5. CHEMICAL AND PHYSICAL PROPERTIES OF INDUCERS

I have previously reviewed many of the chemical and physical properties of the various inducers (Darden, 1970), and only a brief summary is included here.

All inducers which have been tested are inactivated by the proteolytic enzyme pronase and are unaffected by DNase and RNase. They are all susceptible to heat inactivation, but some are much more resistant than others. Partially purified inducers are more heat sensitive (Starr, 1970).

Partial purification of inducers has been achieved using lyophylization, ion-exchange chromatography, electrophoresis, and molecular sieve chromatography (Darden and Yarbrough, 1969; Starr, 1970; McCracken and Starr, 1970; Vande Berg and Starr, 1971; Ely and Darden, 1971, 1972).

The molecular weight of the inducers has been estimated using Sephadex chromatography (Darden, 1970; Starr, 1970) and sucrose gradient centrifugation (Holaday, 1969). Some inducers are apparently heterogeneous in size, and molecular weight estimates range from 10×10^3 for the small component in *V. carteri* f. *weismannia* (Kochert, 1968) up to 2×10^6 for the largest component in *V. aureus* M5 (Ely and Darden, 1971). It is not known whether the apparent heterogeneity of some inducers is due to different inducers, aggregates, or inducers bound to some other cell component.

C. Biochemistry of Development

Biochemical changes which accompany development in *Volvox* are poorly understood. Of the few studies which have been done, most involve the synthesis of nucleic acids. The nucleic acids of *V. carteri* f. *weismannia* have been isolated and characterized (Kochert, 1971; Kochert and Sansing, 1971), and we have studied the course of nucleic

acid synthesis during asexual development in V. *aureus* M5 (Tucker
and Darden, 1972). We found that nucleic acid synthesis is maximal
during the 36- to 48-hour stage of development, and that there was
little synthesis of DNA or RNA during gonidial enlargement. This may
well not be the case in species where gonidial enlargement is much
more extensive than it is in V. *aureus* (Starr, 1970).

Hutt and Kochert (1971) found that certain protein and nucleic acid
inhibitors prevent the formation of the fertilization pore in the NB-7
and KA-1 strains of V. *carteri* f. *nagariensis.* Their data suggest that
the synthesis of a digestive enzyme is required for the formation of
the pore and therefore for the subsequent entrance of the sperm into
the female colony.

A new system of male induction has recently been described in V.
aureus M5 (Darden, 1971). In this system the differentiation of male
colonies can be induced by a combination of a *Volvox* extract and calf
thymus histones. It is hoped that a study of this system will yield additional information concerning the biochemistry of induction.

D. Variations in Developmental Patterns

The most striking developmental variations among the five species
discussed here include the number and type of reproductive cells formed,
the time at which they are differentiated and their ultimate fate (Table
II).

1. *Volvox carteri*

The two forms of V. *carteri* which have been intensively investigated
(*nagariensis* and *weismannia*) are heterothallic. Asexual reproduction
in both male and female strains is accomplished by division of gonidia.
New gonidia are differentiated by an unequal cleavage of certain cells
of the developing embryo (Kochert, 1968; Starr, 1969). This results
in the formation of large gonidial initials and small somatic initials.
Gonidia are differentiated at the division of the 16-celled embryo in
V. *carteri* f. *weismannia* (Kochert, 1968) and at the division of the
32-celled stage in V. *carteri* f. *nagariensis* (Starr, 1969).

When young enlarging gonidia of the female strains are exposed to
inducer, the unequal cleavage is delayed by one division, and the large
cells which are formed differentiate into eggs.

Induction of young gonidia of the male strain of V. *carteri* f. *nagariensis* delays the unequal cleavage until the division of the 256-celled stage.
In this case no further divisions occur and the large cells divide to
produce packets of 64 or 128 sperm (Starr, 1969, 1970).

TABLE II
DEVELOPMENTAL VARIATIONS IN *Volvox*

Species	Morphological appearance of reproductive cells	Fate of induced gonidia	Stage of maximum susceptibility to induction	
			Hours from release	Stage of development
V. aureus M5	Just prior to inversion	Male colonies; then sperm	48	Preinversion
V. carteri f. *nagariensis*				
Vegetative	At division of 32-celled embryo	—	—	—
Male	At division of 256-celled embryo	Male colonies; then sperm	0–24	Gonidial enlargement
Female	At division of 64-celled embryo	Female colonies; then sperm	0–24	Gonidial enlargement
V. carteri f. *weismannia*				
Vegetative	At division of 16-celled embryo	—	—	—
Male	At division of 32-celled embryo	Male colonies; then sperm	Unknown	Unknown
Female	At division of 32-celled embryo	Female colonies; then eggs	0–24	Gonidial enlargement
V. rousseletii f. *grizuaensis*				
Vegetative	Just after inversion	—	—	—
Male	Just after inversion	Sperm directly	0–9	Gonidial enlargement
Female	Just after inversion	Eggs directly	0–9	Gonidial enlargement
V. gigas				
Vegetative	Just before inversion	—	—	—
Male	Just before inversion	Sperm directly	0–24	Gonidial enlargement
Female	Just before inversion	Eggs directly	0–24	Gonidial enlargement

Sperm-bearing colonies appear "spontaneously" in *V. carteri* f. *weismannia* and in early experiments could not be induced (Kochert, 1968). Later experiments have shown that under certain conditions induction of male colonies is possible (Kochert, personal communication). The unequal cleavage which results in the differentiation of androgonidia occurs at the division of the 32-celled embryo (Kochert, 1968). The androgonidia subsequently divide and produce packets of 64 or 128 sperm.

2. *Volvox aureus* M5

Volvox aureus M5 is homothallic and dioecious (sexual reproduction occurs within a clonal culture but male and female reproductive structures are formed on separate individuals).

It is unclear as to the exact developmental stage at which gonidia are differentiated, but they are first morphologically discernible in young colonies shortly before their release from parental coenobia. *Volvox aureus* M5 does not form special eggs but rather undivided gonidia function as female gametes.

The exposure of young vegetative colonies to inducer results in the division of their gonidia to form male colonies in the next generation (Darden, 1966). Young male colonies are indistinguishable from vegetative colonies except that they lack gonidia, and all cells resemble somatic cells. The cells in the posterior three-fourths of the colony enlarge and function as androgonidia. The androgonidia divide and differentiate into packets of 32 biflagellate sperm (Darden, 1966).

The most obvious consequence of induction in this form is that the differentiation of gonidia is prevented. The absence of gonidia is accompanied by division of cells (androgonidia) which are morphologically indistinguishable from somatic cells.

It is also possible to induce parthenospores (structures resembling zygotes but formed without fertilization) in some strains of *V. aureus* (Darden, 1968; Starr, 1968; Darden and Sayers, 1969).

3. *Volvox rousseletii* AND *Volvox gigas*

Three types of cells are found in vegetative colonies of the heterothallic species, *V. rousseletii* and *V. gigas;* large reproductive cells, small reproductive cells and somatic cells. Asexual reproduction in male and female clones of both species is similar. Usually only the large cells divide to form new colonies, the smaller reproductive cells developing as somatic cells (McCracken and Starr, 1970; Vande Berg and Starr, 1971). The precise time of differentiation of new gonidia is unclear but they are first evident in *V. gigas* just prior to inversion and in *V. rousseletii* just after inversion.

Unlike the situation in *V. aureus* and *V. carteri* where induction results in the development of gonidia into male or female colonies which in turn produce male gametes, in *V. rousseletii* and *V. gigas* induction leads to gamete formation directly from induced reproductive cells. Egg induction in both forms must occur during early enlargement of gonidia and before division has occurred (McCracken and Starr, 1970; Vande Berg and Starr, 1971). When properly induced, reproductive cells do not divide but enlarge and differentiate into eggs. Although only the large reproductive cells function in asexual reproduction, both types become eggs when female strains are induced. Therefore, as a consequence of induction, not only is division of reproductive cells prevented but also small reproductive cells, which normally remain vegetative, are induced to differentiate into eggs.

Exposure of the male strains of *V. rousseletii* and *V. gigas* to inducer results in the division of large reproductive cells to form globoids, each containing 256 or 512 sperm. Subsequent to the division of these large reproductive cells, the small reproductive cells of *V. rousseletii* and some of the potential somatic cells of *V. gigas* also divide to form smaller packets of sperm. It is interesting that the development of these smaller cells is delayed and does not occur until the division of the larger ones is well underway or complete (McCracken and Starr, 1970; Vande Berg and Starr, 1971).

It should be noted that only 512 sperm are formed from a single large reproductive cell. This is in contrast with asexual division in which $20–30 \times 10^3$ cells are formed in *V. rousseletii* (McCracken and Starr, 1970) and some 2×10^3 in *V. gigas* (Vande Berg and Starr, 1971). Induction thus reduces the number of divisions of a reproductive cell, prevents the differentiation of new gonidia and results in the formation of sperm.

E. Susceptibility to Induction

Early experiments with *V. aureus* (Darden, 1966) and later with other species (Kochert, 1968; McCracken and Starr, 1970; Vande Berg and Starr, 1971) showed that not all developmental stages are equally susceptible to induction. Table II indicates that maximum suceptibility to induction occurs during gonidial enlargement, except in *V. aureus* M5.

Experiments have also been done in which newly released colonies were placed into fresh medium and allowed to develop in the absence of inducer. A sample of these colonies was removed periodically and placed into inducer for various periods of time. Table III summarizes data collected from the female strain of *V. gigas* (Vande Berg and Starr, 1971) and the male strain of *V. rousseletii* (McCracken and Starr, 1970). These data again indicate that very early stages (0–24 hours in *V. gigas* and 0–9 hours in *V. rousseletii*) are most susceptible to induction. It

TABLE III

SUSCEPTIBILITY OF VARIOUS DEVELOPMENTAL STAGES TO INDUCTION

Volvox gigas (female)		Volvox rousseletii (male)	
Stage exposed to inducer (hours)	% Female	Stage exposed to inducer (hours)	% Male
0–120	100	0–15	98
12–120	100	9–24	98
24–120	100	12–27	82
36–120	79	15–72	51
48–120	30	18–72	32
60–120	0	21–72	8
72–120	0	24–72	0

is equally clear however, that the critical developmental events occur after 24 hours in V. *gigas* and after 9 hours in V. *rousseletii*. This is evident from the fact that full induction is possible beginning with 24- and 9-hour colonies which have had no previous exposure to inducer. It is possible that induction affects the synthesis of materials in the undivided gonidium which are active only at later developmental stages.

F. Summary, Hypothesis, and Speculation

Development in *Volvox* is obviously a complex process involving a number of diverse biochemical processes including genetic repression and derepression. It has been proposed, and I believe correctly, that induction in *Volvox* simply initiates a series of developmental events leading to the ultimate differentiation of gametes (Starr, 1970). Starr (1970) has also suggested that the various induction systems in *Volvox* may be basically quite different. This idea, which is based on the diversity of developmental patterns found in the genus, may prove to be correct. However, in my view, it is more attractive to postulate that the fundamental mechanisms underlying cellular differentiation in all *Volvox* species are basically similar. This implies that differences found in the genus are due to modifications of a basic developmental theme rather than to the presence of multiple themes.

A number of model systems could be proposed which would reduce these seemingly diverse developmental patterns to a common basis. This section presents two such models and points out similarities in the development of the various *Volvox* species. The models, although consistent with most of the presently known facts, are purely hypothetical. They are by no means the only models possible and are presented solely in the hope of stimulating discussion and of focusing attention upon possible unifying developmental principles which may exist in the genus.

1. DEVELOPMENTAL SIMILARITIES IN THE GENUS *Volvox*

a. Equivalence of Cell Types. Meiosis in *Volvox* presumably occurs during zygote germination with three of the products degenerating so that a single functional haploid nucleus remains. All cells of a *Volvox* colony (or clonal culture) are believed to be derived from this single haploid product. Therefore, it is to be expected that all cells of a colony, somatic and reproductive, possess the same genetic potential. For example, cells of male colonies should all possess the potential to form sperm and those of females to form eggs. Evidence substantiating this view comes from *V. aureus* M5 and *V. gigas* where apparent somatic cells may also function as androgonidia (Darden, 1966; Vande Berg and Starr, 1971) and from *V. carteri* in which "eggs" may divide to form new colonies (Kochert, 1968; Starr, 1969, 1970). A number of developmental mutants, many due to single gene changes, have been isolated from *V. carteri* f. *nagariensis* which are also consistent with this interpretation (Starr, 1970). In one such mutant, most of the cells of a colony form sperm packs or egg-like cells; other mutants show a decrease in number of androgonidia with a concomitant increase in the number of somatic cells.

b. Asexual Reproductive Cells. Asexual development in male and female strains is quite similar. The first morphological indication of differentiation into reproductive and somatic cell lines occurs in the young embryo where a few cells differentiate as gonidia, the rest developing as somatic cells. The nature of this switch from somatic to reproductive cells is unknown although Kochert and Yates (1970) have shown that the undivided gonidia of *V. carteri* f. *weismannia* apparently contain a UV-labile morphogenetic substance which is ultimately responsible for the differentiation of gonidia.

C. Female Induction

The most striking similarity found in female-inducing systems is that it is apparently impossible to induce reproductive cells to form eggs (or egg-bearing colonies) once they have divided. It appears that division of reproductive cells, or irreversible commitment to division, precludes egg formation.

A primary result of induction in female strains is the failure of reproductive cells to divide. This inhibition of division may occur directly in the induced cells (*V. gigas, V. rousseletii*) or in reproductive cells of the next generation (*V. carteri*). Reproductive cells in which division is prevented usually differentiate into eggs. This is clearly the case in *V. gigas* and *V. rousseletii* where the same reproductive cells can develop either as asexual embryos or as eggs. The situation is somewhat more complicated in *V. carteri* where induction delays the unequal division which produces reproductive cells (Kochert, 1968; Starr, 1969). Therefore, in this case, special eggs are formed. It is clear however, that these eggs are not completely unlike asexual reproductive cells because under certain conditions they can divide to produce asexual colonies (Kochert, 1968; Starr, 1969). Thus even in *V. carteri*, a major effect of induction is to prevent the division of reproductive cells which clearly possess the potential to do so. It seems clear that two developmental pathways are possible for gonidia of female colonies; one pathway leads to the formation of new vegetative colonies while the second results in the ultimate differentiation of female gametes.

2. The Female Model

It could be proposed that inhibition of the asexual division of reproductive cells is the first step leading to their differentiation as eggs. The female model postulates that the function of inducer in all female systems is to prevent the asexual division of reproductive cells which in turn results in their differentiating as eggs. The model proposes that the initiation of asexual division requires a threshold concentration of some initiating substance (designated x). This substance, possibly a gene product, is synthesized during early stages of gonidial enlargement. Induction of females would repress the synthesis of x and prevent the initiation of asexual division.

Variation in developmental patterns could arise due to different concentrations of the initiating substance already present in young reproductive cells at the time of their release (Fig. 1). Young, unenlarged gonidia of the *V. carteri* type are postulated to contain a threshold concentration of x, and induction would not prevent their division. Induction during the enlargement of such gonidia would however, prevent the synthesis of additional x; therefore, the reproductive cells of the next generation would be deficient in x and would differentiate as eggs.

Young reproductive cells of the *V. gigas* type, on the other hand, presumably do not contain a threshold concentration of x and if induced would differentiate directly into eggs (Fig. 1).

The model suggests that removal of inducer could lead to a renewed synthesis of x and the subsequent division of "eggs." Such dedifferentiation is found in *V. carteri* where unfertilized eggs may resume division

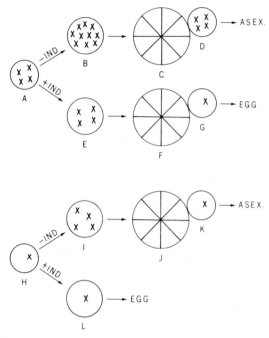

Fig. 1. The female model. Hypothetical scheme for development of female gonidia of the *Volvox carteri* type (A–G) and of the *Volvox gigas* type (H–L). Unenlarged gonidia of *V. carteri* containing threshold concentraton of x (A), and of the *V. gigas* without a threshold concentration of x (H). Additional x is synthesized during gonidial enlargement without inducer (B and I) but not in the presence of inducer (E and L). (C), Embryo before inversion with new gonidium (D) which will develop asexually. (F), Embryo with gonidium which will differentiate as an egg in the presence of inducer. (J), Embryo with gonidium (K) which will develop asexually in the absence of inducer. (L), Enlarged gonidium of *V. gigas* which will differentiate as an egg in the presence of inducer.

and form new colonies (Kochert, 1968; Starr, 1969, 1970). There is also evidence from *V. carteri* f. *nagariensis* that continual contact with inducer prevents this reversion to an asexual state (Starr, 1970). It should be noted that unfertilized eggs of other species have not been observed to divide but careful studies involving removal of inducer have not been conducted.

It is to be expected that any mutation affecting the synthesis of x would have a pronounced effect upon the pattern of differentiation. As has been previously noted, egg differentiation in *V. carteri* f. *nagariensis* usually does not occur in the absence of the inducer produced by the male strain. Starr (1970) has reported the isolation of "spontaneous females" which arose in the female strain in the absence of inducer.

The altered developmental pattern of at least one of these females appears to be due to a gene mutation which is linked to the sex locus. Unlike wild-type females, unfertilized eggs of the mutant strain do not divide when removed from inducer unless the new medium to which they are transferred is supplemented with sodium acetate (Starr, 1970). Some of the clones established from spontaneous females continued to produce only eggs. In such clones, the eggs enlarge and divide to produce new colonies in which eggs are formed and the cycle is repeated in subsequent generations. These strains which continue to produce only eggs could be due to a reduced rate of synthesis of x so that the threshold concentration required for the division of new reproductive cells is not reached during the gonidial enlargement stage, but only much later in the already differentiated egg.

3. THE MALE MODEL

When the reproductive cells of induced male colonies divide, normal gonidia are not differentiated. This effect is common to all of the male-inducing systems described here. Induction may result in the induced reproductive cell functioning directly as an androgonidium (*V. gigas* and *V. rousseletii*) or androgonidia may form in the next generation (*V. carteri* f. *nagariensis* and *V. aureus* M5). In either case, when androgonidia divide, only sperm are produced, somatic and reproductive cells are not formed. Androgonidia could be viewed as reproductive cells which enlarge and divide but do not differentiate new gonidia. On this basis, developing sperm packets would be analogous to rudimentary embryos in which gonidial differentiation is prevented. Concomitant with or as a consequence of the failure of gonidia to differentiate, the cells of the "androembryo" differentiate into sperm. It is worthy of note in this connection that sperm packets, like asexual embryos, undergo an inversion process.

There is some indirect evidence which supports the view that the differentiation of gonidia, either directly or indirectly, results in the suppression of the development of other cells in a colony. Apparent somatic cells of male *V. gigas* and *V. aureus* M5 divide under conditions of induction when gonidial differentiation does not occur. The small reproductive cells of *V. rousseletii* and *V. gigas* do not develop during asexual reproduction; however, they function as eggs or androgonidia under conditions of induction.

On the basis of the male model, induction could not occur once gonidal differentiation had taken place (gonidial differentiation in this sense refers to irreversible commitment to differentiation as well as to the morphological manifestations). Therefore, the developmental stage at which gonidial differentiation occurs should have a profound effect upon the stages at which induction is possible. Forms in which gonidial differentiation occurs early in development would be susceptible to induction only during very early stages before commitment to gonidial formation had been made. Species in which gonidial differentiation occurs later would be susceptible to induction over a longer period of time. Table II indicates that this prediction is satisfied by the species being considered. The differentiation of gonidia in V. *carteri*, V. *rousseletii*, and V. *gigas* apparently occurs quite early in development and induction is possible only during gonidial enlargement (Table II). In V. *aureus* M5 where gonidial differentiation is apparently somewhat later (gonidia not evident until shortly before release) maximum induction is possible up to 48 hours of development.

Variation as to whether sperm are formed directly from induced reproductive cells or in the next generation could be explained in a fashion analogous to the female model. The gonidia of some species presumably are already committed somewhat to asexual division (V. *aureus* and V. *carteri*), while others are not (V. *gigas* and V. *rousseletii*). Under optimum conditions the gonidia of none of these species are already committed to the differentiation of new gonidia. Possible exceptions to this generalization may be found in V. *carteri* under suboptimal conditions (Kochert, 1968).

In final summary, the genus *Volvox* seems to offer excellent material for the study of cellular differentiation. Two of the most promising lines for future investigation appear to be studies of the biochemical changes accompanying development and the genetic analysis of developmental mutants. It is hoped that through such investigations, developmental principles will be elucidated which are applicable not only to *Volvox* but to other forms as well.

REFERENCES

Barksdale, A. W. (1969). *Science* **166**, 831–837.
Berthold, G. (1881). *Mitt. Zool. Stat. Neapel* **2**, 401–413.
Bonner, J. T. (1971). *Annu. Rev. Microbiol.* **25**, 75–92.
Brandham, P. E. (1967). *Can. J. Bot.* **45**, 483–496.
Coleman, A. W. (1959). *J. Protozool.* **6**, 249–264.
Cook, A. H., and Elvidge, J. A. (1951). *Proc. Roy. Soc.* **B138**, 97–114.
Cook, A. H., Elvidge, J. A., and Heilborn, J. (1948). *Proc. Roy. Soc.* **B135**, 293–301.
Darden, W. H. (1966). *J. Protozool.* **13**, 239–255.
Darden, W. H. (1968). *J. Protozool.* **15**, 412–414.
Darden, W. H. (1970). *Ann. N.Y. Acad. Sci.* **175**, (Art. 2), 757–763.
Darden, W. H. (1971). *Biochem. Biophys. Res. Commun.* **45**, 1205–1211.

Darden, W. H., and Sayers, E. R. (1969). *Microbios* **2**, 171–176.

Darden, W. H., and Sayers, E. R. (1971). *Microbios* **3**, 209–214.

Darden, W. H., and Yarbrough, J. D. (1969). *Int. Bot. Congr. Abstr., 11th* p. 41.

Deason, T. R., Darden, W. H., and Ely, S. (1969). *J. Ultrastruct. Res.* **26**, 85–94.

Deason, T. R., Darden, W. H., and Ely, S. (1971). *In* "Contributions in Phycology" (B. C. Parker and R. M. Brown, eds.), pp. 67–79. Allen Press, Lawrence, Kansas.

Diwald, K. (1938). *Flora (Jena)* **132**, 174–192.

Ely, T. H., and Darden, W. H. (1971). *ASB Bull.* **18**, 32.

Ely, T. H., and Darden, W. H. (1972). *Microbios* **5**, 51–56.

Förster, H., and Wiese, L. (1954). *Z. Naturforsch.* **9B**, 548–550.

Förster, H., and Wiese, L. (1955). *Z. Naturforsch.* **10B**, 91–92.

Förster, H., Wiese, L., and Braunitzer, G. (1956). *Z. Naturforsch.* **11B**, 315–317.

Geitler, L. (1931). *Biol. Zentralbl.* **51**, 173–187.

Goldstein, M., and Morrall, S. (1970). *Ann. N.Y. Acad. Sci.* **175** (Art. 2), 600–678.

Green, P. B. (1969). *Annu. Rev. Plant Physiol.* **20**, 365–418.

Green, P. B., Erickson, R. O., and Richmond, P. A. (1970). *Ann. N.Y. Acad. Sci.* **175** (Art. 2), 712–731.

Hartmann, M. (1950). *Pubbl. Staz. Zool. Napoli* **22**, 1–9.

Hlubucek, J. R., Hora, J., Toube, T. P., and Weedon, B. C. L. (1970). *Tetrahedron Lett.* **57**, 5163–5164.

Hoffman, L. (1960). *Southwest. Natur.* **5**, 111–116.

Holaday, J. W. (1969). Masters Thesis, Univ. of Alabama.

Hutt, W., and Kochert, G. (1971). *J. Phycol.* **7**, 316–320.

Jaenicke, L., Akintobi, T., and Müller, D. G. (1971). *Angew. Chem. Int. Ed.* **10**, 492–493.

Kochert, G. (1968). *J. Protozool.* **15**, 438–452.

Kochert, G. (1971). *Arch. Biochem. Biophys.* **147**, 318–322.

Kochert, G., and Sansing, N. (1971). *Biochem. Bophys. Acta* **238**, 397–405.

Kochert, G., and Yates, I. (1970). *Develop. Biol.* **23**, 128–135.

Kunieda, H., and Suto, S. (1938). *Bot. Mag. Tokyo* **52**, 539–546.

Lerche, W. (1937). *Arch. Protistenk.* **88**, 236–268.

Mainx, F. (1931). *Arch. Protistenk.* **75**, 502–516.

McCracken, M. D., and Starr, R. C. (1970). *Arch. Protistenk.* **112**, 262–282.

Müller, D. G. (1967). *Planta* **75**, 39–54.

Müller, D. G. (1968). *Planta* **81**, 160–168.

Müller, D. G., Jaenicke, L., Donike, M., and Akintobi, T. (1971). *Science* **171**, 815.

Pascher, A. (1931). *Jahr. Wiss. Bot.* **75**, 551–580.

Pettus, J. A., and Moore, R. E. (1970). *J. Chem. Soc. D* 1093–1094.

Powers, J. H. (1908). *Trans. Amer. Microsc. Soc.* **28**, 141–175.

Provasoli, L., and Pintner, I. J. (1959). *In* "The Ecology of Algae" (C. A. Tryon and R. T. Hartman, eds.), pp. 84–96. Spec. Publ. No. 2 Pymatung Lab. of Field Biol., Univ. of Pittsburgh.

Rawitscher-Kunkel, E., and Machlis, L. (1962). *Amer. J. Bot.* **49**, 177–183.

Smith, G. M. (1944). *Amer. Microsc. Soc.* **63**, 265–310.

Starr, R. C. (1968). *Proc. Nat. Acad. Sci. U.S.* **59**, 1082–1088.

Starr, R. C. (1969). *Arch. Protistenk.* **111**, 204–222.

Starr, R. C. (1970). *Develop. Biol. Suppl.* **4**, 59–100.

Tucker, R. G., and Darden, W. H. (1972). *Arch. Mikrobiol.* **84**, 87–94.
Tschermak-Woess, E. (1959). *Planta* **52**, 606–622.
Tschermak-Woess, E. (1962). *Planta* **59**, 68–76.
Tsubo, Y. (1957). *Bot. Mag. Tokyo* **70**, 327–334.
Tsubo, Y. (1961). *J. Protozool.* **8**, 114–121.
Vande Berg, W. J., and Starr, R. C. (1971). *Arch. Protistenk.* **113**, 195–219.
Wiese, L. (1961). *Fortschr. Zool.* **13**, 119–145.
Wiese, L. (1969). *In* "Fertilization, Comparative Morphology, Biochemistry, and Immunology" (C. B. Metz and A. Monroy, eds.), pp. 138–188. Academic Press, New York.
Wiese, L., and Metz, C. B. (1969). *Biol. Bul.* **136**, 483–493.
Wiese, L., and Shoemaker, D. W. (1970). *Biol. Bul.* **138**, 88–95.
Wiese, L. S., and Wiese, L. (1971). *J. Phycol. Suppl.* **7**, 12.
Zeigler, J. R., and Kingsbury, J. M. (1964). *Phycologia* **4**, 105–116.
Zimmerman, W. (1921). *Jahrb. Wiss. Bot.* **60**, 256–294.

HORMONAL CONTROL OF ROOT GROWTH

D. Mertz

  I. Introduction 121
 II. Root Organization 122
     Apical Meristem 126
III. Nutritional Requirements of Root Growth 125
 IV. Development of the Hormone Concept 127
     A. Cholodny–Went Theory 127
     B. Geotropism 130
     C. Root Elongation 131
  V. Vascular Differentiation 137
     A. Control of the Vascular Arrangements 137
     B. Secondary Vascularization 139
     C. Primary Vascularization 140
 VI. Concluding Remarks 142
     References 143

I. INTRODUCTION

Root development has never been unequivocally demonstrated to be under the control of endogenous growth regulators supplied by the root apex; therefore, I accepted the assignment of writing a chapter on the chemical control of root development with some trepidation. Furthermore, a perusal of some of the recent treatises on plant growth and development leads one to conclude that a great deal of the early work concerning the participation of the root apex as a source of auxins in controlling root growth and tropic responses has fallen into disrepute.

Nevertheless, very little new information is available to entirely justify such an attitude. With the recognition in recent years that the root tip synthesizes gibberellins and cytokinins, in addition to auxins, there is some justification to reexamine some of the early studies.

The present chapter is not intended to be a comprehensive analysis of all aspects of root development. Such an analysis was essentially made recently by Torrey (1965). This chapter will deal principally with the areas of growth which tends to support the concept that hormone relations exist between the root tip and the remainder of the root and between the aerial portion of the plant and the root proper.

II. ROOT ORGANIZATION

Of the various plant organs the root has been selected most frequently for developmental studies. The principal reasons for selecting the root are that it is a heterotropic organ in which its terminal development lacks complicating lateral appendages and stages of differentiation are aligned in a linear sequence from the apex to the base.

Although the root has been extensively described in morphological and anatomic terms (Esau, 1965), a brief consideration of the development will be presented. The terminology used here is essentially that used by Esau (1965).

Plants differ from animals in possessing formative tissues, the meristems, which add new cells to the plant body, and at the same time perpetuating themselves as such. The root meristem originates early in embryogeny with the differentiation at the suspensor end of the embryo axis of an apical meristem. A few cells in the apical meristem divide regularly and repeatedly; these cells are the apical initials of the meristem. Cells which are produced distal to the apical initials differentiate rapidly into the root cap while others remain meristematic. Cells which are produced proximal to the initials, elongate, enlarge radially and add mature tissue to the root.

The immediate derivatives of the meristematic cells enlarge and divide for various lengths of time during the differentiation of tissues; as a consequence, it is difficult to delineate the meristem proper from its most recent derivatives. Because of this difficulty, the initiating cells and their most recent derivatives are often termed the *promeristem*. The subadjacent meristematic tissues are then segregated according to the tissue systems that are derived from them. The *protoderm* differentiates into the epidermis. The *procambium* gives rise to the primary vascular cylinder or central stele, which consists of the pericycle and the primary xylem and phloem. The *ground meristem* is the precursor of the cortical tissue system which consists of the exodermis, if present, cortical parenchyma and endodermis. The *protoderm, ground meristem,* and *procambium* constitutes what are termed the *primary meristems* and therefore give rise to *primary tissues.*

The sequence of differentiation of the primary tissues have been carefully studied in a number of roots (Heimsch, 1951; Torrey, 1953; Esau, 1940, 1941, 1943). While the relative sequence of differentiation is in an acropetal direction, the rate of differentiation and maturation of tissues differs widely depending upon the rate at which the root is growing (Heimsch, 1951). In rapidly growing roots the mature primary tissues may not have fully developed until 10–15 mm behind the apex while in slow growing roots mature tissue may be found within a few millimeters of the root tip.

The first cells proximal to the apical meristem to exhibit signs of differentiation are the cells of the procambium. Cells begin to vacuolate in a radial pattern proceeding centrifically and proximally producing poles of the protoxylem leaving arcs of densely protoplasmic cell between the poles. The arcs of cells are the future phloem regions. In the xylem the first cells to complete their differentiation are the outer ones, the protoxylem, while the inner ones are the first to show the beginning of differentiation. Xylem maturation therefore proceeds in a centripetal direction. In the phloem region the first cells to vacuolate are the protophloem sieve tubes near the outside of the stele followed by further vacuolation toward the center. The sieve tubes mature rapidly after vacuolation starts and closer to the apex than do vessel elements. Mature protophloem may differentiate between 200–800 μm from the apical meristem while mature protoxylem matures much further behind the meristem. It is in this general area of protoxylem maturation that root hairs of the epidermal cells develop, arising after cell elongation has ceased. Near or at about this same region the mature endodermis is evident, possessing the suberized Casparian strips on the radial and transverse walls. Moving further back from the tip the first branching roots appear, originating from the pericycle generally opposite the xylem strands (see Mallory *et al.,* 1970). The vascular cambium, if present, will be found still further back.

Apical Meristem

Root apices lack leaf primordia and other lateral appendages; therefore, unlike the shoot apex, there is no periodic swelling and contraction, and the root is not divided into nodes and internodes. As a consequence,

the cellular pattern of organization and behavior of the root meristem
is better understood than the shoot meristem. Among the lower vascular
plants, the pteridophytes, there are species which have a single apical
initial such as *Marsilea* and *Azolla* (Clowes, 1961). Other pteridophytes
and all of the seed bearing plants have multicellular promeristems. In
the multicellular meristem a great deal of interest has centered around
the question of the number of apical initial cells. Over the years two
concepts have evolved (see Clowes, 1961; Esau, 1965). According to
one concept the apical initials comprise a very small number or, frequently, only one cell from which all primary tissues develop. According
to a second concept the apical meristem is made up of many initials
with different orientations and arrangements. The latter concept arose
when it was recognized that neither the root nor shoot of angiosperms
possessed a prominent apical cell. The recognition that primary tissues
arise from different initials had earlier led Hanstein (1868) to formulate
the concept of three cell layers or histogens. The plerome, periblem
and the dermatogen were terms applied to designate the meristematic
tissues of the procambium, cortex, and epidermis respectively. Later
Janczewski (1874) proposed a forth histogen, the calyptrogen, which
gives rise to the root cap.

Brumfield (1943) attempted to resolve the differences between the
two concepts. By irradiating germinating seeds of *Vicia faba* and *Crepis*
with sublethal doses of X-ray, he was able to induce chromosomes ab-
berations in the apical meristem. By following subsequent development
of cells with abberant chromosomes he concluded that all primary tissue
developed from possibly as few as three initials. Brumfield, therefore,
concluded that the histogen concept was incorrect.

More recently Clowes (1961) concluded from a series of studies on
the pattern and planes of division of cells within the apical meristem
that a large part of the center of the apical meristem was not truly
meristematic. The center consisted of a hemispherical group of cells
that divided infrequently, which he termed the *quiescent center*. Such
a radical change in the concept of the apical meristem led to a series
of investigations of the turnover rate of DNA in the apical meristem.
In studies using tritiated thymidine and autoradiography it was very
clearly demonstrated that the cells within the quiescent center of *Zea
mays* incorporated virtually no radioactive thymidine during a 24-hour
exposure (Clowes, 1959). Only cells on the periphery of the quiescent
zone incorporated thymidine in the synthesis of DNA during mitosis.
The boundary between the quiescent center and the dividing cells was
less well defined on the proximal side than on the distal side. On the
distal side the boundary corresponded with the boundary between the

root cap initials and the rest of the root. The irregularity of the boundary on the proximal side indicated that there are numerous cells contributing to the development of the primary tissues.

The quiescent center has been confirmed in root apices of *Vicia faba* and *Allium cepa* by an analysis of the mitotic frequency and DNA distribution (Jensen, 1958; Jensen and Kavaljian, 1958). Clowes (1971) has further characterized the quiescent center in *Zea mays* by a careful analysis of cell divisions by pulse labeling with tritiated thymidine. The quiescent center in the maize root contains approximately 600 cells, of these cells about 270 divide; however, only 100 divide once in approximately 40 hours. Mitosis, therefore, contributes only some 60 cells per day from the quiescent center at 23°C. Two hundred micrometers proximal to the quiescent center encompass the rest of the meristem which contains approximately 125,000, of which 100,000 have a high rate of division of once every 14 hours. Thus this area of the meristem contributes the vast majority of the cells to the basal portions of the root. The root cap meristem just distal to the quiescent center contains 6600 cells of which 4400 divide every 10.4 hours, thus the root cap is replaced every day. The root cap in dicotyledenous plants appears to be replaced at a much slower rate (Phillips and Torrey, 1971). The discovery of the quiescent center has led to a general acceptance of a meristem made up of many initials and is therefore more in keeping with Hanstein's histogen theory.

Although the presence of a quiescent center has been confirmed in a number of different plants (Byrne and Heimsch, 1970), very little is known of its regulatory role in controlling tissue differentiation. Yet as will be seen later this region of the root plays an important role in regulating tissue pattern formation and the synthesis of plant growth regulators.

III. NUTRITIONAL REQUIREMENTS OF ROOT GROWTH

For the most part roots are terrestrial and nonphotosynthetic; therefore, the root is dependent upon the aerial photosynthetic portion of the plant for an energy supply, generally in the form of sucrose (Street, 1957). In addition to an energy source the root also requires all the known macronutrients (N, P, K, Ca, etc.) and micronutrients (Mg, Fe, Mn, Mo, Zn, B, etc.) essential to the plant. These are supplied either by way of the root hairs in the maturing region of the root or possibly they are absorbed directly by the tissues near the tip (Torrey, 1965).

The root is not only heterotrophic for an energy supply but it also is usually unable to synthesize several vitamins or other growth substances for continuous and normal growth. White (1934) was the first to demonstrate that tomato roots could be isolated from the parent plant and continuously grown in a nutrient medium containing inorganic salts, sucrose and extracts of brewer's yeast.

Once it was recognized that the root could be successfully grown apart from the parent plant a method was available for an analysis of those factors supplied by the aerial portion of the plant. Bonner (1937) showed that vitamin B_1 or thiamine, a component of yeast extract, was essential for growth of isolated pea roots. It was also found that B_1 was required for the growth of tomato roots (Robbins and Bartley, 1937; White, 1937). In the absence of thiamine, cell divisions in the apical region of the root ceases (Addicott, 1939) thus the root is dependent upon a continuous supply of the vitamin from the above ground, green portion of the plant. In a classical sense vitamins supplied by the aerial portion of the plant can be considered as growth hormones.

Certain roots have, however, been reported to be able to synthesize thiamine (McClary, 1940; Street, 1957). Recently Neales (1967) reported that excised roots of a strain of *Arabidopsis thaliana* did not require thiamine.

Soon after the recognition of the requirement of B_1 for excised pea and tomato roots, nicotinic acid and pyridoxine (vitamin B_6) were identified as growth factors for isolated pea and tomato roots (Addicott and Bonner, 1938; Robbins and Schmidt, 1939). Boll (1954) demonstrated that all three pyridoxine homologs—pyridoxine, pyridoxal, and pyridoxamine—could be utilized by tomato roots.

Our understanding of the nutritional requirements for isolated monocotyledenous roots is rather limited since continuous and unlimited growth has been successfully achieved with roots of only one plant—the rye plant (Roberts and Street, 1955; Almestrand, 1957). Continuous growth was obtained by supplying excised roots with yeast extract. Roberts and Street (1955) demonstrated that 80% of the growth promoting activity of yeast extract was caused by the content of L-tryptophan. The failure to obtain continuous growth of isolated monocotyledenous roots results from the inability of the meristem (promeristem) to continue to divide; although cell elongation may continue for some time after excision from the parent plant.

The vitamin requirement for cultured monocotyledenous roots is far from clear. Almestrand (1957) found that thiamine had no effect upon growth of barley, oats, wheat, and rye roots. Wheat roots, however, responded to pyridoxine by promoting cell division for a limited period of time. Neither barley nor oats responded to pyridoxine. Almestrand was unable to detect any biotin response in any of the roots studied. Bonner and Bonner (1948) had previously noted the unresponsiveness of roots to biotin.

In most cases where roots have been continuously cultured over an extended period of time, the roots exhibit essentially the same pattern of tissue differentiation as roots developing on seedlings. Thus isolation of the root from the parent plant does not significantly alter the genetic pattern of development.

IV. DEVELOPMENT OF THE HORMONE CONCEPT

A. Cholodny–Went Theory

The fundamental concept that plant growth is controlled by chemical agents or hormones stems from the classical work of Charles Darwin in 1880. He noted that the tips of both roots and shoots were the site of geotropic and phototropic perception and that the stimulus was transmitted basipetally where it induced a growth response. Over the next few years a large number of investigators characterized these two tropic responses which led to the isolation of the stimulus or hormone. The tropic studies culminated in the classical Cholodny–Went theory around 1928.* This theory simply stated that growth curvatures induced either by internal or external factors were caused by an asymmetrical distribution of the growth substance (auxin†) between the two sides of the curving organ. The opposite response of the root and shoot to a gravitational stimulus was considered to be caused by a supraoptimal concentration of the growth hormone on the lower side of the root causing growth inhibition on that side. Elongating cells on the upper side of the root caused a positive geotropic response.

The isolation and identification of indole-3-acetic acid (IAA) as one of the native hormones has led to an overwhelming body of evidence to support the Cholodny–Went theory when the tropic responses in stems are considered (Wilkins, 1966; McCready, 1966). However, in spite of a span of over 40 years there is still some uncertainty as to the precise role IAA plays in the geotropic response in roots.

---

* Early references relating to the development of the Cholodny–Went Theory are found in the book "Phytohormones" by Went and Thimann (1937).

† Auxins are organic substances which at low concentration (less than 0.001 M) promote growth along the longitudinal axis, when applied to shoots of plants freed as far as possible from their own inherent growth promoting substance but normally inhibit root elongation (Thimann, 1969).

At the close of the last century several investigators noted that decapitation of roots did not significantly influence their subsequent rate of
growth. But Cholodny (see Went and Thimann, 1937) was able to demonstrate an acceleration of root growth following decapitation in *Zea mays*
and *Lupinus angustifolius* if precautions were taken to prevent water loss
by spraying the roots during the experimental period. He subsequently
demonstrated that if the decapitated root was capped with a root tip
or coleoptile, tip growth was inhibited. He concluded that both root
and shoot tips secreted a substance which inhibited growth. This sensitivity of root growth to auxin supplied by the tip was interpreted to
indicate that the root tip supplied auxin to the elongating cells in amount
supraoptimal for root elongation.

Several studies followed that supported the Cholodny–Went theory.
Keeble *et al.* (1931) demonstrated that when tips from geotropically
stimulated maize roots were stuck on to the stump of decapitated, unstimulated roots, geotropic curvature results. Hawker (1932) found that
geotropically stimulated roots yield more auxin by diffusion into gelatine
blocks from the lower half of the root tip than from the upper half.
She further demonstrated that if the agar block was placed asymmetri-
cally upon the stump of unstimulated roots a curvature resulted toward
the agar block, thus indicating a basipetal transport. A study by Boysen-
Jensen (1933) confirmed Hawker's findings. Cholodny (1934) subsequently demonstrated that when corn coleoptile tips were applied to
the apical or basal cut surface of decapitated roots and geotropically
stimulated, the stumps with tips on their apical ends curved geotropically
while the basal application did not cause curvature.

Although the early studies based on biological assays clearly indicated
basipetal transport of auxin, more recent studies have resulted in conflicting views. Hertel and Leopold (1963) demonstrated that acropetal
transport of [^{14}C]IAA-1 in 2-mm maize root segments was less than
one-half the basipetal transport. The transport was partially inhibited
by 2,3,5-triiodobenzoic acid, a metabolic inhibitory of auxin transport.
Nagao and Ohwaki (1968) using intact seedlings of *Vicia faba* demonstrated that by ringing the root with lanolin containing IAA (50 μg/gm)
growth was always inhibited basipetal to the site of application, thus
indicating a basipetal transport of auxin.

Yoemans and Audus (1964) were unable to demonstrate a basipetal
transport. They applied [^{14}C]IAA-1 in agar donor blocks to the apical
and basal ends of 4-mm segments cut 4 mm from the root tip of *Vicia
faba*. After 1–6 hours, agar receivers and transverse sections of the segments were assayed for auxin. Essentially no radioactivity was detected
in the receiver blocks; however, the radioactivity in the segments indi-

cated a preferential transport toward the apex. They found no evidence for an active transport. In similar studies with *Lens culinaris* roots, it was found that after 4 hours twice as much ^{14}C had accumulated in the receivers at the apical end of segments as in receivers at their base (Pilet, 1964). Kirk and Jacobs (1968) noted a much stronger acropetal than basipetal transport stream in *Lens* and *Phaseolus* roots.

In a series of studies with 6- and 12-mm root segments excised 1 mm below the apex a group at the University of Nottingham in Great Britain also demonstrated a preferential transport of [^{14}C]IAA (Wilkins and Scott, 1968a,b; Scott and Wilkins, 1968; and Wilkins and Cane, 1970). The acropetal transport was sensitive to anoxic condition and sodium fluoride and was therefore metabolically controlled. They were able to demonstrate a slight basipetal flux of auxin only at low temperature; however, even at low temperature basipetal flux was considerably less than acropetal flux.

The failure to determine the extent to which IAA is lost by oxidation or immobilization during transport may be a partial explanation of the conflicting reports of auxin transport in roots. Goldsmith and Thimann (1962), stressed the destruction and immobilization of IAA as a source of error in transport studies. Iversen and Aasheim (1970), and Aasheim and Iversen (1971) have clearly shown an enzymatic destruction of [^{14}C]IAA-1 at the cut surfaces during transport in sunflower and cabbage roots. Although they demonstrated a preferential acropetal translocation of IAA, they were able to show a significant increase in the basipetal flux in sunflower root segments cut 5 mm from the apex when $^{14}CO_2$ was trapped as $Ba(CO_3)_2$ in the receiver blocks. When decarboxylation of [^{14}C]IAA-1 was taken into account the ratio of acropetal to basipetal flux was between 1.3 and 2.4. Wilkins and Scott (1968a) found a ratio of 9.0 in sunflower root segments excised 1 mm behind the apex. Although they were not dealing with precisely the same segments it nevertheless appears that they recovered only a small fraction of the basipetally transported auxin.

Wilkins and Scott (1968b) reported an increase in basipetal transport under anoxic condition. This increase may indicate that an IAA oxidase was inhibited, particularly at the basal end of the segment, thus permitting a greater accumulation in the receiver block. In actual measurements they have calculated that the area of the basal cut surface of the 6-mm segments in maize was almost 3 times greater than the apical surface. The weak basipetal transport could be related to a greater destruction of IAA associated with the greater surface area.

Hillman and Phillips (1970) studying transport in 5-mm root segments of *Pisum sativum* excised from 2-day-old dark grown seedlings also demonstrated a preferential acropetal movement; however, a considerable basipetal flux was recorded. They also noted that up to 70% of the IAA supplied to the donor block was taken up by the root segments but only 2–3% was recovered in the receiver block. They concluded that during passage through the tissue the majority of the auxin was metabolized. Kirk and Jacobs (1968) noted that only a small percentage (0.30) of total auxin added to the donor block reached the receiver block in *Lens* roots.

Aside from the question of auxin destruction and immobilization in transport studies much of the controversy centering around the polarity of auxin movement in roots may have resulted in part from the failure to delineate the region of the root studied. Although there is no question that there is a strong acropetal transport of auxin, particularly in the older more mature regions of the root (Bonnett and Torrey, 1965; Morris *et al.*, 1969), a weak basipetal transport is still evident.

B. Geotropism

It has been known for 100 years that the removal of the very tip of the root prevented the geotropic response (Ciesielski, 1872); however, it has only recently been unequivocally established that the cells of the columella—central cells of the root caps—are able to perceive geotropic stimuli. The amyloplasts within the cells of the columella serve as the geosensors. The surgical removal of the cap from the roots of *Pisum sativum* (Konings, 1968) and *Zea mays* (Juniper *et al.*, 1966), or the digestion of starch grains from cap cells of *Lepidium sativum* (Iversen, 1969) induced by hormone treatment prevented the perception of gravitational stimuli. The removal of the root cap did not affect the growth in length of the roots regardless of the orientation of the root. Only after regeneration of the major part of the root cap did response to gravity return. These findings indicate that the root cap is either the site of the gravi-perception mechanism, the source of growth-regulating substances involved in a geotropic response, or both.

Gibbons and Wilkins (1970) have demonstrated that removing half of the root cap resulted in curvature toward the remaining half of the cap regardless of the orientation of the root with respect to gravity. A half-cap on the lower side of a horizontally oriented root, however, gave rise to a much greater curvature (approximately doubled) than a half-cap on the upper side of a horizontal root. Several possibilities are suggested from these observations. A horizontally placed root cap produces an inhibitor which is released more from the lower side than from the upper side, a preferential lateral transport of the inhibitor in response to gravity, or an increase in the sensitivity of the cell on the lower side over the cells of the upper side of the root. Although it is not possible to distinguish between these possibilities it is of some interest to note that it has been demonstrated that geotropically stimulated cells of *Helianthus* hypocotyles are more responsive to and are able to absorb more exogenously applied auxin than unstimulated cells (Brauner and Diemer, 1971). It is, therefore, conceivable that the cells of the lower side of the root are more sensitive to the influence of the inhibitor.

It was not demonstrated by Gibbons and Wilkins whether a cap-inhibitor concentration gradient was established between the upper and lower halves of a horizontal root; however, the concentration of plasmodesmata on the transverse wall of cells of the columella and the junction between the cap initials and the quiescent zone could provide a longitudinal channel of communication from the cells of the columella where the stimulus is perceived to the elongation zone of the root proper (Juniper and Barlow, 1969; Juniper and French, 1970).

That auxin is one of the factors responsible for geotropism of roots is suggested by the findings that low levels of IAA inhibit root growth by inducing the formation of the growth regulator ethylene (Chadwick and Burg, 1967). Chadwick and Burg (1970) have concluded that the geotropic curvature is related to an IAA-dependent ethylene production on the lower side of the root.

In what way the root cap-inhibitor participates in geotropism is not known; however, it is possible that the cap-inhibitor could regulate auxin synthesis in the meristem. If the cap-inhibitor is laterally and basipetally displaced following geotropic stimulation it is conceivable that it could increase auxin synthesis on the lower side of the root. Increased auxin levels on the lower side could induce ethylene biosynthesis resulting in the inhibition of cell elongation, thus resulting in geotropic curvature.

Until which time it has been unequivocally established that there is no basipetal transport of auxin in the apical few millimeters of root tips, the Cholodny–Went theory is still the most plausible explanation for geotropism in the root. As will be seen later a preferential acropetal movement is however undoubtedly one of the controlling factors in acropetal polar differentiation.

C. Root Elongation

Root elongation measurements have been the most frequently made because of the ease and speed of the determination. However, as has been pointed out by Torrey (1956) only in rare cases is it possible

to measure directly root growth which results wholly from cell enlargement. Rather, in the usual situation, mitosis, cytokinesis, cell elongation,
and cellular differentiation are inextricably interrelated. Although few
studies over the years have delineated what phase of root growth is
specifically affected by plant growth regulators, it seems to be reasonably
clear that auxins, gibberellins, and cytokinins are required for root
growth and development. It is also clear that these same growth regulators are synthesized in the root tip.

1. AUXIN

Thimann (1937) was the first to suggest that root, bud, and stems
all behave in a comparable way, their growth being inhibited by relatively high concentrations and promoted by relatively low concentrations
of auxin. The difference between the three organs was considered to
be of a quantitative rather than a qualitative nature. On this basis the
response of different organs to auxin, a series of optimal curves of similar
shapes, was presented. Root elongation was promoted in the neighbor-
hood of 10^{-11}–10^{-10} M concentration. Van Overbeek (1939) provided
evidence that auxin was produced in isolated pea roots cultivated *in
vitro* under sterile conditions. The presence of auxin in root tips has
now been firmly established; however, there is still some doubt whether
these always include IAA. In *Vicia faba* the main auxins appeared to
be substances different from IAA (Lahiri and Audus, 1960; Burnett
et al., 1965). Street (1969) concluded that IAA was not present in the
tomato roots. He was unable to detect IAA in sterile root cultures and
concluded that IAA does not occur; however, a number of substances
with activity in conventional auxin bioassays were present, several of
which were indolic compounds. Recently, Dullaart (1970) has demonstrated that IAA is the main auxin produced in root nodule tissue in
yellow lupine infected with *Rhizobium*. He concluded that a substantial
part of the IAA produced was synthesized by the plant enzymes as
a result of a metabolic alteration induced by the rhizobial infection.

Robbins and Hervey (1969, 1971) demonstrated that excised roots
of *Bryophyllum calycinum* required both auxin and cytokinin for continuous growth. Two millimeter root tips grew very poorly in a basal
medium of mineral salts, sucrose, and vitamins supplemented with either
an auxin or cytokinin; however, in presence of both hormones significantly greater growth resulted. Since large inocula could be successfully
grown in the absence of both hormones, it was concluded that both
auxin and cytokinin were synthesized in the root. The poor growth

of 2-mm root tip segments was considered to be caused by too little hormone synthesis or the hormones were leached into the culture medium resulting in an unfavorable hormone balance.

In view of the fact that auxin must ultimately act on the cell wall, one of the most extensively studied effects has been the change in the physical properties of the cell wall following auxin treatment (Heyne, 1940; Ray, 1969; Celand, 1971). The major action of auxin in facilitating cell expansion is to increase the elasticity and plasticity of the cell wall which is followed by water uptake resulting from a decrease in the water potential of the vacuolar sap (Ray and Ruesink, 1963). The first phase of cell elongation starts with auxin-stimulated dissolution of the cell wall which becomes elastically and subsequently plastically extensible. Elasticity is a reversible expansion while plasticity is a nonreversible wall deformation thought to be caused by the breaking of crosslinks between the cellulose microfibrils of the cell wall (Lockhart, 1965). The first phase of cell elongation is followed by water uptake, wall expansion, and a deposition of new wall material by intussusception of new cellulose microfibrils. The extreme sensitivity of root elongation to exogenously applied auxin is considered to be a consequence of the inhibition of the duration of cell elongation rather than the inhibition of the initiation of cell elongation (Burström, 1969; Morré and Bonner, 1965), possibly through auxin-induced synthesis of ethylene (Chadwick and Burg, 1967, 1970). Recently, Burström *et al.* (1970) have shown by means of a resonance frequency instrument measuring Young's modulus of elasticity (E) that IAA causes a decrease in elasticity (decrease in E means an increase in extensibility) 2–3 minutes before cell elongation in pea internode sections. Because of technical difficulties they were unable to measure quantitatively the change in elasticity of the pea root; however, the addition of auxin caused a decrease in elasticity within 15 minutes in a region where IAA was assumed to initiate elongation. Root segments excised 4 and 6 mm from the tip that were undergoing cell elongation exhibited no measurable effect to auxin as far as the elastic modulus was concerned.

Nissl and Zenk (1969) have demonstrated that the lag period associated with cell elongation of oat coleoptiles following auxin treatment can be shortened to zero by increasing the temperature and auxin concentration. It therefore appears that the primary site of auxin action causing the initial cell elongation is associated with a *preformed enzyme* system which mediates cleavage and reformation of wall polymers. These observations suggest that IAA initiates cell elongation in both root and stems via the same mechanism.

Although IAA appears to be required for the initiation of cell elongation in root cells and under normal conditions seems to limit the duration of elongation, much less is known concerning the effect of IAA upon organized apical meristems, particularly in the root. It is well known that IAA is essential for cell division in lateral meristems and in unorganized callus growth; however, it is generally impossible under normal conditions to increase cell multiplication in the root apex by the addition of auxin (Burström, 1968). In a study concerning the effect of the herbicide, Barban (4-Cl-2-butynyl-*n*-3-Cl-phenylcarbamate) on the growth of wheat root, Burstrom has shown that the herbicide causes a rapid inhibition of cytokinesis and cell elongation. The inhibitions of both cytokinesis and cell elongation spontaneously reversed itself after 24–48 hours. The reversion of the meristem inhibition was enhanced however by both thymidylic acid and IAA. The reversal of cell elongation was promoted by IAA and 6-furfurylaminopurine (kinetin). Since the primary effects of Barban was to block cell division at metaphase, it was suggested that blocking of cell division leads to a reduction in IAA, kinetin, and possibly other metabolites. Burström concluded that the initiation of cell elongation requires both auxin and kinetin supplied by the meristem or produced in the cells as a consequence of mitosis preceding the onset of elongation.

Barlow (1969) has shown that IAA in the presence of hydroxyurea—an inhibitor of DNA synthesis—induced a sharp rise in the growth rate of the cells of the quiescent center and in protoxylem cells immediately proximal to the quiescent zone in the maize root tips. These data, although indirect, suggest that auxin is synthesized in the apical meristem possibly as a sequel to cell divisions. By contrast, however, the cells of the cap initial showed no response to IAA. Whether these cells require IAA for elongation and differentiation is not known.

2. GIBBERELLINS

Since the report of Brian *et al.* (1954) that cress roots showed no growth response to GA_3, relatively few studies have appeared that were directly concerned with gibberellin-induced root growth. Kato (1955) reported that gibberellin at 0.1–5 mg/liter resulted in a slight growth stimulation of roots of seedlings of *Allium fistulosom*. Whaley and Kephart (1957a, b) reported that GA_3 stimulated the growth of excised maize roots with maximum growth obtained by a combination of IAA and GA_3. Butcher and Street (1960a) demonstrated that GA_3 had rather profound effects upon growth in length of tomato roots. At appropriate concentrations both GA_3 and NAA enhanced the main axis growth of tomato root grown in culture medium containing sucrose at concentrations below 1%. Both GA and NAA increased the number of emergent laterals at low sucrose content.

That gibberellins are involved in root growth finds further support in the work of Richardson (1958). He noted that 5 mg of GA_3 per liter applied to seeds of *Pseudotsuga menziesii* caused a significant increase in root elongation. Burström (1960) found that GA_3 decreased cell multiplication and elongation in wheat roots grown in the dark, while in the light multiplication was only slightly decreased but the rate of elongation was strongly increased. Raghavan and Torrey (1964) noted that GA_3 was more effective than auxin in promoting elongation of roots in light and dark grown embryos of *Capsella bursa-pastoris*. Paleg *et al.* (1964) showed that gibberellins promoted root growth of lettuce seedlings. Especially active was GA_6, which at a concentration of 1 ppm caused roughly a 40% increase in root elongation. Mertz (1966) demonstrated that isolated 10-mm root tips of a gibberellin-deficient dwarf mutant of maize (d_1d_1) cultured in White's medium increased in growth by 80% over the control following GA_3 treatment while their normal sibling (Dd_1) was much less responsive.

It has been unequivocally established that the root tip produces gibberellins. Jones and Phillips, (1966, 1967) have shown by diffusion techniques that the apical 3–4 mm of the sunflower root synthesizes gibberellin. The gibberellin content of root apices treated with the plant growth retardant (2-chloroethyl) trimethylammonium chloride (CCC) was found to be significantly reduced as compared to the control. Wylie *et al.* (1970) demonstrated that pea root tips incubated in the presence of N-dimethylaminosuccinic acid (Alar), 4-hydroxyl-5-isopropyl-2-methylphenyltrimethylammonium chloride, 1-piperidine carboxylate (Amo 1618), or CCC all inhibitors of GA biosynthesis resulted in the accumulation of several intermediates involved in the biosynthesis of the gibberellins. Growth analyses were not reported by Wylie *et al.*, but several studies have been reported that indicate that Amo 1618 inhibits root elongation, (Mertz, 1966; McComb and McComb, 1970). Mertz (1966) reported that Amo 1618-inhibited root growth could be overcome by exogenously applied GA_3.

A number of studies have appeared which reported that gibberellins are inhibitory or are unresponsive to root growth (see Mertz, 1966). However, if native gibberellins are not limiting factors in growth then their application would not stimulate growth and, like the auxins, they may be inhibitory at supraoptimal concentrations. Furthermore, in view of the number of structurally different gibberellins identified from higher plants the failure to apply the native hormone must be considered (Lang, 1970).

3. CYTOKININS

Our knowledge of the cytokinins and their possible function in root growth is rather limited. Generally the exogenous application of cytokinins induces growth inhibition of intact seedling roots as well as root tips cultivated *in vitro*. However, Butcher and Street (1960b) have defined condition under which kinetin will stimulate or inhibit root growth of excised tomato roots. Kinetin was inhibitory in culture medium containing 1.5% or less of sucrose. At higher concentrations, kinetin promoted linear growth by extending the period of the high growth rate. The inhibitory effect of kinetin at low sucrose concentration was caused by decreasing the production of new cells in the meristematic zone. Under conditions where kinetin enhanced growth, roots exhibited a higher rate of production of new cells. In culture media containing low concentrations of sucrose, the inhibition of growth by kinetin was intensified by the addition of GA_3 or naphthalineacetic acid.

Recently Gaspar and Xhauflaire (1967) and Darimont *et al.* (1971) have demonstrated that *Lens* root tips grown in the presence of inhibitory concentrations of kinetin possessed more peroxidase and destroyed more IAA than control roots. The reversal of inhibition by auxin, and conversely the reversal of auxin inhibition by kinetin has led them to propose that root growth inhibition by cytokinins might be caused by an enhanced auxin catabolism leading to a lower auxin content. Yang and Dodson (1970) have also noted that the inhibitory effect of 0.1 mg/liter of IAA could be largely overcome by 0.01 mg/liter kinetin.

Kende (1965) has clearly shown that the root tip synthesizes cytokinins. Burström (1968) has demonstrated that the inhibition of cytokinesis in wheat roots led to a cessation of cell elongation. Both IAA and kinetin reversed the inhibition of cell elongation. It therefore appears that in the normal organization of the root tip, the meristem may produce both kinetin and auxin which are required for the initiation of cell elongation.

4. ABSCISIC ACID

Virtually nothing is known of the role abscisic acid (ABA) plays in root growth; however, ABA has been reported in pea roots (Tietz, 1971) and to significantly reduce root elongation, particularly the S-isomer in germinating barley seeds (Sondheimer *et al.*, 1971). In roots of *Lens culinaria* ABA has been shown to be antagonistic to inhibitory concentrations of IAA (Pilet, 1970) while in barley seedlings kinetin and ABA were reported to cause synergistic inhibition of root growth.

Although the fundamental question of how hormones control root elongation has not been answered, it seems safe to conclude that auxins, gibberellins, cytokinins, ethylene, and possibly abscisic acid are all synthesized in and participate in elongation.

V. VASCULAR DIFFERENTIATION

A. Control of the Vascular Arrangements

As already noted, a portion of the apical meristem is occupied by the quiescent center, a zone made up of cells which are relatively inert metabolically. It has been shown in both monocotyledenous and dicotyledenous roots that the cells of the quiescent center are low in protein, RNA, DNA, carbohydrates and oxygen uptake (Brown and Broadbent, 1950; Jensen, 1955; Jensen and Kavaljian, 1958). Although this region of the apical meristem is metabolically very sluggish it has a very profound effect upon the patterns of vascular arrangement of the root.

Reinhard (1954) was one of the first to show that small pieces of pea root tips (0.7 mm) could be grown in culture to normal roots if the tip included the apical meristem and adjacent procambial tissue. Torrey (1954) also demonstrated that the apical 0.5 mm of pea roots, which included the root cap and roughly 200 μm of the meristematic region behind the root cap could be grown into organized roots provided a number of components of the nutrient medium were increased in concentration. The 0.5 mm root tips included the region of the apical initials and immediate derivatives, but no mature tissues. It had been previously noted that the presence of mature tissues in isolated 10 mm root tips grown in culture influenced the activity of the meristem possibly by supplying factors, such as vitamins, and micronutrients missing when extremely small tips were cultured (Brown and Wightman, 1952). The apical meristem of the root, if provided with the materials normally supplied from the shoot system is capable of developing into a normal organized root.

One of the most characteristic patterns established in a seedling during the ontogeny of the plant is the alternate and radial arrangement of the primary vascular tissue within the primary root (Esau, 1965).* In many species this radial arrangement with its characteristic number of vascular strands may persist for the life of the primary root. The number of alternate and radially arranged xylem and phloem strands in any given plant may be quite characteristic of genus, species, or even family.

---

* Depending upon the number of protoxylem poles, one, two, three, or more, the root or vascular patterns are called monarch, diarch, triarch, and so on.

Torrey (1955) noted that the vascular arrangement derived from isolated pea root tips cultured 1 week in sterile medium showed a uniform
pattern of vascular tissue differentiation. The earliest visible pattern
of vascular differentiation following the delimitation of the procambium
was the enlargement of the future metaxylem elements, beginning
around 175 μm proximal to the apical initials and resulting in a blocking
out of a triarch primary xylem pattern. The division of the protophloem
mother cells situated on alternate radii occur at about 260 μm proximal
to the apical initials and mature sieve tube elements were apparent
within 340 μm of the apical meristem. Torrey excised 0.5 mm root tips
from week-old root cultures, the root tips containing no mature vascular
tissue were grown in sterile defined medium for 1 week, and then analyzed histologically. Approximately 20% of the tips showed a reduction
from the normal triarch radial pattern to a symmetrical diarch or monarch arrangement of the vascular tissue. The abnormal patterns which
resulted were believed to have resulted from a partial or complete destruction of the preexisted triarch pattern in the procambium at the
level of excision. On continued culture the symmetrical monarch or
diarch roots underwent a transition returning to the original triarch
arrangement. From these studies it is apparent that the primary vascular
tissue pattern does not result from an influence of the older mature vascular tissue but rather the vascular pattern formation is under the immediate control of the rather inert quiescent center of the apical meristem.

Additional information concerning the control of vascular pattern formation by the apical meristem of pea roots has been derived from experiments on regeneration of the root meristem. Torrey (1957) studied the
regeneration of meristems under two different conditions, one in which
the decapitated root was allowed to grow in controlled nutrient medium
lacking IAA and in a medium in which IAA was added at a concentration inhibitory to root elongation. In the auxin-free medium two types
of regeneration were apparent. In one type, regeneration of the tip pro-
ceeded from tissues of the central stele wherein the vascular pattern
between the triarch base and the new root showed some morphological
discontinuity; however, in other roots there was found almost complete
continuity of the triarch vascular pattern between the old and the new
tissue. In other roots the vascular pattern of the newly formed tissue
was triarch, but showed no continuity with the preexisted triarch vascular pattern. Thus the new vascular arrangement formed was unrelated
to the old pattern. In the second type of regeneration the meristem
was formed from the pericycle usually opposite the protoxylem points
of a decapitated root in a manner analogous to the formation of lateral
roots.

In sharp contrast to the regeneration phenomena which occurred in the decapitated roots grown in the controlled auxin-free medium, was the type of regeneration which occurred when decapitated roots were cultured in media supplied with IAA. In the presence of IAA at 10^{-5} M regeneration was very precise and regular. It resembled somewhat the type described above, wherein a new meristem was organized from the procambial cylinder with cell division principally from the pericycle contributing to the meristem. The regenerated root elongated very little because of the inhibition of cell elongation by auxin in the medium; however, a strikingly different vascular pattern within the newly regenerated root was found. Instead of the normal triarch vascular arrangement of the root, a symmetrical hexarch vascular pattern developed. The original triarch strands were evident in essentially their original positions but three new strands of xylem had been interposed and six phloem bundles now alternated with the six xylem strands. As long as the newly developed root continued its slow growth in the auxin medium it maintained its hexarch pattern; however, when the root was transferred to a medium free of auxin the meristem changed through a gradual sequence to the production of pentarch, tetrarch and finally to the original triarch pattern.

Although the normal triarch root is genetically determined, the external auxin concentration has an overiding effect upon the organization and activities of the apical meristem. It is tempting to suggest that auxin supplied by the apex provides the proper hormone balance and thus the genetic integrity of the procambium; however, this has never been established.

B. Secondary Vascularization

Isolated roots grown in culture do not usually form secondary tissues; however, the application of sugars and hormones via the cut root base has led to a partial understanding of the hormonal control of the initiation of secondary vascular tissues (Torrey, 1963).

Torrey and Shigemura (1957) reported that root tip excised from 48-hour pea seedlings grown in an 0.5% agar culture containing 10^{-5} M IAA caused the initiation of a vascular cambium from periclinal divisions of cells in the area between the primary phloem bundles and the primary xylem. Digby and Wanngerman (1965) noted that cambial activity in the pea root was stimulated by auxin coming from the shoot rather than from auxin supplied from the root apex. Removal of the root apex had virtually no effect upon xylem or cambium development, while removal of the shoot apex greatly reduced both. Loomis and

Torrey (1964) and Torrey and Loomis (1967) have demonstrated an absolute requirement for auxin and the cytokinin benzylaminopurine to initiate cambium development in the root of the radish *Raphanus sativum*. The hormones along with myoinositol and 8% sucrose were supplied to the root by way of the cut basal end. The addition of inositol, while not an absolute requirement for cambium initiation increased the magnitude of the response markedly.

Sorokin *et al.* (1962) have also shown a requirement for auxin and kinetin for cambial initiation. Using excised pea epicotyls it was shown that auxin caused the activation of a fasicular cambium with some interfasicular cambium which resulted in the production of large amounts of xylem. Application of both auxin and kinetin caused the formation of a much more active cambium over the entire circumference of the internode resulting in the conversion of the epicotyl to essentially a woody habit of growth.

Very little is known of the role gibberellins play in secondary vascularization of roots; however, it is rather clear from studies of woody stems that gibberellins participate in vascular differentiation (Roberts, 1969). The relative levels of IAA and GA are important in determining whether mainly xylem or phloem tissue is produced. High IAA to low GA concentrations favor xylem formation whereas low IAA to high GA concentrations favor phloem production (see Wareing and Phillips, 1970). Gibberellins have been shown to also promote the maturation of immature sieve cells from dormant branches of *Pinus strobus* L. (DeMaggio, 1966) and to promote the initiation of xylem closer to the apex in GA-treated bean roots (Odhnoff, 1963).

C. Primary Vascularization

Only indirect evidence is available concerning the hormonal control of primary vascularization in the root. Evidence that primary vascular tissue formation is hormonally controlled, stems principally from studies of wound-vessel differentiation and vascularization in callus cultures. Jacobs (1952, 1954) has demonstrated that auxin limits xylem regeneration around stem wounds in *Coleus*, and that increases in the number of mature xylem elements formed from the procambium in the leaf are coincident with the maximum rate of auxin production in that leaf.

The importance of both auxin and sugar for vascular tissue formation in callus tissue is well known. Experimental induction of vascular tissue has been achieved in callus tissue derived from a number of different species of plants. Auxin and sugars, particularly sucrose, are necessary for the complete differentiation of xylem and phloem in callus tissue.

Concentration of sugars has been shown to alter the proportion of xylem to phloem. Wetmore and Rier, (1963) have shown that, at a fixed auxin concentration, low concentrations of sucrose (1.5–2.5%) favored xylem formation, high concentrations (3–4%) favored phloem formation while, intermediate concentrations (2–3%) favored the presence of both xylem and phloem with a cambium in between. Sucrose appears not only to supply raw materials for the synthesis of lignin and cellulose but to also play a regulatory role in differentiation. Jeff and Northcote (1967) have shown that neither fructose nor glucose could replace the requirement for sucrose in the differentiation of vascular tissue derived from *Phaseolus vulgaris*. Xylem elements increased progressively with increasing concentrations of sucrose in the medium. In the absence of exogenously applied IAA the wall structure of regenerated xylem cells was found to depend in part upon the concentration of sucrose provided. Annular and sclariform elements were produced at low sucrose levels while reticulate elements were formed at high levels. Sclariform and reticulate members were produced at intermediate levels. The interaction of auxin with the intact sucrose molecules may regulate the quantity as well as quality of vascular differentiation (Beslow and Rier, 1969).

Foskett and Torrey (1969) have demonstrated in soybean callus tissue derived from cotyledons that auxin and cytokinin in the presence of sugar are required for tracheary element formation. In the presence of 10^{-5} M NAA cell number increased as the kinetin concentration was increased between 10^{-9} and 10^{-6} M. However, tracheary element formation was not initiated until the kinetin concentration was 5×10^{-8} M or above. Neither auxin nor kinetin at any concentration tested stimulated tracheary element formation in the absence of an effective level of the other hormone. However, 2,4-D at 10^{-7} or 10^{-6} M promoted both cell proliferation and tracheary element differentiation in the absence of an exogenous cytokinin. It was suggested by Foskett and Torrey that 2,4-D may function as a weak cytokinin as well as an auxin, or 2,4-D may cause the tissue to produce its own cytokinin.

More convincing evidence that both cytokinin and auxin are required for xylogenesis comes from the work of Torrey and Foskett (1970). When culturing 1-mm thick segments cut 10–11 mm proximal to the root tip of germinating seeds of *Pisum sativum,* they noted that in the presence of auxin pericycle proliferation occurred. The cortical cells did not divide and were sluffed off as a callus tissue of diploid cells was formed. In the presence of kinetin (0.1–1.0 ppm) the cortical cells were stimulated to divide and underwent DNA synthesis prior to division. Following cell division the polyploid daughter cell rapidly underwent cytodifferentiation to form mature tracheary elements.

In the majority of the cases involving vascular tissue differentiation cell division normally precedes differentiation. It is generally considered that cell division is necessary for the initiation of xylogenesis and that mitosis must occur in a particular hormonal environment for the derivatives to appear as tracheary elements (Foskett, 1968; Foskett and Torrey, 1969). However, in studying the influence of auxin on vascular strand formation from cortical cells of pea roots, Sachs (1968) found that strand formation occurred in parenchyma tissue without any pronounced growth or cell division. The tracheary elements formed following the application of auxin to cortical root tissue from which the vascular cylinder had been removed had essentially the same shape as the cortical parenchyma cells. It should be noted that Sachs' interpretations were based upon free hand sections of the cortical tissue, thus he may have failed to detect cell division.

The redifferentiation of parenchyma to tracheary elements was demonstrated to involve stimuli, presumably auxin, coming from the shoot and not from the root tip. Sachs (1969) separated a shoot-root graft zone with a foreign tissue so that the shoot to root polarity of the foreign tissue was at different angles to the grafted plant. New xylem always formed from parenchyma cells following the original shoot to root polarity of the cells of the foreign tissue.

It appears that both auxin and cytokinin are required for tracheary element formation. Both hormones are synthesized in the root and may function in vascular differentiation *in situ;* however, the strong preferential acropetal transport of auxin into the root, particularly in the older more mature region, suggests that auxin is supplied by the aerial portion of the plant. This polar movement into the root may be one of the determining factors in polar vascularization. Auxin supplied by the root apex may regulate cell elongation and possibly tissue pattern formation.

VI. CONCLUDING REMARKS

The chemical control of root growth and development is still one of the most poorly understood areas of plant morphogenesis. Yet, it is becoming increasingly apparent that growth of the root is under the influence of both the root and shoot apex. The strong acropetal transport of auxin into the root from the aerial portion of the plant coupled with the weak basipetal movement from the root apex suggests that two transport streams of auxin are involved in root development. Unfortunately very little is known of the polar transport of the other growth regulators, particularly the cytokinins and gibberellins; however, they are both synthesized in the root tip and are, along with auxin, involved in the regulating cell division in the apical meristem and participate in vascular differentiation.

The participation of indoleacetic acid and gibberellins in controlling cell elongation in the shoot is well established; however, in roots the evidence for a similar hormonal control of cell enlargement is not nearly as clear. The extreme sensitivity of roots to exogenously applied auxins has led some investigators to conclude that auxins play no role in root growth. This is a rather extreme point of view, since auxins have been shown to induce root elongation at extremely low concentrations, to stimulate lateral root initiation, to be involved, along with cytokinins, in cambial initiation and primary vascularization.

The effect of gibberellins on root elongation has not been extensively studied; nevertheless, dwarf mutants of maize, such as d_1d_1, respond to exogenously applied gibberellin by increased root elongation. Such dwarfs are deficient in endogenous gibberellins and possess shorter and larger diameter roots than normal seedlings. Inhibitors of gibberellin biosynthesis, such as Amo 1618, have been shown to inhibit root elongation with the exogenous application of gibberellin overcoming the inhibition.

To what extent the two growth inhibitors, ethylene and abscisic acid, play a role in normal root growth and development is unknown. However, it is becoming increasingly apparent that the chemical control of growth is dependent upon interactions, both synergistic and inhibitory. In future studies these inhibitors will have to be carefully considered along with the interaction of the other growth regulators.

REFERENCES

Aasheim, T., and Iversen, T. (1971). *Physiol. Plant.* **24**, 325.
Almestrand, A. (1957). *Physiol. Plant.* **10**, 521.
Addicott, F. T. (1939). *Bot Gaz.* **100**, 836.
Addicott, F. T., and Bonner, J. (1938). *Science* **88**, 577.
Barlow, P. W. (1969). *Planta* **88**, 215.
Beslow, D. T., and Rier, J. P. (1969). *Plant Cell Physiol.* **10**, 69.
Boll, W. G. (1954). *Plant Physiol.* **29**, 325.
Bonner, J. (1937). *Science* **85**, 183.
Bonner, J., and Bonner, H. (1948). *Vitam. Horm.* **6**, 225.
Bonnett, H. T., and Torrey, J. G. (1965). *Plant Physiol.* **40**, 813.
Boysen-Jensen, P. (1933). *Planta* **20**, 688.
Brauner, L., and Diemer, R. (1971). *Planta* **97**, 337.
Brian, P. W., Elson, G. W., Hemming, H. G., and Radley, M. (1954). *J. Sci. Food. Agr.* **5**, 602.

Brown, R., and Broadbent, D. (1950). *J. Exp. Bot.* 1, 249.

Brown, R., and Wightman, F. (1952). *J. Exp. Bot.* 3, 253.

Brumfield, R. T. (1943). *Amer. J. Bot.* 30, 101.

Burnett, D., Audus, L. J., and Zinsmeister, H. D. (1965). *Phytochemistry* 4, 891.

Burström, H. (1960). *Physiol. Plant.* 13, 597.

Burström, H. (1968). *Physiol. Plant.* 21, 1137.

Burström, H. (1969). *Amer. J. Bot.* 56, 679.

Burström, H., Uhrstrom, I., and Olausson, B. (1970). *Physiol. Plant.* 23, 1223.

Butcher, D. N., and Street, H. E. (1960a). *J. Exp. Bot.* 11, 206.

Butcher, D. N., and Street, H. E. (1960b). *Physiol. Plant.* 13, 46.

Byrne, J. M., and Heimsch, C. (1970). *Amer. J. Bot.* 57, 1179.

Celand, R. (1971). *Ann. Rev. Plant Physiol.* 22, 197.

Chadwick, A. V., and Burg, S. P. (1967). *Plant Physiol.* 42, 415.

Chadwick, A. V., and Burg, S. P. (1970). *Plant Physiol.* 45, 192.

Cholodny, N. G. (1934). *Planta* 21, 511.

Ciesielski, T. (1872). *Ber. Biol. Pflugers* 1, 30.

Clowes, F. A. L. (1959). *Biol. Rev.* 34, 501.

Clowes, F. A. L. (1961). "Apical Meristem." Davis, Philadelphia, Pennsylvania.

Clowes, F. A. L. (1971). *Ann. Bot.* 35, 249.

Darimont, E., Gaspar, T., and Hofinger, M. (1971). *Z. Pflanzen Physiol.* 64, 232.

DeMaggio, A. E. (1966). *Science* 152, 370.

Digby, J., and Wangermann, E. (1965). *New Phytol.* 64, 168.

Dullaart, J. (1970). *Acta Bot. Neerl.* 19, 573.

Esau, K. (1940). *Hilgardia* 13, 175.

Esau, K. (1941). *Hilgardia* 13, 437.

Esau, K. (1943). *Hilgardia* 15, 299.

Esau, K. (1965). "Plant Anatomy." Wiley, New York.

Foskett, D. E. (1968). *Proc. Nat. Acad. Sci.* 59, 1089.

Foskett, D. E., and Torrey, J. G. (1969). *Plant Physiol.* 44, 871.

Gaspar, T., and Xhaufflaire, A. (1967). *Planta* 72, 252.

Gibbons, G. S. B., and Wilkins, M. B. (1970). *Nature (London)* 226, 558.

Goldsmith, M. H. M., and Thimann, K. V. (1962). *Plant Physiol.* 37, 492.

Hanstein (1868). *Festschr. Niederrhein. Ges. Natur. Heilkunde.*, p. 109–143.

Hawker, L. E. (1932). *New Phytol.* 31, 321.

Heimsch, C. (1951). *Amer. J. Bot.* 38, 523.

Hertel, R., and Leopold, A. C. (1963). *Planta* 59, 535.

Heyne, A. N. J. (1940). *Bot. Rev.* 6, 515.

Hillman, S. K., and Phillips, I. D. J. (1970). *J. Exp. Bot.* 21, 959.

Iversen, T. (1969). *Physiol. Plant.* 22, 1251.

Iversen, T., and Aasheim, T. (1970). *Planta* 93, 354.

Jacobs, W. P. (1952). *Amer. J. Bot.* 39, 245.

Jacobs, W. P. (1954). *Amer. Natur.* 88, 327.

Janczewski, E. von (1874). *Bot. Z.* 32, 113.

Jeff, R. A., and Northcote, D. H. (1967). *J. Cell Sci.* 2, 77.

Jensen, W. A. (1955). *Exp. Cell Res.* 8, 506.

Jensen, W. A. (1958). *Exp. Cell Res.* 14, 575.

Jensen, W. A., and Kavaljian, L. G. (1958). *Amer. J. Bot.* 45, 365.

Jones, R. L., and Phillips, I. D. J. (1966). *Plant Physiol.* 41, 1381.

Jones, R. L., and Phillips, I. D. J. (1967). *Planta* 72, 53.

Juniper, B. E., and Barlow, P. W. (1969). *Planta* 89, 352.

Juniper, B. E., and French, A. (1970). *Planta* **95**, 314.
Juniper, B. E., Groves, S., Schachar, B. L., and Audus, L. J. (1966). *Nature (London)* **209**, 93.
Kato, Y. (1955). *Bot. Gaz. (Chicago)* **117**, 16.
Keeble, F., Nelson, M. G., and Snow, R. (1931). *Proc. Roy. Soc. B.* **108**, 537.
Kende, H. (1965). *Proc. Nat. Acad. Sci.* **53**, 1307.
Khan, A. A., and Downing, R. D. (1968). *Physiol. Plant.* **21**, 1301.
Kirk, S. C., and Jacobs, W. P. (1968). *Plant Physiol.* **43**, 675.
Konings, H. (1968). *Acta Bot. Neerl.* **17**, 203.
Lahiri, A. N., and Audus, L. J. (1960). *J. Exp. Bot.* **11**, 341.
Lang, A. (1970). *Ann. Rev. Plant. Physiol.* **21**, 537.
Lockhart, J. A. (1965). *In* "Plant Biochemistry" (J. Bonner and J. E. Varner, eds.), pp. 826–849. Academic Press, New York.
Loomis, R. S., and Torrey, J. G. (1964). *Proc. Nat. Acad. Sci.* **52**, 3.
Mallory, T. E., Chiang, S., Aetter, E. G., and Gifford, E. M. (1970). *Amer. J. Bot.* **57**, 800.
McClary, J. E. (1940). *Proc. Nat. Acad. Sci.* **26**, 581.
McComb, A. J., and McComb, J. A. (1970). *Planta* **91**, 235.
McCready, C. C. (1966). *Ann. Rev. Plant. Physiol.* **17**, 283.
Mertz, D. (1966). *Plant Cell Physiol.* **7**, 125.
Mooré, D. J., and Bonner, J. (1965). *Physiol. Plant.* **18**, 635.
Morris, D. A., Briant, R. E., and Thomson, P. G. (1969). *Planta* **89**, 178.
Nagao, M., and Ohwaki, Y. (1968). *Bot. Mag. (Tokyo)* **81**, 44.
Neales, T. F. (1967). *New Phytol.* **67**, 159.
Nissl, D., and Zenk, M. H. (1969). *Planta* **89**, 323.
Odhnoff, C. (1963). *Physiol. Plant.* **16**, 474.
Overbeek, J. van (1939). *Bot. Gaz. (Chicago)* **101**, 450.
Paleg, L., Aspinall, D., Coombe, B., and Nicholls, P. (1964). *Plant Physiol.* **39**, 286.
Phillips, H. L., and Torrey, J. G. (1971). *Plant Physiol.* **48**, 213.
Pilet, P. E. (1964). *Nature (London)* **204**, 561.
Pilet, P. E. (1970). *J. Exp. Bot.* **21**, 446.
Raghavan, V., and Torrey, J. G. (1964). *Plant Physiol.* **39**, 691.
Ray, P. M. (1969). *Develop. Biol. Suppl.* **3**, 172.
Ray, P. M., and Ruesink, A. W. (1963). *J. Gen. Physiol.* **47**, 83.
Reinhard, E. (1954). *Z. Bot.* **42**, 353.
Richardson, S. D. (1958). *Nature (London)* **181**, 429.
Robbins, W. J., and Bartley' M. A.*(1937). *Proc. Nat. Acad. Sci.* **23**, 385.
Robbins, W. J., and Hervey, A. (1969). *Proc. Nat. Acad. Sci.* **64**, 495.
Robbins, W. J., and Hervey, A. (1971). *Proc. Nat. Acad. Sci.* **68**, 347.
Robbins, W. J., and Schmidt, M. B. (1939). *Proc. Nat. Acad. Sci.* **25**, 1.
Roberts, E. H., and Street, H. E. (1955). *Physiol. Plant.* **8**, 238.
Roberts, L. W. (1969). *Bot. Rev.* **35**, 201.
Sachs, T. (1968). *Ann. Bot.* **32**, 391.
Sachs, T. (1969). *Ann. Bot.* **33**, 263.
Scott, T. K., and Wilkins, M. B. (1968). *Planta* **83**, 323.
Sondheimer, E., Galson, E. C., Chang, Y. P., and Walton, D. C. (1971). *Science* **174**, 829.
Sorokin, H. P., Mathur, S. N., and Thimann, K. V. (1962). *Amer. J. Bot.* **49**, 444.
Street, H. E. (1957). *Biol. Rev.* **32**, 117.

Street, H. E. (1969). *In* "Root Growth" (W. J. Whittington, ed.), pp. 20–41. Butterworths, London and Washington, D.C.

Thimann, K. V. (1937). *Amer. J. Bot.* **24**, 407.

Thimann, K. V. (1969). *In* "Physiology of Plant Growth and Development" (M. B. Wilkins, ed.), pp. 3–37. McGraw-Hill, New York.

Tietz, A. (1971). *Planta* **96**, 93.

Torrey, J. G. (1953). *Amer. J. Bot.* **40**, 525.

Torrey, J. G. (1954). *Plant Physiol.* **29**, 279.

Torrey, J. G. (1955). *Amer. J. Bot.* **42**, 183.

Torrey, J. G. (1956). *Ann. Rev. Plant Physiol.* **7**, 237.

Torrey, J. G. (1957). *Amer. J. Bot.* **44**, 859.

Torrey, J. G. (1963). *Symp. Soc. Exp. Biol.* **17**, 288.

Torrey, J. G. (1965). *In* "Encyclopedia of Plant Physiology" (W. Ruhland, ed.), Vol. XV, pp. 1256–1327. Springer-Verlag, Berlin.

Torrey, J. G., and Fosket, D. E. (1970). *Amer. J. Bot.* **57**, 1072.

Torrey, J. G., and Loomis, R. S. (1967). *Amer. J. Bot.* **54**, 1098.

Torrey, J. G., and Shigemura, Y. (1957). *Amer. J. Bot.* **44**, 334.

Wareing, P. E., and Phillips, I. D. J. (1970). "The Control of Growth and Differentiation in Plants." Pergamon, Oxford.

Went, F. W., and Thimann, K. V. (1937). "Phytohormones." MacMillan, New York.

Wetmore, R. H., and Rier, J. P. (1963). *Amer. J. Bot.* **50**, 418.

Whaley, W. G., and Kephart, J. (1957a). *Plant Physiol.* **32**, xxxiii.

Whaley, W. G., and Kephart, J. (1957b). *Science* **125**, 234.

White, P. R. (1934). *Plant Physiol.* **9**, 585.

White, P. R. (1937). *Plant Physiol.* **12**, 803.

Wilkins, M. B. (1966). *Ann. Rev. Plant Physiol.* **17**, 379.

Wilkins, M. B., and Cane, A. R. (1970). *J. Exp. Bot.* **21**, 195.

Wilkins, M. B., and Scott, T. K. (1968a). *Nature (London)* **219**, 1388.

Wilkins, M. B., and Scott, T. K. (1968b). *Planta* **83**, 335.

Wylie, A. W., Ryugo, K., and Sachs, R. M. (1970). *J. Amer. Soc. Hort. Sci.* **95**, 627.

Yeomans, L. M., and Audus, L. J. (1964). *Nature (London)* **204**, 559.

Yang, D., and Dodson, E. O. (1970). *Can. J. Bot.* **48**, 19.

III

ADDITIONAL ASPECTS

7

CHEMICAL SIGNALS BETWEEN ANIMALS: ALLOMONES AND PHEROMONES

George M. Happ

I.	Introduction	149
II.	Characteristics of Chemical Signals	151
III.	Methods of Study	154
IV.	Allomones	155
	A. Allomones for Defense	156
	B. Allomones Promoting Symbiotic Associations	161
V.	Pheromones	163
	A. Alarm Substances	164
	B. Recruitment Pheromones	166
	C. Sex Pheromones	168
	D. Territoriality and Recognition Scents	172
	E. Primer Pheromones	173
	F. Pheromone Perception	175
VI.	Exocrines and Endocrines	178
	References	181

I. INTRODUCTION

Air and water transmit a multitude of chemical signals, many of which are not by-products of human "civilization". A great variety of these air- and water-borne molecules, present at very low concentrations, carry specific information from one animal to another. Sensory receptors of the recipient animal may be astoundingly sensitive: only one molecule need hit certain sense cells to trigger an action potential (Kaissling and Priesner, 1970). Chemical ecology, the study of these signals and the interactions they mediate, is now a rapidly expanding field (Sondheimer and Simeone, 1970; Whittaker and Feeney, 1971).

Chemical interactions are both diverse and complex. As a whole, the chemical signals which act between organisms are termed *semiochemicals* (Law and Regnier, 1971). In multicellular organisms, the distinction between semiochemicals and hormones is usually clear: semiochemicals are exocrine secretions, produced by one individual and acting upon another. The signal (semiochemical) is the central element in a system consisting of producer–signal–recipient. When producer and recipient are of the same species, communication is intraspecific and the signal is known as a *pheromone* (Karlson and Lüscher, 1959; Karlson and Butenandt, 1959). When the signal acts between two different species, it is called an *alleochemic* (Whittaker, 1970; Whittaker and Feeney, 1971). Alleochemics may benefit the producer (e.g., by repelling a predator), or may be disastrous for the producer (e.g., by attracting a predator), or they may benefit both producer and recipient (e.g., floral scents which attract pollinating insects to nectar). Alleochemics which are adaptive for the producer are called *allomones* and those which are of adaptive advantage to the recipient are known as *kairomones* (Brown, 1968; Brown *et al.*, 1970).

This chapter is concerned with interspecific allomones and intraspecific pheromones, namely those chemical signals which have clear adaptive value to the animals producing them. In one species, several chemical signals may serve quite diverse functions, as illustrated by the mealworm beetle, *Tenebrio molitor*. Scent communication plays at least four roles in the reproduction of *Tenebrio*. The female produces a scent which attracts and sexually excites the male (Valentine, 1931; Tschinkel *et al.*, 1967, Happ and Wheeler, 1969). The male produces a scent which attracts females (Happ, 1969; August, 1971). A male which has "smelled" a female emits a scent which inhibits other males, i.e., makes them less responsive to the female attractant (Happ, 1969). Mature males, and to a lesser extent mature females, produce a scent which accelerates reproductive maturation in younger adult females (Happ *et al.*, 1970). Finally, adult *Tenebrio* possess glands (Roth, 1945) which may repel predators by means of defensive allomones, *p*-benzoquinones (Schildknecht, 1963).

Chemical signalling systems have been found in many phyla, as Wilson (1970) has noted, "they continue to turn up regularly in species when a deliberate search is made for them." Within the last 25 years, over 100 allomones and half as many pheromones have been chemically identified; the bulk of this research has involved insect material (Weatherston and Percy, 1970; Law and Regnier, 1971). Insect studies have progressed especially rapidly for three principal reasons: (1) the dramatic improvements in analytic chemical techniques which now allow characterization of organic molecules present in microgram quantities, (2) the fact that chemical cues play a predominant role in many aspects of insect biology, and (3) the attractive possibility that natural chemical signals could be used to manipulate natural populations and thus contribute to control of insect pests. Yet insects are not unique; a vast number of observations on behavior and natural history indicates the importance of chemical signals in many animals, notably vertebrates see Wilson and Bossert, 1963, various papers in Seboek, 1968, Sondheimer and Simeone, 1970, and Johnston *et al.*, 1970, for references). Except for the studies of perfume chemists on animal musks (Lederer, 1950), relatively little information has been available on mammalian exocrine secretions. The present review will focus on two groups of animals: the terrestrial arthropods, where much biological and chemical data are available, and the vertebrates, where the importance of the chemical signals are acknowledged but the chemical data are, as yet, rather sparse. Of necessity, the choice of examples from the literature has been somewhat arbitrary, but I have attempted to indicate the diversity in both signal molecules and their functions.

II. CHARACTERISTICS OF CHEMICAL SIGNALS

In contrast to auditory or visual signals, chemical signals physically occupy space and are relatively persistent. Within the last 8 years, E. O. Wilson and W. H. Bossert have published a series of fascinating papers which elucidate many of the general features of chemical signal transmission (Wilson and Bossert, 1963; Bossert and Wilson, 1963; Bossert, 1968; Wilson, 1968, 1970; Wilson *et al.*, 1969). Crucial to their conclusions is a mathematical model of the system. I will attempt to describe its major features below, but the interested reader should consult the original papers (especially Bossert and Wilson, 1963) for fuller mathematical development.

Consider a simple system: a stationary animal on a flat surface begins to emit a pheromone into still air. Assume that the response to this signal is all-or-none, and thus a recipient will respond only when the pheromone concentration in his own vicinity exceeds threshold—this response threshold is designated as K (molecules/cm^3). As the pheromone diffuses out from the stationary emitter into ever-increasing volume of surrounding air, a concentration gradient appears, declining away from the emitter. At some distance from the emitter, pheromone concentration will be less than K, and no potential recipients will respond. However, there will be a certain volume (in the vicinity of the emitter) where pheromone concentration is at least K or exceeds K, and a response occurs. This volume is designated as the *active space,* and as Wilson (1970) has noted, "the signal is the active space."

The shape of the active space varies with three factors: (1) the position of the emitter, (2) air movement, and (3) the behavior of the emitter. If the emitter is at the top of a tall tree, the active space may be almost spherical, but when the emitter is on a flat surface, the active space will be essentially hemispheric, at least in still air. When the air is not still, the shape will be modified: in a constant wind, the hemispheric active space is smeared into a semiellipsoid. Finally, if the *emitter* is moving and marking the substrate with liquid pheromone, he leaves a trail of active space behind him which persists until the "marks" have evaporated.

The volume and life-span of any active space will vary with four parameters: (1) rate of pheromone emission into the air, designated Q and expressed in molecules per second, (2) the rate of diffusion characteristic of each molecular species, (3) the response threshold K, and (4) temporal factors, namely elapsed time since emission began and the duration of the emission. By a refinement of the diffusion equation, Bossert and Wilson (1963) have shown that these parameters are interrelated according to the following equation (for animals on a flat nonabsorbent surface).

$$ K = \frac{Q}{2D\pi r} \, efrc \, \frac{r}{\sqrt{4Dt}} $$

where Q, D, and K are emission rate, diffusion coefficient, and threshold concentration, respectively, and where r is the "radius" of the active space (cm), t is the time from the beginning of emission (seconds) and where $efrc(x)$ is the complementary error function.

An increase in Q (emission rate) or a decrease in K (threshold concentration) will lead to a greater maximum volume for the active space. Wilson and Bossert (1963) have shown that the ratio between these two parameters

$$ \frac{Q}{K} = \frac{\text{molecules emitted/second}}{\text{molecules/cm}^3 \text{ at threshold}} $$

effectively describes not only the maximum volume but also the temporal characteristics of the signal. The Q/K ratio predicts both the lag-time required for expansion of the active space to maximum volume (which is equivalent to lag-time for signal transmission at maximum range) and the fade-out time after emission ceases. Wilson (1970) has calculated the volumes and temporal characteristics for a substance diffusing ($D = 0.1$ cm^2/second) into still air (Table I).

It should be emphasized that *optimum* volume and duration of the active space are not synonomous with *maximum* volume and duration. Sex attractants emitted by female moths operate well over many kilometers (and the male moth can fly that distance upwind) but trail markers of ants must keep the following ants close to the original trail. In the case of the moth, Q/K is large and the calling female remains stationary. For the trail substance, the opposite holds. Natural selection has operated to maximize the efficiency of each signal system, including the properties of the molecular signal itself.

Neither nitrogen gas not glycogen would be effective as an air-borne chemical signal: such a signal must be volatile (which glycogen is not) and must be different from environmental noise (which nitrogen is not). A chemical signal must be of low molecular weight to be volatile, and yet of sufficient structural complexity to convey a signal of some specificity. There are an infinite number of organic compounds and a multitude of structural, geometric, and optical isomers of many. An upper limit on size is imposed by the volatility requirement and a lower limit by specificity. Wilson and Bossert (1963) propose a judicious compromise: air-borne pheromones will have 5–20 carbons and molecular weights from 80 to 400. As subsequent chemical studies have shown (see Sections IV,V) their predictions were amazingly accurate.

Given this variety of molecular signals, can the prospective recipient distinguish one from another? As noted below, discrimination is extremely precise in some cases. And at least in mammals, a very large number of distinct odors are recognized. Even in man, a notoriously

TABLE I

Q/K	Maximum radius (cm)	Time to reach maximum radius (sec)	Fade-out time (sec)
1	0.6	0.4	1
100	2	8	20
10,000	10	150	500
1,000,000	60	40,000	10,000

microsomatic species, Hainer *et al.* (1954) estimate that 10,000 odorants can be distinguished.

Transmission of a signal through aqueous media possess essentially the same sorts of problems, except that the rate of diffusion is usually much lower and one would expect the signal molecule to be rather polar in order to be water soluble. Wilson (1970) has made the calculations for such an aqueous medium: on the assumption that $D = 10^{-5}$ cm²/second, he finds that for any given Q/K ratio, the volume of the maximum active space is similar to that in air but the time for expansion to maximum and time for fade-out are much longer.

III. METHODS OF STUDY

The human nose is something of a handicap in studies of chemical signals which act in natural populations. Whereas in visual signalling, we can usually *see* something (ultraviolet and infrared cues excepted) and then record it on film, and in auditory signalling we can usually *hear* something (except ultrasonic signals) and record it on tape for analysis, when chemical signals are employed we often smell nothing. Defensive allomones are obvious exceptions. Formic acid is unpleasant to man, as it is to most animals. Commonly, a study of chemical signals begins with a chance observation that an animal responds in the absence of noticeable stimuli. The tasks then become: to determine whether the covert stimuli are chemical, to search for the molecule(s) responsible, and to define precisely their role in the behavioral or physiological changes of the recipient.

It may be relatively easy to demonstrate that a *chemical* signal triggers certain behavior. Elimination of auditory and visual cues is often not too difficult, and the recipient may then respond when exposed to a scented airstream or a liquid sample. It is sometimes reassuring to demonstrate that ablation of putative chemoreceptors or their axons, i.e., clipping off segments of an insect antenna or sectioning the olfactory bulb of a vertebrate, abolishes the response to the chemical stimuli; but this surgical insult may affect many other facets of nervous function, and the results must be interpreted with caution. In any case, for chemical isolation and characterization of the signal, it is necessary to have a suitable bioassay, preferably one which utilizes behavioral responses of intact organisms in reasonably natural situations.

Electrophysiological techniques, for example monitoring the electroantennograms of insects, have been employed profitably for bioassays, but eventually the purified chemical must be tested on intact animals. The major requisite is that the assay must be quantitative. To cite an example: in 1967, Wheeler and I isolated an homogeneous substance which attracted and excited male mealworm beetles, and we rather thought that we had found the natural sex pheromone. However, quantitative bioassay revealed that our "homogeneous substance" was 100 times *less* active than a partially purified extract of females. In fact, the "homogeneous substance" was dibutyl phthlate, a volatile contaminant introduced by Tygon tubing in our collection of female scent (Happ and Wheeler, 1969).

Given the quantitative bioassay, what usually follows is chemical drudgery—repeated purification and repurification, and biological ennui—bioassay of each fraction generated by the chemistry. (These procedures may be cleverly avoided in some cases by direct assay of a battery of known chemicals as demonstrated by W. Roelofs and his co-workers at Cornell). But in the usual routine, when the chromatography finally yields an apparently pure substance of high potency in the bioassay, modern spectral techniques (mass, infrared, nuclear magnetic resonance, and ultraviolet spectrometry) often allow identification. The power of these techniques cannot be overestimated, for they are sensitive to micro- and nanogram amounts. The structure determination must then be confirmed by careful comparison with an authentic sample. Especially with pheromones, which are often present in very small amounts, the authentic samples must be very rigorously purified, as a very minor contaminant may account for the biological activity. If the authentic sample is not available, it must be synthesized. Premature publication of pheromone structure on the basis of reasonable structural evidence only (but no synthesis) has led to unfortunate errors.

IV. ALLOMONES

In their recent review of alleochemics, Whittaker and Feeney (1971) define allomones as "chemical agents of adaptive value to the organism producing them" which do not act between members of the same species. Included within this broad definition are a host of chemical agents which will not be considered within the present chapter. Among those excluded will be counteractants (such as antibodies), venoms which are used to kill prey, escape substances (such as cephalapod inks), and substances which are primarily nutritive (such as vitamins provided by symbiotic bacteria). The following discussion will be limited to two

classes of chemical *signals* which act upon another species: the repellents used for defense and the substances which regulate symbiotic
interactions.

A. Allomones for Defense

Predators do not always succeed in capturing their potential prey:
the presumptive meal may hide, flee, or counterattack. Among the most
dramatic defensive weapons are the chemical ones: the notoriety of
the skunk is founded on the overwhelming persuasiveness of its repellent
spray. The millipede *Apheloria* responds to attack by emitting a mixture
of benzaldehyde and hydrogen cyanide (Eisner *et al.*, 1963). Both the
skunk and this millipede are distinctively colored, presumably so that
in future encounters those predators which can learn (especially vertebrates) will heed the warning which these colors present. Chemical
repellents are also found in fish, newts, frogs, and snakes, but among
terrestrial and freshwater animals, arthropods have the most diverse
chemical defenses (Roth and Eisner, 1962; Schildknecht, 1963; Cavill
and Robertson, 1965; Eisner and Meinwald, 1966; Weatherston, 1966;
Weatherston and Percy, 1970; Eisner, 1970; Schildknecht, 1970).

1. SMALL REPELLENT MOLECULES

In 1670, Fisher examined a distillate of formicine ants and detected
an organic acid which was subsequently characterized as formic acid.
As one of the simplest repellents, formic acid can serve to illustrate
many of the characteristics of the substances found in defensive secretions. Because of its high acid strength ($pK_a = 3.77$), formic acid causes
protein coagulation and is therefore cytotoxic. Ants squirt a finely dispersed stream of liquid acid at attackers, thus maximizing the chance
that sufficient numbers of toxic molecules will be delivered to the assail-
ant (see Eisner, 1970, for a dramatic photograph). Formic acid is of
course a volatile substance (boiling at about 100°C) with a pungent
odor; it is repellent in both the liquid and the gas phase.

Representative defensive allomones are shown in Fig. 1. Aliphatic
acids and aldehydes are common; usually the acids (I–III, V) are short
chain (C_1–C_5) while the aldehydes (IV, VI) are of intermediate length
(C_4–C_8) and often α- β-unsaturated. Aliphatic alcohols are rare, but
their sulfur analogs (VII, mercaptans) are utilized by mustelid mammals. Cyclic molecules include benzoquinones (VIII), phenols, benzaldehyde derivatives (IX, X), and an unusual chlorinated hydrocarbon
(XI) apparently derived from injested herbicide. A number of defensive
allomones are terpenoid (XII, XIII).

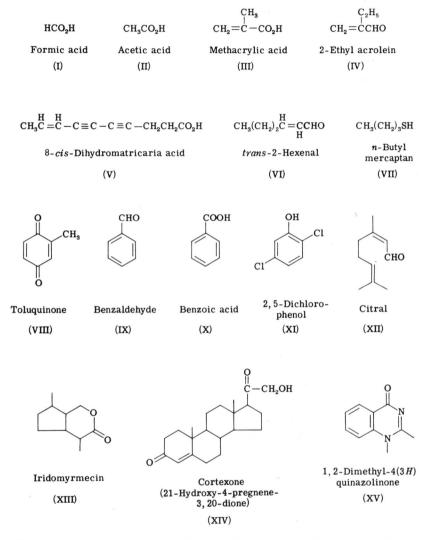

Fig. 1. Representative defensive allomones. I, Ants, caterpillars, beetles; II, whip scorpion, bugs; III, carabid beetles; IV, cockroaches; V, cantharid beetle; VI, cockroaches, bugs; VII, mustelid mammals (Lederer, 1950); IX, millipede, ant (Eisner *et al.*, 1961; Blum *et al.*, 1969); X, dytiscid beetles; XI, grasshopper (Eisner *et al.*, 1971); XII, ants, bees, beetles; XIII, dolichoderine ants; XIV, dytiscid beetles; XV, glomerid millipedes. Unless otherwise indicated, references in Roth and Eisner, 1962; Schildknecht, 1963, 1970; Eisner and Meinwald, 1966; Weatherston and Percy, 1970; Eisner, 1970.

The acids, aldehydes, quinones, and aromatic compounds are all reactive general toxicants which are effective against a broad spectrum of predators. Albeit by rather different mechanisms, both the aldehydes and the acids are protein fixatives. Whereas the acids are nonadditive fixatives in the sense of Baker (1958), the aldehydes can be regarded as additives which bond to nucleophilic groups (amines, thiols, etc.) of proteins. The α,β unsaturation in both the aldehydes and quinones would favor such addition reactions.

Acids such as formic and acetic, which are relatively polar, often must penetrate layers of integumental lipids before they can exert cytotoxic effects. A variety of interesting adaptations ensure penetration of these acids. In the whip scorpion, *Mastigoproctus giganteus*, the defensive glands produce a mixture of 84% acetic acid, 5% caprylic (octanoic) acid, and 11% water (Eisner *et al.*, 1961). Eisner and his co-workers nicely demonstrated that the caprylic acid disrupts the lipoid epicuticle of arthropods and thus allows the more irritating and cytotoxic acetic acid to penetrate. An analogous mixture, namely formic, methacrylic, or tiglic acids accompanied by a hydrocarbon is found in the defensive secretions of many ground beetles (Schildknecht, 1970). The ant *Acanthomyops claviger* uses both mechanical and chemical means to disrupt epicuticle; its mandibles scratch the attacker; its mandibular gland secretion (citral and citronellal) is applied, and formic acid is sprayed on this site (Ghent, 1961).

Many defensive secretions are mixtures, apparently because intermediate polarity properties of the secretion as a whole are optimum for broad-spectrum repellency. The defensive secretion of the tenebrionid beetle *Eleodes longicollis* has at least seven components in the nonpolar phase including three benzoquinones, caprylic acid, and three hydrocarbons (Eisner and Meinwald, 1966). That of the bug *Nezara viridula* is a mixture of 20 molecular species, but mostly aliphatic aldehydes (Gilby and Waterhouse, 1965). When one component predominates, it is usually reactive, of intermediate polarity, and often has detergent action. Phenolic compounds usually have alkyl chains which increase their toxicity over that of phenol itself (Sexton, 1963). The bacteriostatic and fungistatic activities of phenols are due in large part to their physiochemical characteristics, especially their tendency to "orient at an oil–water interface and so perhaps by this means to interrupt life processes" (Sexton, 1963).

The general toxicity of many of these allomones raises an interesting problem: How can a living system produce and store such potent toxicants without poisoning itself in the process? For some animals, the answer lies in spatial separation of the final steps in the biosynthesis of the toxic end product—an adaptation that Eisner (1970) has called the "reactor gland." In the defensive glands of *Apheloria* (which emit hydrogen cyanide and benzaldehyde) and of *Brachinus* (which ejects a mixture of benzoquinones and hydrocarbons at 100°C), the living secretory cells produce a relatively nontoxic precursor which is stored in a glandular reservoir. When the precursors pass into an outer cuticular compartment, they are enzymatically converted into the defensive toxicant (Schildknecht and Holoubek, 1961; Eisner *et al.*, 1963; Schildknecht *et al.*, 1968; Aneshausley *et al.*, 1969; Schildknecht *et al.*, 1970). Most defensive glands lack such large cuticular reaction compartments, but the same general strategy may be employed within each secretory cell unit. Each secretory cell is drained by a fine efferent cuticular ductule in many glands, and as Eisner *et al.* (1964) suggested, the ductule and cavities associated with it could serve as serially arranged reaction chambers. Histochemical evidence on quinone production in tenebrionid defensive glands supports Eisner's suggestion (Happ, 1968). Even where such a system of cuticular ductules is lacking, as in the defensive gland of the osmeteria of papilionid caterpillars, the cuticle over the secretory cells is complex and riddled with fine labyrinthine channels (Crossley and Waterhouse, 1969) and these many canaliculi could function as micro reaction compartments.

A considerable variety of predators have been tested in the laboratory and in the field in order to establish clearly the effectiveness of defensive secretions. Many are antimicrobial as well as antipredator. For references to some of these studies, see Eisner (1970).

Topical application of the defensive allomone to the predator is most often the case. An acrid mist may be ejected a distance of several feet and aimed directly at the site of attack, as in the whip scorpion (Eisner *et al.*, 1961) or the secretion may ooze from a gland orifice and then be wiped directly onto the point of attack, as in a soldier beetle (Eisner, 1970). The blind snake *Leptotypholops* coats itself with an emulsion of fatty acids and glycoprotein from its cloacal sac which effectively repels attacking army ants (Gehlbach *et al.*, 1968; Watkins *et al.*, 1969; Blum *et al.*, 1971b).

A topically applied blatant chemical insult obviously deters predators, but the effect of the defensive allomone may be more subtle. Fatty acids produced by the blind snake and citral produced by ants are scarcely general toxicants, yet they do repel predators. They may function by acting primarily on chemoreceptors to mask the effects of food stimuli. Dethier and Chadwick (references in Dethier, 1963) have shown that a sugar solution may be rendered unpalatable to blowflies by addition of various small organic molecules. An increase in length of the alkyl chain increased inhibitory efficiency. No one has done a systematic study of known defensive allomones (and closely related molecules) to determine which predators might be discouraged by this sort of sensory (as opposed to cytotoxic) mechanism.

Defensive allomones may also act in the gas phase, forming a repellent active space as the volatile molecules diffuse into the surrounding air. In certain bugs, a zone of specialized "fuzzy" cuticle lies adjacent to the orifice of the defensive gland, and secretion trapped within this space slowly evaporates, creating a small active space which moves with the bug as it leaves the site of an attack (Remold, 1962). Even if the concentration is quite low, the active space could still bias predator behavior, making attack less likely. Who would fail to detour around the active space left by a frightened skunk?

2. Steroids and Alkaloids

Steroids and alkaloids found within defensive secretions are usually nonvolatile pharmacologically active agents. In the skin secretions of Amphibia are the most potent of neurotoxins, including salamandrin (Habermehl, 1966), bufotalin (Meyer, 1952), tetrodotoxin (Woodward, 1964) and batracotoxin (Tokuyama *et al.*, 1969). These defensive allomones apparently serve to teach vertebrate predators to avoid the aposematically colored urodeles and anurans which produce them. Their chemical structures and pharmacological effects have been the subject of recent reviews (Bücherl *et al.*, 1968–71).

The steroids expelled from the prothoracic glands of dytiscid beetles, and the quinazolinones in the defense secretions of glomerid millipedes, are less understood from a pharmacological point of view. Most of the dytiscid steroids (including testosterone) are pregnane derivatives (XIV) and they are present in surprising quantities: a Mexican *Cybister* can store as much as 1 mg of 12-hydroxy-4,6-pregnadiene-3,20-dione. Both the crude prothoracic secretions and the purified steroids are anaesthetic and sometimes lethal to ,fish and amphibians (Schildknecht, 1970). Glomerid millipedes expel a proteinaceous fluid which contains 1-methyl-2-ethyl-4(3H)-quinazolinone and the related 1,2-dimethyl derivative (XV). The quinazolinones have a bitter taste to man, and a delayed general toxicity to birds, mice, and spiders (Schildknecht *et al.*, 1966, 1967; Y. C. Meinwald *et al.*, 1966; Eisner, 1970).

Perhaps the most surprising defensive substance to be reported is colymbetin, a small nucleoprotein (MW > 700), produced in the defensive glands of a water beetle (*Colymbetes fuscus*). When colymbetin was injected into rats, a drastic reduction on blood pressure occurred (Schildknecht and Tacheci, 1971).

B. Allomones Promoting Symbiotic Associations

Symbiotic associations have long fascinated biologists and the respective contributions of each partner to the success of the association have been cataloged by many ingenious investigations (see Henry, 1967). In a few cases, the allomones which regulate these associations have been identified, but this vast field still awaits systematic biochemical study.

Probably the most dramatic demonstration of an allomone acting between symbiotic partners comes from the classic work of L. R. Cleveland on the wood roach *Cryptocercus* and its intestinal flagellates. The cellulose-digesting flagellates are obligate anaerobes in the hindgut of the cockroach, whose cuticle (including that of the hindgut) is shed at each molt. Before the molt, the flagellates encyst and thus survive exposure to oxygen. The signal for encystment is a chemical (ecdysone) which is also the molting hormone of the insect. In this case ecdysone is both a hormone and an allomone (Cleveland *et al.*, 1960).

Nowhere are the symbioses between two insect species more diverse and richly documented than within insect societies. In his comprehensive and very readable monograph, E. O. Wilson (1971) has graphically described the many gradations which occur between species of social insects ranging from casual association to parasitism. Many of these must involve, at the very least, a tolerance or habituation of one species to the pheromones of another. Even more bizarre are the nonsocial species which exploit the insect societies at some point in their own life cycles. These social parasites placate their hosts with secretions, which Wilson has termed "appeasement substances." The appeasement substances are often produced by a special set of epidermal glands. In *Termitella*, a staphylinid beetle which lives in the nests of nasute termites, Pasteels (1968) has painstakingly described two glandular systems: the "primary" which is common in free-living and symbiotic staphylinids, and the "secondary" which is unique to the social parasites and produces the appeasement substances. Hölldobler (1971) describes analogous glandular specialization in *Atemeles*, a staphylinid found in ant nests. When an ant worker approaches an *Atemeles*, the beetle bends the tip of its abdomen toward the ant and the ant feeds on the appeasement secretion. Next, the ant licks the "adoption gland" on the lateral margins of the beetle's abdomen, and finally, the ant carries the beetle into its nest.

Attine ants, certain termites, and many scolytid beetles may transport, culture, and consume ectosymbiotic fungi. For each insect species there is a characteristic microflora, usually consisting of only one or two species

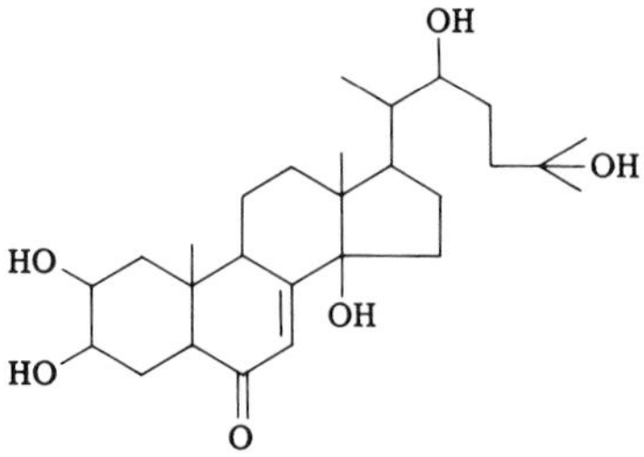

α-Ecdysone
(XVI)

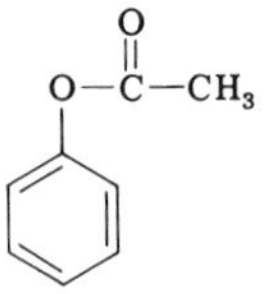

Phenylacetic acid
(XVII)

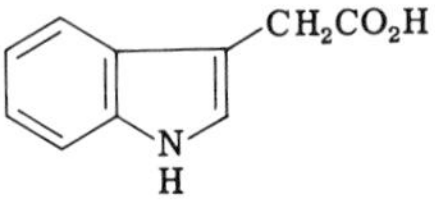

Indolylacetic acid
(XVIII)

OH
|
CH₃(CH₂)₆—CH—CH₂—CO₂H

Myrmicacin
(β-hydroxydecanoic acid)
(XIX)

Fig. 2. Allomones promoting symbiosis. XVI, Woodroach, Cleveland *et al.*, 1960; XVII, XVIII, and XIX ants, Maschwitz *et al.*, 1970; Schildknecht and Koob, 1971.

of fungi. The purity of the fungal culture maintained by these insects is intriguing, and at least for attine ants, an explanation is at hand. The ants mechanically remove alien spores from the fungus garden (Weber, 1955), provide proteolytic enzymes and amino acids which the symbiotic fungi require for optimum growth (Martin, 1970) and, according to a recent report, regulate the growth of microflora by allomones. The metapleural glands of *Atta* produce phenyl acetic acid, β-indolyl acetic acid, and β-hydroxydecanoic acid (Fig. 2) (Maschwitz *et al.*, 1970; Schildknecht and Koob, 1971). Schildknecht and his colleagues envision the role of each compound as follows: Phenylacetic acid is an antibiotic which prevents the growth of bacteria and some fungi in the fungus garden. Indoleacetic acid is, of course, a plant hormone and it is thought to promote mycelial growth of the symbiotic crop. "Myrmicacin" (β-hydroxydecanoic acid) is an inhibitor which prevents the germination of extraneous spores in the fungus garden. Growth of the symbiotic fungi is unaffected by either phenylacetic acid or myrmicacin. Myrmicacin is also found in several other ant genera, notably a harvester ant (*Messor*) which collects and stores large numbers of seeds in special chambers of its nest. These seeds do not sprout during storage, apparently because Myrmicacin prevents germination.

Many scolytid beetles transport fungi within a special cuticular compartment, the mycangium (Francke-Grossman, 1967). In many species of beetle, the mycangium is a selective culture chamber; even though weed fungi can be readily isolated from the body cuticle, only the symbiotic species proliferate within the mycangium (Barras and Perry, 1972). The growth of fungi is apparently regulated by secretory cells which surround the mycangial lumen. In *Gnathotrichus* a distinct cycle of secretory activity is correlated with the period of fungal proliferation (Schneider and Rudinsky, 1969). There are two morphological types of secretory cells in the southern pine beetle, and it may be that one cell type nourishes the symbiotic crop while the other inhibits the growth of alien contaminants (Barras and Perry, 1971; Happ *et al.*, 1971).

V. PHEROMONES

Pheromones act within a defined context: between individuals of a single species. On the basis of their modes of action, pheromones can be divided into two general classes: (1) *releasers* which trigger a relatively rapid behavioral response and (2) *primers* which produce a more gradual and prolonged shift in the physiology of the recipient (Wilson and Bossert, 1963). Releasers act mainly through the nervous system, while the mechanism of primer action usually involves the endocrine system as well. Although some authors (e.g., Bronson, 1968) have objected to the ethological implications of the word "releaser," it seems to me that such a minor semantic shortcoming hardly justifies the substitution of a new term for one which is already in general use. As treated in the present chapter, releasers affect behavior, either by evoking an immediate stereotyped response or they act in concert with other stimuli to bias the behavior of the recipient.

Communication systems are found in all animal species, although they are most highly developed in social forms. Within insect societies and within local populations of vertebrates, odor signals play many specific roles; at least a score of situations can be listed for mammals (Mykytowycz, 1970). For the purposes of this chapter, releaser pheromones will be discussed under four general headings: (1) alarm, (2) recruiting, (3) reproductive, and (4) recognition. In each case, selection has influenced both the rate of emission of the signal and the response threshold of the target organisms to favor optimum efficiency of the signal system.

A. Alarm Substances

Alarm substances communicate the presence of danger. The classic demonstration of such a chemical signal stems from the work of Karl von Frisch on the minnow *Phoxinus* (von Frisch, 1941). Von Frisch removed one minnow from a normal school, and after slightly injuring the minnow, returned it to the school. The school promptly dispersed. In a series of experiments, von Frisch demonstrated that the stimulus for dispersal was a water-borne chemical substance (Schreckstoff), and that the intensity of the response to this signal varied with its concentration and also with the physiological state of the recipient. Analogous alarm substances have been found in many fish species (Bardach and Todd, 1970). In each case, they are liberated after injury, and Pfeiffer (1962, 1963) has argued that the likely sources in some species are the club cells of the epidermis. Aside from von Frisch's (1941) data which indicate that Schreckstoff retains its potency after 5 minutes of boiling but is partially inactivated by longer boiling, almost nothing is known of the chemical nature of these substances.

Alarm substances (variously known as fright substances, warning substances, fear substances, etc.) have been found in other vertebrate groups. Tadpoles of the toad *Bufo vulgaris* produce a substance (perhaps steroidal) which elicits the fright reaction in other *Bufo* tadpoles (Eibl-Eibesfeldt, 1949; Hrbacek, 1950). Cloacal secretions may evoke an alarm reaction in snakes (Burghardt, 1970). When red foxes are alarmed, they release a mixture of short-chain carboxylic acid from their anal glands (Albone and Fox, 1971). Among rodents, alarm substances have been demonstrated in the urine of traumatized house mice (Müller-Velten, 1966), laboratory rats (Valenta and Rigby, 1968), and golden hamsters (Sherman, unpublished). It is surprising that few chemical characterizations of vertebrate alarm substances have been attempted, for since the evasive behavior is repeatedly and easily evoked, a quantitative bioassay for monitoring chemical fractions during purification could certainly be devised.

Among social insects, alarm substances are widely used to signal the presence of an intruder into the nest (Maschwitz, 1964, 1966; Butler, 1967; Blum, 1969; Gabba and Pavan, 1970; Stuart, 1970; Wilson, 1971). In at least some ant species, the alarm substance serves two roles: at low concentrations it attracts other workers while at higher concentrations it produces a state of high excitement and releases attack behavior (Wilson, 1958; Moser, 1970). Thus the alarm substance acts not only to alert other workers to the presence of danger but also to recruit other workers for a collective defense effort. Alarm substances are often

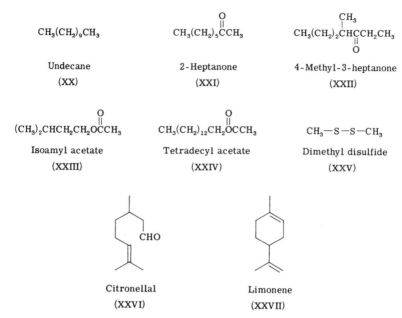

$CH_3(CH_2)_9CH_3$

Undecane

(XX)

$CH_3(CH_2)_5\overset{\displaystyle O}{\overset{\|}{C}}CH_3$

2-Heptanone

(XXI)

$CH_3(CH_2)_2\overset{\displaystyle CH_3}{\underset{\underset{O}{\|}}{CH}}CCH_2CH_3$

4-Methyl-3-heptanone

(XXII)

$(CH_3)_2CHCH_2CH_2O\overset{\displaystyle O}{\overset{\|}{C}}CH_3$

Isoamyl acetate

(XXIII)

$CH_3(CH_2)_{12}CH_2O\overset{\displaystyle O}{\overset{\|}{C}}CH_3$

Tetradecyl acetate

(XXIV)

$CH_3-S-S-CH_3$

Dimethyl disulfide

(XXV)

Citronellal

(XXVI)

Limonene

(XXVII)

Fig. 3. Representative alarm pheromones. XX, Formicine ants; XXI, bees, dolichoderine ants; XXII, myrmicine ants; XXIII, honeybees; XXIV, formicine ants (Bergstrom and Löfqvist, 1968; Regnier and Wilson, 1971); XXV, ponerine ants; XXVI, formicine ants; XXVII, termites. Unless otherwise indicated, references in Blum, 1969; Gabba and Pavan, 1970; Wilson, 1971; Law and Regnier, 1971.

emitted concurrently with defensive allomones, and may facilitate the penetration of toxicants through cuticular barriers (Ghent, 1961).

In order to effectively communicate danger and to accurately pinpoint its locus, an alarm substance must be volatile and the active space should be moderately small. Wilson and Bossert (1963) predicted that most alarm substances would have molecular weights of 100–200 and their Q/K ratios would be intermediate (10^2–10^4) so that the active space is about 10 cm in radius and that it expands and fades out within minutes. At least twenty alarm substances of insects have been identified (Blum, 1969) (see Fig. 3). Many appear to meet the criteria set forth by Wilson and Bossert (1963). C_6–C_8 ketones (XXI, XXII) or esters (XXIII, XXIV) predominate in the social Hymenoptera while termites utilize terpenoid hydrocarbons (XXVII). Often a mixture of several substances, produced by several glands, individually or collectively evoke alarm behavior (Regnier and Wilson, 1968).

Alarm pheromones are sometimes exploited for offense. Robber bees

(*Lestrimelitta limao*) obtain their protein by plundering the nests of other stingless bees (*Trigona* spp.) (Moure *et al.*, 1958). Citral (Fig. 1, XII), produced by the mandibular glands of *Lestrimelitta* workers when they attack a nest of *Trigona*, attracts other *Lestrimelitta* and pervades the nest of *Trigona*. Citral also causes disorientation and dispersal of the *Trigona*, and thus allows *Lestrimelitta* to plunder the nest without serious opposition (Blum, 1966). An analogous situation is found in slave-maker ants (*Formica sanguinea*). As *F. sanguinea* raid other colonies to obtain slave workers, the raiders expel their alarm pheromones (a mixture of decyl, dodecyl, and tetradecyl acetates). The mixture, dubbed "propaganda substances," produces apparent panic in the workers of the colony under attack and thus any organized defense is precluded (Regnier and Wilson, 1971).

Danger to one species is often danger to many. The alarm vocalizations of many species of passerine birds are almost identical, and thus different species alert one another (Marler, 1959). Alarm pheromones also act interspecifically: Maschwitz (1964) has shown that common alarm substances are often found throughout genera or even subfamilies of social Hymenoptera. Rodent alarm pheromones have yet to be analyzed in this regard. Only three species of rodents have been studied and the experiments were confined to artificial laboratory situations. In the wild, alarm pheromones most probably affect dispersal and therefore they are density-regulating. If species-specificity is low, then the sharing of a chemical signal must influence interspecies interactions as well. The possibility deserves investigation.

B. Recruitment Pheromones

Chemical signals are employed by worker castes of social insects to guide their nest mates to a food source. These signals are of two kinds: stationary scent marks at the site of the food and chemical trails which lead to it (Gabba and Pavan, 1970; Blum, 1970; Moser, 1970; Wilson, 1971; Stuart, 1970).

Trail substances are common in ants and termites. The trail is laid by workers returning to the nest from a food source. The details of the trail laying behavior vary with the particular biology of each species. In fire ants the gland producing the trail substance is associated with the sting (Wilson, 1962) while in termites, the trail substance is produced by sternal glands (see Stuart, 1970 for references). The trail substance left behind the returning forager forms an active space which guides outgoing foragers. At least in fire ants, the Q/K ratio is quite small so that the active space is narrow and, unless other returning

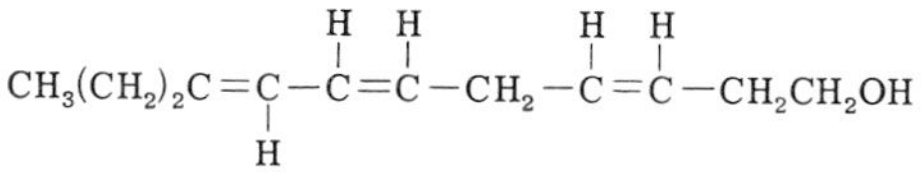

$$CH_3(CH_2)_2C=\overset{\displaystyle H}{\underset{\displaystyle H}{C}}-\overset{\displaystyle H}{C}=\overset{\displaystyle H}{C}-CH_2-\overset{\displaystyle H}{C}=\overset{\displaystyle H}{C}-CH_2CH_2OH$$

$$CH_3(CH_2)_4CO_2H$$

cis-3, *cis*-6, *trans*-8-Dodecatriene-1-ol

Hexanoic acid

(XXVIII)

(XXIX)

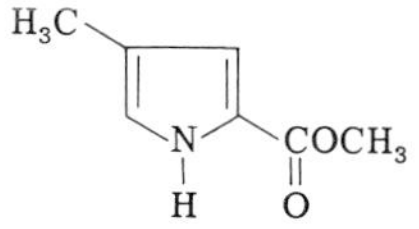

H_3C⧵
        ⧵
         ⧸N⧵COCH_3
         | ‖
         H   O

4-Methylpyrrole-
2-carboxylate

(XXX)

Fig. 4. Trail substances. XXVIII, Termite (Matsumura *et al.*, 1968); XXIX, termite (Hummel and Karlson, 1968); XXX, attine ants (Tumlinson *et al.*, 1971).

foragers reinforce the original trail, the trail evaporates within 100 seconds (Wilson, 1962). Hangartner (1967) showed that outgoing foragers of the ant *Lasius fuliginosus* actually weave from side to side within the active space of the trail and apparently turn back toward the original trail whenever the concentration of the pheromone falls below *K*.

Very few trail substances have been identified (Fig. 4). In termites, Hummel and Karlson (1968) have implicated a non-terpenoid hydrocarbon ($C_{11}H_2O$) and hexanoic acid (XXIX) as the trail substance for *Zootermopsis* and Moore (1966) belives a diterpenoid hydrocarbon ($C_{20}H_{32}$) plays this role for *Nasuititermes*. Matsumura *et al.* (1968) have shown that *cis*-3, *cis*-6, *trans*-8 dodecatrienol (XXVIII) is most probably the trail substance of *Reticulitermes*. Very recently Tumlinson *et al.* (1971) have identified 4-methyl-pyrrole-2-carboxylate (XXX) as the trail substance of the leaf-cutting ant *Atta texana*. Ant workers will follow a trail which contains less than 10^{-13} gm/cm (3.48×10^8 molecules/cm). One-third milligram is sufficient to lay a trail around the earth! Unlike the short-lived trails of *Solenopsis* (Wilson, 1962), those of *Atta* persist for several days (Moser, 1970).

Although trails are rare in flying insects, certain South American bees (Meliponi) are exceptions. As elegantly shown by Lindauer and Kerr (1958, 1960), foragers returning from a food source pause every 2 or 3 meters to mark tufts of vegetation with droplets of secretion from their mandibular glands. Outgoing foragers follow the aerial odor trail, and returning recruits reinforce the scent marks. Blum and associates (cited in Blum, 1970) have recently identified some of these secretions. In *Trigona postica,* ten methyl ketones, benzaldehyde, and two hydrocarbons are present. In *T. tubiba,* eight of the methyl ketones are lacking. It is quite possible that the collective effect of several molecules allows workers to distinguish trails of their own species.

Are other trail pheromones species specific? The answer is unclear and is complicated by the fact that results in the laboratory (where many species follow one anothers' trails) seem in conflict with the field data, which suggest species and even nest specificity. It may be that in addition to the primary trail pheromone, each nest has its own dialect due to minor components (Blum, 1970).

Although the dance language of honeybees seems to be the major communication system used for recruitment over some distance (von Frisch, 1967), pheromones play a role in short-range attraction. When a worker honeybee has located a good food source, she often exposes her abdominal Nasanov gland and fans her wings. The acylic terpenes emitted from this gland, which include geraniol, citral (both isomers), geranic acid, and nerolic acid, attract other foragers (Boch and Shearer, 1962; 1964; Butler and Calam, 1969). According to Butler and Calam (1969), citral is the most attractive constituent.

C. Sex Pheromones

Scents influence reproductive behavior in many species. The chemical signals may act either as attractants which bring the sexes together or aphrodisiacs which trigger specific aspects of precopulatory or copulatory behavior.

The most studied communication systems are those of nocturnal Lepidoptera. In a typical situation, a stationary female exposes a gland in her abdomen from which attractant molecules diffuse into the surrounding air. Air movements cause the active space to form a scent plume extending downwind from the female. When pheromone concentration exceeds the threshold for the males, they fly to the female. It seems unlikely that the males find the female by following a concentration gradient per se, but rather they orient by anemotaxis, i.e., merely flying upwind. By a very convincing series of experiments with flightless male silkmoths which ran along a surface to find "calling" females, Schwinck (1958) showed that the orientation to females is a two-step process: at low concentrations of pheromone, the males run upwind, and at high concentrations they search randomly. Thus anemotaxis accounts for long-distance orientation while random search apparently suffices when females are nearby.

Females emit their signals only in the proper context: females call only in the early hours of the evening and female *Polyphemus* moths call only in the presence of 2-hexenal, a volatile constituent of their food plant (Riddiford, 1967). The majority of the attractants which have been isolated are long-chain alcohols, esters, or acids (Fig. 5). Most often, a single molecular species appears to constitute the signal and the differences between species may appear slight. For example, two species of gelechid moths utilize 9-tetradecenyl acetate (XXXIII) as a sex attractant, but one species produces the *trans*-isomer and the other the *cis*-isomer. Furthermore, each isomer attracts only conspecific males and inhibits the response of the other species to its own isomer (Roelofs and Comeau, 1969).

Although one might expect that every species would emit its own unique molecular signal, such is not the case. Often, signal ambiguity is avoided in spite of chemical overlap because the species are temporally, geographically, or ecologically isolated. In other species, minor synergistic chemicals apparently prevent signal ambiguity (Brady *et al.*, 1971).

Upon reaching the female, the male may release an aphrodisiac which induces the female to mate. In noctuid moths, the aphrodisiac is liberated from special scent brushes everted from the male's abdomen (Birch, 1970). In noctuids the molecules are small and highly volatile, for example, butyric acid or benzaldehyde (Alpin and Birch, 1970). Birch (1970) showed that after surgical removal of the brushes, the females were unresponsive and mating was unsuccessful.

For many day-flying butterflies, visual cues mediate long-distance attraction of the sexes and many specific stages in the courtship sequence. However, for the queen, the monarch, and other danaid butterflies, an aphrodisiac pheromone is required for successful copulation. The aphrodisiac is produced in the eversible hair pencils of the male and dusted onto the antennae of the female (Brower *et al.*, 1965; Pliske and Eisner, 1969). The molecule which seduces the female queen is a ketonic pyrrolizidine (methyl 1,2,3-dihydro-1*H*-pyrrolizidin-1-one, XL) (Meinwald *et al.*, 1969; Pliske and Eisner 1969).

Sex attractants of several coleopterans have been chemically characterized. The female black carpet beetle (*Attagenus*) attracts the male by emitting megatomic acid (XXXII) (Silverstein *et al.*, 1967), a substance rather like those produced by many female Lepidoptera. Four attractant substances are produced by female *Trogoderma*, two of which have been identified as a C_{16} alcohol and a methyl ester of the corresponding C_{16} acid (Rodin *et al.*, 1969). The male boll weevil (*Anthomonas*) produces a mixture of two terpenoid alcohols (XXXVIII, XXXIX) and

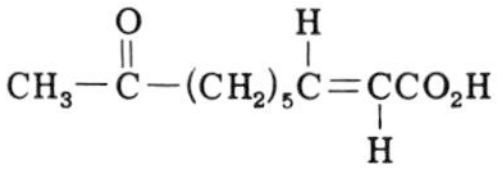

$$CH_3-\overset{\overset{\displaystyle O}{\|}}{C}-(CH_2)_5\overset{\overset{\displaystyle H}{|}}{C}=C\overset{}{CO_2H}$$
$$\underset{H}{|}$$

Honeybee queen substance
(*trans*-9-keto-2-decanoic acid)

(**XXXI**)

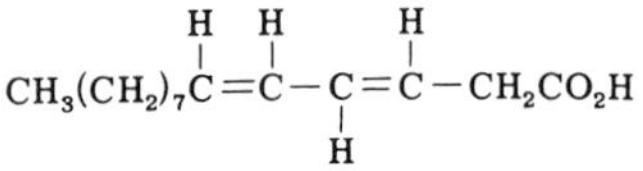

$$CH_3(CH_2)_7\overset{\overset{\displaystyle H}{|}}{C}=\overset{\overset{\displaystyle H}{|}}{C}-\overset{}{C}=\overset{\overset{\displaystyle H}{|}}{C}-CH_2CO_2H$$
$$\underset{H}{|}$$

Megatomic acid (*trans*- 3, *cis*-
5-tetradecadienoic acid)

(**XXXII**)

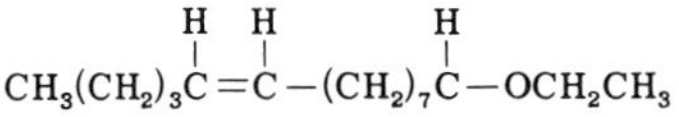

$$CH_3(CH_2)_3\overset{\overset{\displaystyle H}{|}}{C}=\overset{\overset{\displaystyle H}{|}}{C}-(CH_2)_7\overset{\overset{\displaystyle H}{|}}{C}-OCH_2CH_3$$

cis-9-Tetradecyl acetate

(**XXXIII**)

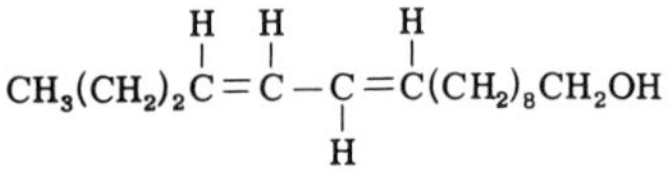

$$CH_3(CH_2)_2\overset{\overset{\displaystyle H}{|}}{C}=\overset{\overset{\displaystyle H}{|}}{C}-\overset{}{C}=\overset{\overset{\displaystyle H}{|}}{C}(CH_2)_8CH_2OH$$
$$\underset{H}{|}$$

Bombykol (*trans*-10, *cis*-
12-hexadecadien-1-ol)

(**XXXIV**)

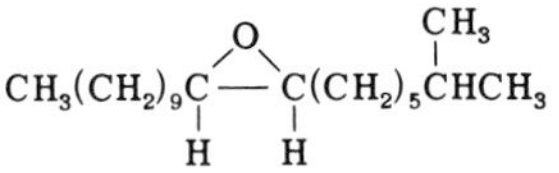

$$CH_3(CH_2)_9\overset{}{C}\overset{\overset{\displaystyle O}{\diagup\!\diagdown}}{}\overset{}{C}(CH_2)_5\overset{\overset{\displaystyle CH_3}{|}}{C}HCH_3$$
$$\underset{H}{|}\underset{H}{|}$$

Disparlure (*cis*-7, 8-epoxy-
2-methyl octadecane)

(**XXXV**)

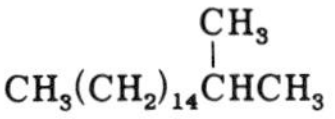

$$CH_3(CH_2)_{14}\overset{\overset{\displaystyle CH_3}{|}}{C}HCH_3$$

2-Methyl heptadecane

(**XXXVI**)

Figs. 5 and 6. Representative sex pheromones. XXXI, Honeybee queen; XXXII, female black carpet beetle; XXXIII, female gelecid moth; XXXIV, female silkmoth; XXXV, female gypsy moth; XXXVI, female arctiid moths (Roelofs and Cardé, 1971); XXXVII, female housefly (Carlson *et al.*, 1971); XXXVIII and XXXIX, male boll weevil; XL, male danaid butterflies; XLI, male muskdeer (Lederer, 1950). Unless otherwise indicated, references in Law and Regnier (1971).

two aldehydes which act synergistically to attract the females (Tumlinson *et al.*, 1969) (Fig. 6). In the scolytid beetle, *Ips confusus*, male feces contain a mixture of three terpene alcohols which attract both sexes and all three are required for full biological activity (Silverstein *et al.*, 1966). Although Vite (1967) has pointed out that these substances are not true sex attractants (since they primarily promote aggregation), mating occurs within these aggregations and I will therefore include them in this section. In other scolytids of the genus *Dendroctonus*, a series of terpenoid compounds, produced by males or females, promote the aggregations and thus allow mating (see Silverstein, 1970 for references). The most interesting feature of these coleopteran pheromones is that, at least for several species, the signal is a medley of several substances.

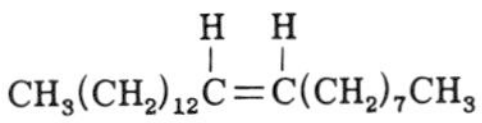

$$CH_3(CH_2)_{12}\overset{\overset{H}{|}}{C}=\overset{\overset{H}{|}}{C}(CH_2)_7CH_3$$

cis-9-**Tricosene**

(**XXXVII**)

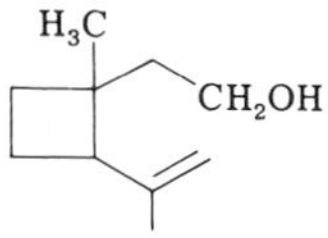

cis-2-Isopropenyl-1-
methylcyclobutane ethanol

(**XXXVIII**)

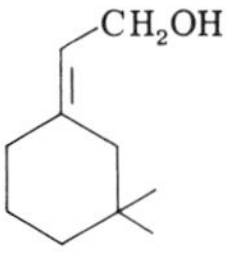

cis-3, 3-Dimethyl-$n^{1,\,\beta}$
cyclohexane ethanol

(**XXXIX**)

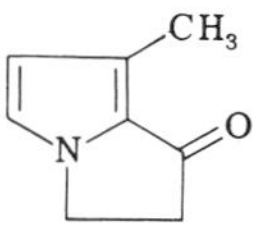

2, 3-Dihydro-7-methyl-
1H-pyrrolizin-1-one

(**XL**)

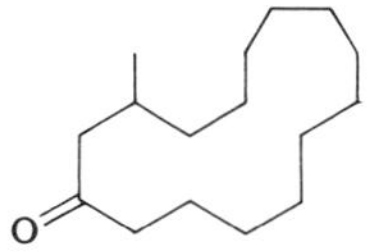

Muskone

(**XLI**)

See legend on page 170.

Sex pheromones have been identified for many other insects, including houseflies (*cis*-9-tricosene (XXXVII); Carlson *et al.*, 1971) and honeybees (9-keto-2-decenoic acid (XXXI); Gary, 1962). Assembly scents (esters, alkanones, alcohols, hydrocarbons) which attract conspecifics of both sexes are produced by the mandibular glands of male bumblebees (Kullenberg *et al.*, 1970). Aphrodisiacs have been reported for numerous insect species, for example "seducin" produced by some male cockroaches (*Nauphoeta cinerea*) (Roth and Dateo, 1966).

Although it is known that many vertebrates possess sex pheromones (principally aphrodisiacs), relatively little chemical information is available. A number of fish (Bardach and Todd, 1970; also Losey, 1969; Rossi, 1969; Gandolfi, 1969) produce aphrodisiacs. Newts apparently possess sex attractants (Twitty, 1955). The musk glands of male alligators produce yacarol, a mixture of compounds including citronellal (Lederer, 1950), which may be attractive to females (Burghardt, 1970). The complex mammalian secretions used by the perfume industry, including muskone from male musk deer (XLI), civetone from male civet cats, and beaver castoreum (Lederer, 1950) may well be sex pheromones or territorial markers. Most of the putative mammalian sex pheromones have a musky odor and are large cyclic compounds, either steroids (such as the "boar taint substance") or cycloketones (muskone and civetone). The proposition that man may have similar pheromones has been delightfully argued by Comfort (1971a, b). Unfortunately, there is little experimentally derived data on the exact roles of these musky scents in the reproduction of any mammal.

Many female mammals indicate their physiological readiness to mate by emitting characteristic scents (Gleason and Reynierse, 1969; Le Magnen, 1970; Mykytowycz, 1970). Recently, Michael, Keverne, and their co-workers have shown that vaginal secretions of receptive female rhesus monkeys contain an aphrodisiac pheromone which they call "copulin." Copulin production is estrogen-dependent, and thus copulins are not present in ovariectomized females (Michael and Keverne, 1970). Topical application of estrogen-stimulated vaginal secretions on to the sexual skin of ovariectomized females renders these females attractive to males which respond by mounting, ejaculation, or masturbation. Ovariectomized females were used routinely for bioassay of gas-chromatographic fractions from vaginal secretions, and the copulins have been identified. They comprise a mixture of a short-chain acids, namely acetic, propionic, isobutyric, isovaleric, and isocaproic (Michael *et al.*, 1971; Curtis *et al.*, 1971).

D. Territoriality and Recognition Scents

Pheromones are widely used to mark territories in mammals (Gleason and Reynierse, 1969; Mykytowyctz, 1970; Ralls, 1971). With the exception of the musky scents exploited by perfumery (Lederer, 1950), little is known of their chemical nature. These scents may be deposited in dung or urine or they may be produced by special glands, for example, the chin glands of rabbits (Mykytowycz, 1970).

It is often difficult to distinguish between scents which label a territory, scents which signal social status, and scents which allow individual recognition. In mammals, a medley of exocrine products may play all three roles.

The "colony odors" of ants, bees, and wasps have long been recognized but have resisted precise chemical characterization to date. Perhaps this

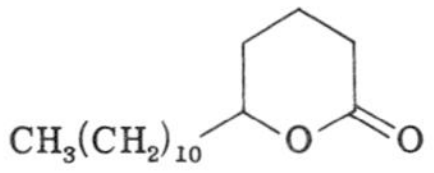

5-Hydroxydecanoic
acid lactone

(XLII)

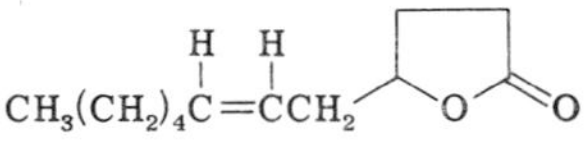

cis-4-Hydroxydodec-
6-enoic acid lactone

(XLIII)

Fig. 7. Recognition pheromones. XLII, Oriental hornet queen (Ikan *et al.*, 1969); XLIII, male black-tailed deer, (Brownlee *et al.*, 1969).

is because these odors are actually medleys, and the scent of each colony is influenced by diet and microhabitat (Wilson, 1971). Recognition of social status is also mediated by pheromones; 9-keto-2-decanoic acid and 9-hydroxy-2-decanoic acid produced by queen honeybees causes settling of worker swarms (Butler and Simpson, 1967), while δ-*n*-hexadecalactone (Fig. 7, XLII) produced by a queen hornet (*Vespa orientalis*) attracts workers and stimulates them to build new queen cells at the appropriate season (Ikan *et al.*, 1969).

In many vertebrates, mother-young recognition is dependent on chemical cues (Gleason and Reynierse, 1969). Within a local population, social status is often correlated with urination; for example, the urine of male mice contains a pheromone that increases aggression in other males. The urine from dominant male mice is more potent than that from submissive ones (Mugford and Nowell, 1970).

Only one known vertebrate recognition scent has been chemically identified: *cis*-4-hydroxydodec-6-enoic acid lactone (XLIII) produced by the tarsal glands of male black-tailed deer (Brownlee *et al.*, 1969; Müller-Schwarze, 1969). The female deer lick and nuzzle the tuft of hair associated with the tarsal gland and apparently distinguish between individuals on the basis of the tarsal scent. Several components are present in the natural secretion, and the lactone is merely one. For a maximum response from the female, at least four components are necessary (Müller-Schwarze, 1969).

E. Primer Pheromones

The actions of primer pheromones are more covert than are those of the releasers discussed above, for primers regulate physiological functions of a recipient. Primers may inhibit, accelerate, or synchronize reproductive maturation and/or reproductive cycles in the target organisms. Primers may be produced by either or both sexes and may operate on males or females. For the desert locust (*Schistocerca gregaria*), mature males give off a scent which accelerates reproductive maturation of young males (Loher, 1961), and mature mealworm beetles of both sexes emit a scent which increases the rate of ovarian growth in young females (Happ *et al.*, 1971).

The development of the various castes in insect societies is largely regulated by pheromones. The mandibular glands of queen honeybees produce 9-keto-2-decanoic acid and 9-hydroxy-2-decanoic acid which inhibit both the growth of ovaries of worker bees and the building of queen cells by the workers (Butler and Fairey, 1963; Butler and Callow, 1968). In a series of elegant experiments, Lüscher and his co-workers have shown that several pheromones regulate differentiation of worker termites into reproductives in *Kalotermes flavicollis*. In the termite colony, there is normally only one functional reproductive of each sex. Each royal male or female produces an inhibitory pheromone which prevents reproductive maturation of pseudergate workers of the same sex. In the absence of the appropriate inhibitor, a pseudergate molts several times and transforms into a replacement reproductive. In addition, the royal male produces a pheromone which accelerates transformation of pseudergate females. The pheromones are passed by contact, from the royal pair to pseudergates and thence between pseudergates (Lüscher, 1961).

Although primer pheromones are suspected in many mammalian species, they have been clearly demonstrated in only one order, Rodentia. At least four distinct roles are played by primer pheromones in mice (Wilson, 1970; Whitten and Bronson, 1970). The *Lee–Boot effect* is a suppression of estrous and the development of pseudopregnancy in over 50% of the females when four or more females are grouped together and a male is not present (Lee and Boot, 1955). The *Whitten effect* is the induction or acceleration of the estrous cycle in the female mouse when she is exposed to an odor from male urine. This effect is most clearly seen in groups of females (after the Lee–Boot effect), (Whitten, 1958). The *Bruce effect* is a failure of implantation and rapid return to estrous in a female mouse which has been exposed to the odor of a strange male whose odor is unlike that of her stud (Bruce, 1960). The *Ropartz effect* describes the adrenal hypertrophy which occurs when isolated mice are exposed to the odor of other mice (Ropartz, 1966, 1968). Most of these investigations have utilized laboratory mice, and demonstration of similar effects in laboratory rats has been difficult. However, many of the experiments have been repeated with deermice (*Peromyscus*) and thus the phenomena may be widespread among rodents (Bronson and Eleftheriou, 1963; Bronson and Marsden, 1964; Bronson and Dezell, 1968).

The physiological effects of these pheromones are usually statistical, and attempts to develop quantitative bioassay techniques have been discouraging (Whitten and Bronson, 1970). The lack of reliable assays probably accounts for the fact that none of the molecules have been identified. The adaptive significance of these pheromones is unclear, although both the Lee–Boot and Ropartz effects may contribute to the well-known stress syndrome observed in cases of very dense rodent populations (Wilson, 1970; Whitten and Bronson, 1970).

F. Pheromone Perception

Airborne pheromones are detected by primary olfactory sense cells in the antennae of mandibulate arthropods and the nasal epithelia of vertebrates. An analysis of the primary transduction at the receptor and the processing of information in the central nervous system is prerequisite to an understanding of the characteristics of the signal system. The analysis is far from complete.

Both behavioral studies, such as those cited earlier, and the capacity of the appropriately educated human nose attest to the fact that vertebrates can distinguish a large number of specific odorants. The discriminative powers are extremely refined: some pairs of enantiomeric isomers differ in their smell (Russell and Hills, 1971; Friedman and Miller, 1971; Leitereg *et al.*, 1971). Neurophysiological investigations of the vertebrate nasal epithelium and its primary receptor cells provide only slight insight into the olfactory process. In their excellent study of olfaction in the frog, Gesteland *et al.* (1965) recorded from a large number of individual receptors; specific odorants produced shifts in the rather irregular background firing of each unit. Each individual unit is odor-selective, i.e., sensitive to a certain set of molecular species and, within this set, some molecules increase the frequency of action potentials while others inhibit depolarizations. The units are highly heterogeneous, and could not easily be grouped into distinct classes. Downstream, the primary fibers terminate in the glomeruli of the olfactory bulb. On anatomic grounds, one can argue that the dendritic field of the mitral, tufted, and plexiform cells within the bulb are capable of filtering and consolidating the signals from the primary fibers. If a few molecules are to trigger a behavioral response, some sort of signal amplification is probably necessary. Recently, Nicoll (1971) has obtained neurophysiological evidence that recurrent axons from mitral and tufted cells run back into the external plexiform layer and within this layer excite other mitral or tufted cells. Positive feedback via the recurrent axons could be the basis of signal amplification.

In vertebrates, relatively few studies have concerned pheromone reception *per se,* and those few have been primarily concerned with a demonstration that olfaction was involved in the response to the molecules. Anosmic minnows do not react to Schreckstoffe (von Frisch, 1941), nor do anosmic female mice exhibit the Lee–Boot or Whitten effects (Lee and Boot, 1956; Whitten, 1965). Repeated topical application of male urine on to the nostrils of intact female mice produces the Whitten effect (Marsden and Bronson, 1964). Male rhesus monkeys may be rendered reversibly anosmic with noseplugs; when the noseplugs are in place, the males do not resond to the female copulins (Michael and Keverne, 1968).

A recent study by Pfaff and Gregory (1971) employed putative crude pheromone (namely the urine of female rats) in an attempt to analyze coding in the olfactory bulb and medial forebrain bundle of normal and castrated male rats. It had previously been shown that urine from "estrous females" evokes more intense male exploratory behavior than does that from ovariectomized females (Pfaff and Pfaffmann, 1969). Pfaff and Gregory (1971) were unable to detect units in the olfactory bulb or preoptic area which responded *exclusively* to estrous female urine, but 24% of the units in the olfactory bulb and 58% of the units if the preoptic area did respond differentially to urine of estrous and ovariectomized females. Perhaps the difference between the areas reflects the signal amplification suggested by Nicoll (1971).

Electrophysiological studies of pheromone reception in insects have met with considerable success. The neural response to pheromone stimuli has been monitored on three levels: at the individual olfactory sensillum over the antennae as a whole, and in the antennal lobe of the brain. The most extensive studies are those of Dietrich Schneider and his associates on the antennae of the domestic silkmoth *Bombyx mori* (see Schneider, 1969, 1970 for references).

When whole antennae of male *Bombyx* were exposed to bombykol, the sex attractant of the female, Schneider detected a slow potential shift of a few millivolts—termed the electroantennogram (EAG). The magnitude of the EAG is proportional to the logarithm of bombykol concentration and apparently the EAG represents the summed generator potentials of many receptor cells. Microelectrode studies of individual receptor cells revealed that they fall into two classes: odor specialists which react only to the pheromone, and odor generalists which respond to a variety of scents but differ widely from one another. Many odor specialists are present, at least 25,000 on each antenna (Kaissling and Priesner, 1970). By using tritiated bombykol, it has been possible to estimate accurately how many molecules of bombykol are necessary at each receptor for a unit response, and how many units must be affected to trigger a behavioral response. At behavioral threshold, 200–300 molecules of bombykol are available to 25,000 cells. By application of Poisson statistics, it was calculated that 99% of the units were stimulated by only a single molecular hit (Kaissling and Priesner, 1970). EAG's in response to pheromones have been recorded in other Lepidoptera as well in cockroaches and bark beetles (Payne, 1970 and references therein).

To my knowledge, recording from the antennal lobes has been successful with only two species, the American cockroach (Yamada, 1971) and the mealworm beetle (Freundlich and Happ, unpublished). In the male mealworm some of the units (presumably at least second order) in the antennal lobe responded differentially to the scents of male and of female beetles, while certain of the units in the cockroach antennal lobe were specialists for partially purified pheromones.

How specific are the specialist receptor cells? For *Bombyx*, the behavioral threshold concentration for bombykol (hexadeca-10-*trans*, 12-*cis*-dien-1-ol) is at least two orders of magnitude lower than the related *cis-trans* isomers (Schneider, 1963). Theories to explain the mechanisms by which primary sense cells detect odorant molecules are numerous, but the most recent debate has concerned two alternatives: the vibrational theory advocated by Wright (1966) and the stereochemical theory espoused by Amoore (1964). The bulk of the evidence on pheromone perception in insects favors the stereochemical interpretation. Potency of structural analogs is well correlated with the similarity between the size and shape of the analog and that of the natural pheromone (Amoore *et al.*, 1969). Blum *et al.* (1971a) screened 99 ketones $(C_2–C_{13})$ for their effectiveness in releasing the alarm response in the harvester ant *Pogonomyrmex badius*. Two of the 99 were maximally potent: 4-methyl-3-heptanone, the natural pheromone identified by McGurk *et al.* (1966), and 4-methyl-3-hexanone. Structural similarities are obvious. Furthermore, deuteration of these substances, which shifts their vibrational spectra, has no effect on their biological activity, and this fact argues against the Wright theory. Other studies using analogs of alarm substance in honeybees (Boch and Shearer, 1971) and a trail substance of termites (Tai *et al.*, 1971) also support the Amoore explanation.

When both the natural pheromone and an analog are presented simultaneously, the analog can inhibit or act synergistically with the natural pheromone. Roelofs and Comeau (1971a) have demonstrated such interaction between *cis*-11-tetradecenyl acetate and a series of analogs of this attractant for male red-banded leaf rollers (Torticidae: Lepidop-

tera). Both synergists and attractants are similar to the natural pheromone, and in fact, all chemicals which are attractant or attractant-modi-
fying elicit strong EAG's (Roelofs and Comeau, 1971b). It may be that
the modifiers affect the time course of sensory adaptation or central
habituation to the natural pheromone, i.e., inhibitors accelerate habituation while synergists prolong habituation (Roelofs and Comeau, 1971a).

In a fascinating paper, Riddiford (1970) reported that after antennae
of male saturniid moths (*Anthera pernyi*) had been exposed to tritiated
female scent, a saline wash of the antennae contained a radioactively
labeled protein. This protein might serve to convey the attractant
through overlying fluid to the sense cell, or it might be the "receptor
protein" on the surface of the cell membrane.

VI. EXOCRINES AND ENDOCRINES

Chemical regulatory systems are ubiquitous. Semiochemicals occur
not only in metazoans but also in protozoans (Siegel and Cohen, 1962;
Starr, 1968). Most probably, chemical signals between unicellular organisms appeared early in the evolution of living systems, and as Wilson
(1970) has suggested, "pheromones are in a special sense the lineal
ancestors of hormones." Among the lower plants, chemical signals between reproductive cells have been classed as hormones (Raper, 1970)
or pheromones (gamones) (Müller *et al.*, 1971). In these forms, the
distinction is largely a matter of the taste of the experimenter. Both
in their origins and in their interactions, one can see the close relationships between the internal and the external signal systems.

Hormones act within a single organism, pheromones between genetically similar organisms (of the same species), and allomones between
genetically dissimilar organisms. The internal system should be relatively
free of noise, since sender–molecule–milieu–target are all part of a closed
system and are co-adapted for efficient communication. Chemical noise
is inevitable in the external milieu through which pheromones and allomones are transmitted. For pheromones and symbiotic allomones, the
selective pressures operate on both the producer and the target organism
to favor an appropriate diffusible molecule which is emitted at reasonable rates by the producer and is discriminated at optimum distance
by the target. For defensive allomones, selection operates on the producer to favor signal efficiency and on the target to favor mechanisms
which allow the signal to be ignored. Thus the most common defensive
allomones are general toxicants.

Similar carbon chains form the skeletons of the signals which constitute allomones, pheromones or hormones. The same molecule may function in two of the categories, for example ecdysone (a symbiotic allomone and molting hormone of the cochroach *Cryptocercus*) and citral which acts as a defensive allomone, an alarm substance, a recruitment substance, and a propaganda substance in ants. In spite of the almost unlimited structural diversity theoretically possible for chemical signals (see Wilson and Bossert, 1963), only a few classes of molecules are commonly utilized: aliphatics, terpenoids, peptides (hormones), and in a few cases, small cyclic compounds. The explanation of this conservatism, "biochemical parsimony" as Blum (1969) has termed it, is probably two-fold: first, a sort of Hendersonian "fitness" (Henderson, 1958) of those molecules which diffuse properly, and second, the fact that it is easier for cells to make some carbon skeletons than others. It is simpler to modify a pre-existing pathway than to develop an entirely new one. When unusual carbon skeletons are found (e.g., iridomyrmecin) their biogenesis from intermediates in pre-existing pathways can be predicted.

Some generalizations about the molecules employed as allomones and pheromones are possible. In general, the requirements of volatility and reasonable specificity, enunciated so clearly by Wilson and Bossert (1963), have been supported by the chemical identifications over the last decade. For the most part, defensive allomones are distinguished by a functional group (often carbonyl) which renders the molecule reactive and toxic. The more specific pheromones, such as sex attractants, tend to be less reactive and to have a certain structural rigidity. Sex attractants are often terpenoid or *unsaturated* fatty acid derivatives. As Clayton (1970) has suggested, the unsaturation introduces structural rigidity which makes the geometric shapes of these fatty acid derivatives, like that of terpenes, quite highly defined. Such speculation is of course consistent with the stereochemical theory of olfaction (Amoore, 1964), and the discrimination of *cis-trans* isometers from one another (Schneider, 1963). In spite of the fact that only a few chemical classes are utilized, many distinct signals are possible because of the specificity of the biochemical pathways which produce them, and also the specificity of the receptors which detect them.

An increasing number of chemical signals are proving to be medleys of several substances. Theoretically this allows an increase in information if the various molecular species all diffuse at a common rate. One might expect medleys to be more common at close range (copulins of rhesus monkeys or tarsal scent of deer) than at great distances.

Pheromones influence the endocrine system and endocrines influence both pheromone output and receptivity to pheromones. The pheromone regulation of endorcine activity is most dramatically seen in the effects of primer pheromones. Both in social insects (Wilson, 1971) and in mice (Whitten and Bronson, 1970), odors control endocrine gland size and the effects of the pheromone may be prevented by hormone injections. It is also apparent that physiological state, including endocrine activity, affects emission. In cockroaches (Barth, 1961), saturniid moths, (Riddiford and Williams, 1971), mealworm beetles (Menon, 1970), and rhesus monkeys (Michael and Keverne, 1970), a certain endocrine state is prerequisite to pheromone production. Also, an increase in the behavioral response of insects to sex pheromones often accompanies reproductive maturation (Shorey *et al.*, 1968; Happ 1971), and this increased responsiveness apparently stems from shifts in the central nervous system, since the EAG's of mature and immature males are indistinguishable (Payne *et al.*, 1970). A direct correlation between hormone levels and olfactory sensitivity has been demonstrated in man by Le Magnen (1948, 1950). Le Magnen has shown that the synthetic compound exaltolide is odorless to men and children but strongly musky to women, and the sensitivity of women to exaltolide varies with the stage of the menstrual cycle. In addition, estrogen-treated men can smell exaltolide. Vierling and Rock (1967) have confirmed many of Le Magnen's results.

If the odorous steroids found in mammalian urine are functional pheromones, the origins of these steroids, their delivery, and their titer epitomize physiological economy. The steroids may well originate as by-products of circulating hormones; thus little or no special biochemical or cytological machinery is necessary for their production. The steroids are not exported through an independently derived gland, but merely pass into the nephron and are not reabsorbed. If the level of circulating hormones is related to the titer of odorous steroids in urine, then the coordinating link between endocrines and exocrines is built into the system.

Chemical ecology is yet in its infancy. The importance of chemical signals between organisms is increasingly apparent, and the structures of many signals have been established. Many more signals remain to be characterized and many of their roles need more precise definition. The field of chemoreception and subsequent processing of the information is scarcely understood. Much more information is needed on the ways in which external and internal chemical signals interact with each other. Finally, the potential importance of exocrine signals as regulators of population density, acting for example as epideitic pheromones (Corbet, 1971), and the contribution of such signals to the stability of an ecosystem are largely matters of appealing conjecture.

ACKNOWLEDGMENTS

Work carried out in these laboratories and discussed above was supported by Grants in Aid USPHS Grant CC00343, NIH Biomedical Institutional Grants to New York University, and NSF Grant GB 29502X.

REFERENCES

Albone, E. S., and Fox, M. W. (1971). Anal gland secretion of the red fox. *Nature (London)* **233**, 569–570.

Alpin, R. T., and Birch, M. C. (1970). Identification of odorous compounds from male Lepidoptera. *Experientia* **26**, 1193–1194.

Amoore, J. E. (1964). Current status of the steric theory of odor. *Ann. N.Y. Acad. Sci.* **116**, 457–476.

Amoore, J. E., Palmieri, G., Wanke, E., and Blum, M. S. (1969). Ant alarm pheromone activity: correlation with molecular shape by scanning computer. *Science* **165**, 1266–1269.

Aneshansley, D., Eisner, T., Widom, J. M., and Widom, B. (1969). Biochemistry at 100°C: The explosive discharge of bombardier beetles (*Brachinus*) *Science* **165**, 61–63.

August, C. J. (1971). The role of male and female pheromones in the mating behaviour of *Tenebrio molitor*. *J. Insect Physiol.* **17**, 739–751.

Baker, J. R. (1958). "Principles of Biological Microtechnique." Methuen, London.

Bardach, J. E., and Todd, J. H. (1970). Chemical communication in fish. *In* "Communication by Chemical Signals. Advances in Chemoreception" (J. W. Johnston, Jr., D. G. Moulton, and A. Turk, eds.). Vol. I, Appleton, New York.

Barras, S. J., and Perry, T. (1971). Gland cells and fungi associated with the prothoracic mycangium of *Dendroctonus adjunctus*. *Ann. Entomol. Soc. Amer.* **64**, 123–126.

Barras, S. J., and Perry, T. (1972). Fungal symbionts in the prothoracic mycangium of *Dendroctonus frontalis* (Coleoptera: Scolytidae). *Z. Angew. Entomol.* **71**, 95–104.

Barth, R. H. (1961). Hormonal control of sex attractant production in the Cuban cockroach. *Science* **133**, 1598–1599.

Bergström, G., and Löfqvist, J. (1968). Odour similarities between the slave-keeping ants *Formica sanguinea* and *Polyergus rufescens* and their slaves *Formica fusca* and *Formica rufibarbis*. *J. Insect Physiol.* **14**, 995–1011.

Birch, M. (1970). Precourtship use of abdominal brushes by the nocturnal moth, *Phlogophora meticulosa* (L.) (Lepidoptera: Noctuidae). *Anim. Behav.* **18**, 310–316.

Blum, M. S. (1966). Chemical releasers of social behavior—VIII. Citral in the mandibular gland secretion of *Lestrimelitta limao* (Hymenoptera: Apoidea: Melittidae). *Ann. Entomol. Soc. Amer.* **59**, 962–964.

Blum, M. S. (1969). Alarm pheromones. *Annu. Rev. Entomol.* **14**, 57–80.

Blum, M. S. (1970). The chemical basis of insect sociality. *In* "Chemicals Controlling Insect Behavior" (M. Beroza, ed.). Academic Press, New York.

Blum, M. S., Padovani, F., Curley, A., and Hawk, R. E. (1969). Benzaldehyde: Defensive secretion of a harvester ant. *Comp. Biochem. Physiol.* **29**, 461–465.

Blum, M. S., Doolitle R. E., and Beroza, M. (1971a). Alarm pheromones: Utilization in evaluation of olfactory theories. *J. Insect Physiol.* **17**, 2351–2361.

Blum, M. S., Byrd, J. B., Travis, J. R., Watkins, J. F., II, and Gehlbach, F. R. (1971b). Chemistry of the cloacal sac secretion of the blind snake, *Leptotypholops dulcis. Comp. Biochem. Physiol.* **28B**, 103–107.

Boch, R., and Shearer, D. A. (1962). Identification of geraniol as the active component in the Nassanoff pheromone of the honeybee. *Nature (London)* **194**, 704–706.

Boch, R., and Shearer, D. A. (1964). Identification of nerolic and geranic acids in the Nassanoff pheromone of the honeybee. *Nature (London)* **202**, 320–321.

Boch, R., and Shearer, D. A. (1971). Chemical releasers of alarm behaviour in the honey-bee, *Apis mellifera. J. Insect Physiol.* **17**, 2277–2285.

Bossert, W. H. (1968). Temporal patterning in olfactory communication. *J. Theoret. Biol.* **18**, 157–170.

Bossert, W. H., and Wilson, E. O. (1963). The analysis of olfactory communication among animals. *J. Theoret. Biol.* **5**, 443–469.

Brady, U. E., Tumlinson, J. H., Brownlee, R. G., and Silverstein, R. M. (1971). Sex stimulant and attractant in the Indian meal moth and in the almond moth. *Science* **171**, 802–894.

Bronson, F. H. (1968). Pheromonal influences on mammalian reproduction. *In* "Perspectives in Reproduction and Sexual Behavior: a Memorial to Wm. C. Young" (M. Diamond, ed.). Indiana Univ. Press, Bloomington, Indiana.

Bronson, F. H., and Dezell, H. E. (1968). Studies on the estrus-inducing (pheromonal) action of male deermouse urine. *Gen. Comp. Endocrinol.* **10**, 339–343.

Bronson, F. H., and Eleftheriou, B. E. (1963). Influence of strange males on implantation in the deermouse. *Gen. Comp. Endocrinol.* **3**, 515–518.

Bronson, F. H., and Marsden, H. M. (1964). Male-induced synchrony of estrus in deermice. *Gen. Comp. Endocrinol.* **4**, 634–637.

Brower, L. P., Brower, J. V. Z., and Cranston, F. P. (1965). Courtship behavior of the queen butterfly, *Danaus gilippus bernice* (Cramer). *Zoologica* **50**, 1–39.

Brown, W. L., Jr. (1968). An hypothesis concerning the function of the metapleural glands in ants. *Amer. Natur.* **102**, 188–191.

Brown, W. L. Jr., Eisner, T., and Whittaker, R. H. (1970). Allomones and kairomones. Transspecific chemical messengers. *Bioscience* **20**, 21–22.

Brownlee, R. G., Silverstein, R. M., Muller-Schwarze, D., and Singer, A. G. (1969). Isolation, identification, and function of the chief component of male tarsal scent in black-tailed deer. *Nature (London)* **221**, 284–285.

Bruce, H. M. (1960). A block to pregnancy in the mouse caused by proximity of strange males. *J. Reprod. Fertil.* **1**, 96–103.

Bücherl, W., Buckley, E., and Deulofeu, V. (eds.) (1968–71). "Venomous Animals and their Venoms." 3 volumes. Academic Press, New York.

Burghardt, G. M. (1970). Chemical perception in reptiles. *In* "Communication by Chemical Signals. Advances in Chemoreception" (J. W. Johnston, Jr., D. G. Moulton, and A. Turk, eds.). Vol. I. Appleton, New York.

Butler, C. G. (1967). Insect pheromones. *Biol. Rev.* **42**, 42–87.

Butler, C. G., and Calam, D. H. (1969). Pheromones of the honeybee—The secretion of the Nassanoff gland of the worker. *J. Insect Physiol.* **15**, 237–244.

Butler, C. G., and Callow, R. K. (1968). Pheromones of the honeybee (*Apis mellifera L.*): The "inhibiting scent" of the queen. *Proc. Roy. Entomol. Soc. (London)* **A43**, 62.

Butler, C. G., and Fairey, E. M. (1963). The role of the queen in preventing oogenesis in worker honeybees. *J. Apicult. Res.* **2**, 14–18.

Butler, C. G., and Simpson, J. (1967). Pheromones of the queen honeybee (*Apis mellifera L.*) which enable her workers to follow her when swarming. *Proc. Roy. Entomol. Soc.* (*London*) **A42**, 149–154.

Carlson, D. A., Mayer, M. S., Silhacek, D. L., James, J. D., Beroza, M., and Bierl, B. A. (1971). Sex attractant pheromone of the house fly: Isolation, identification and synthesis. *Science* **174**, 76–77.

Cavill, G. W. K., and Robertson, P. L. (1965). Ant venoms, attractants, and repellents. *Science* **149**, 1337–1345.

Clayton, R. B. (1970). The chemistry of non-hormonal interactions: Terpenoid compounds in ecology. *In* "Chemical Ecology" (E. Sondheimer and J. B. Simeone, eds.) Academic Press, New York.

Cleveland, L. R., Burke, A. W., and Karlson, P. (1960). Ecdysone induced modifications in the sexual cycles of the Protozoa of *Cryptocercus. J. Protozool.* **7**, 229–239.

Comfort, A. (1971a). Communication may be odorous. *New Scientist Sci. J.* **49**, 412–414.

Comfort, A. (1971b). Likelihood of human pheromones. *Nature* (*London*) **230**, 432–433.

Corbet, S. A. (1971). Mandibular gland secretion of larvae of the flour moth, *Anagasta kuehniella*, contains an epideictic pheromone and elicits oviposition in a hymenopteran parasite. *Nature* (*London*) **232**, 481–484.

Crossley, A. C., and Waterhouse, D. F. (1969). The ultrastructure of the osmeterium and the nature of its secretion in *Papilio* larvae (Lepidoptera). *Tissue Cell* **1**, 525–554.

Curtis, R. F., Ballantine, J. A., Keverne, E. B., Bonsall, R. W., Michael, R. P. (1971). Sex pheromone—Identification in rhesus monkeys and properties of synthetic attractants. *Nature* (*London*) **232**, 396–398.

Dethier, V. G. (1963). "The Physiology of Insect Senses." Methuen, London.

Eibl-Eibesfeldt, I. (1949). Über das Vorkommen von Schreckstoffen bei den Erdkrötenquappen. *Experienta* **5**, 236.

Eisner, T. (1970). Chemical defense against predation in arthropods. *In* "Chemical Ecology" (E. Sondheimer and J. B. Simeone, eds.). Academic Press, New York.

Eisner, T., and Meinwald, J. (1966). Defensive secretions of arthropods. *Science* **153**, 1341–1350.

Eisner, T., Meinwald, J., Munro, A., and Ghent, R. (1961). Defense mechanisms of arthropods—I. The composition and function of the spray of the whipscorpion, *Mastigoproctus giganteus* (Lucas) (Arachnida, Pedipalpida). *J. Insect Physiol.* **6**, 272–298.

Eisner, T., Eisner, H. E., Hurst, J. J., Kafatos, F. C., and Meinwald, J. (1963). Cyanogenic glandular apparatus of a millipede. *Science* **139**, 1218–1220.

Eisner, T., McHenry, F., and Salpeter, M. M. (1964). Defense mechanisms of arthropods. XV. Morphology of the quinone-producing glands of a tenebrionid beetle (*Eleodes longicollis* Lec.) *J. Morphol.* **115**, 355–399.

Eisner, T., Hendry, L. B., Peakall, D. B., and Meinwald, J. (1971). 2,5-Dichlorophenol (from ingested herbicide?) in defensive secretion of grasshopper. *Science* **172**, 277–278.

Fisher, S. (1670). Reported by J. Wray. Some uncommon observations and experiments made with an acid juice to be found in ants. *Phil. Trans. Roy. Soc.* (*London*) 2063.

Francke-Grosmann, H. (1967). Ectosymbiosis in wood-inhabiting insects. In "Symbiosis" (S. M. Henry, ed.), Vol. II. Academic Press, New York.

Friedman, L., and Miller, J. G. (1971). Odor incongruity and chirality. *Science* 172, 1044–1046.

Frisch, K. von (1941). Über einen Schreckstoff der Fischhaut und seine biologische Bedeutung. Z. *Vergl. Physiol.* 29, 46–145.

Frisch, K. von (1967). "The Dance Language and Orientation of Bees." Belknap Press of Harvard Univ. Press, Cambridge, Massachusetts.

Gabba, A., and Pavan, M. (1970). Researches on trail and alarm substances in ants. In "Communication by Chemical Signals. Advances in Chemoreception" (J. W. Johnston, Jr., D. G. Moulton, and A. Turk, eds.), Vol. I. Appleton, New York.

Gandolfi, G. (1969). A chemical sex attractant in the guppy *Poecilia reticulata* Peters (Pices, Poeciliidae) *Monitore Zool. Ital. (N.S.)* 3, 89–98.

Gary, N. E. (1962). Chemical mating attractants in the queen honeybee. *Science* 136, 773–774.

Gehlbach, F. R., Watkins, J. F., and Reno, H. W. (1968). Blind snake defensive behavior elicited by army ant attacks. *Bioscience* 18, 784–785.

Gesteland, R. C., Lettvin, J. Y., and Pitts, W. H. (1965). Chemical transmission in the nose of the frog. *J. Physiol.* 181, 525–559.

Ghent, R. L. (1961). Adaptive refinements in the chemical defense mechanisms of certain Formicinae. Ph.D. Thesis, Cornell Univ., Ithaca, New York.

Gilby, A. R., and Waterhouse, D. F. (1965). The composition of the scent of the green vegetable bug, *Nezara viridula. Proc. Roy. Soc. (London)* B162, 105–120.

Gleason, K. K., and Reynierse, J. H. (1969). The behavioral significance of pheromones in vertebrates. *Psychol. Bull.* 71, 58–73.

Habermehl, G. (1966). Chemie und Toxikologie der Salamanderalkoide. *Naturwissenschaften* 53, 123–128.

Hainer, R. M., Emslie, A. G., and Jacobson, A. (1954). An information theory of olfaction. *Ann. N.Y. Acad. Sci.* 58, 158–174.

Hangartner, W. (1967). Spezifität und Inaktivierung der Spur-pheromons von *Lasius fuliginosus* Latr. und Orientierung der Arbeiterinnen im Druftfeld. Z. *Vergl. Physiol.* 57, 103–126.

Happ, G. M. (1968). Quinone and hydrocarbon production in the defensive glands of *Eleodes longicollis* and *Tribolium castaneum* (Coleoptera; Tenebrionidae). *J. Insect Physiol.* 14, 1821–1837.

Happ, G. M. (1970). Multiple sex pheromones of the mealworm beetle, *Tenebrio molitor* L. *Nature (London)* 222, 180–181.

Happ, G. M. (1970). Maturation of the response of male *Tenebrio molitor* to the female sex pheromone. *Ann. Entomol. Soc. Amer.* 63, 1782.

Happ, G. M., and Wheeler, J. W. (1969). Bioassay, preliminary purification and effect of age, crowding, and mating on the release of sex pheromone by female *Tenebrio molitor. Ann. Entomol. Soc. Amer.* 62, 846–851.

Happ, G. M., Schroeder, M. E., and Wang, J. C. H. (1970). Effects of male and female scent on reproductive maturation in young female *Tenebrio molitor. J. Insect Physiol.* 16, 1543–1548.

Happ, G. M., Happ, C. M., and Barras, S. J. (1971). Fine structure of the prothoracic mycangium, a chamber for the culture of symbiotic fungi in the southern pine beetle, *Dendroctonus frontalis. Tissue Cell* 3, 295–308.

Henderson, L. J. (1958). "The Fitness of the Environment." Beacon Press, Boston, Massachusetts.

Henry, S. M. (ed.) (1967). "Symbiosis." Academic Press, New York.

Hölldobler, B. (1971). Communication between ants and their guests. *Sci. Amer.* **224**(3), 86–93.

Hrbacek, I. (1950). On the flight reaction of tadpoles of the common toad caused by chemical substances. *Experienta* **6**, 100–101.

Hummel, H., and Karlson, P. (1968). Hexansäure als Bestandteil des Spurpheromons der Termite *Zootermopsis nevadensis* Hagen. *Hoppe-Seyl. Z. Physiol. Chem.* **349**, 725–727.

Ikan, R., Gottlieb, R., and Bergmann, E. D. (1969). The pheromone of the queen of the oriental hornet, *Vespa orientalis. J. Insect Physiol.* **15**, 1709–1712.

Johnston, J. W. Jr., Moulton, D. G., and Turk, A. (eds.) (1970). "Communication by Chemical Signals. Advances in Chemoreception," Vol. I. Appleton-Century-Crofts, New York.

Kaissling, K.-E., and Priesner, E. (1970). Die Reichswelle des Seidenspinners. *Naturwissenschaften* **57**, 23–28.

Karlson, P., and Butenandt, A. (1959). Pheromones (Ectohormones) in insects. *Annu. Rev. Entomol.* **4**, 39–58.

Karlson, P., and Lüscher, M. (1959). "Pheromones": a new term for a class of biological active substances. *Nature (London)* **183**, 55–56.

Kullenberg, B., Bergström, G., Ställberg-Stenhagen, S. (1970). Volatile components of the cephalic marking secretion of male bumble-bees. *Acta Chem. Scand.* **24**, 1481–1483.

Law, J. H., and Regnier, F. E. (1971). Pheromones. *Annu. Rev. Biochem.* **40**, 533–548.

Lederer, E. (1950). Odeurs et parfums des animaux. *Fortschr. Chem. Org. Naturstoffe* **6**, 87–153.

Lee, S., van der, and Boot, L. M. (1955). Spontaneous pseudopregnancy in mice II. *Acta Physiol. Pharm. Neerland.* **5**, 213–215.

Leitereg, T. J., Guardagni, D. G., Harris, J. H., Mon, T. R., and Teranishi, R. (1971). Evidence for the difference between the odours of the optical isomers (+)- and (−)-carvone. *Nature (London)* **230**, 455–456.

LeMagnen, J. (1948). Un cas de sensibilité olfactive se présentant comme un caractère sexuel secondaire féminin. *C. R. Acad. Sci. Paris* **226**, 694–695.

LeMagnen, J. (1950). Nouvelle donées sur le phénomènes de exaltolide. *C. R. Acad. Sci. Paris* **230**, 1103–1105.

LeMagnen, J. (1970). Communication by chemical signals: Conclusion. In "Communication by Chemical Signals. Advances in Chemoreception" (J. W. Johnston, Jr., D. G. Moulton, and A. Turk, eds.), Vol. I. Appleton, New York.

Lindauer, M., and Kerr, W. E. (1958). Die gegenseitige Verständigung bei den stachellosen Bienen. *Z. Vergl. Physiol.* **41**, 405–434.

Lindauer, M., and Kerr, W. E. (1960). Communication between the workers of stingless bees. *Bee World* **41**, 29–41, 65–71.

Loher, W. (1961). The chemical acceleration of the maturation process and its hormonal control in the male of the desert locust. *Proc. Roy. Soc. (London)* **B153**, 380–397.

Losey, G. S. Jr. (1969). Sexual pheromone in some fishes of the genus *Hypsoblennius* Gill. *Science* **163**, 181–183.

Lüscher, M. (1961). Social control of polymorphism in termites. *Sym. Roy. Entomol. Soc. London* 1, 57–67.

Marler, P. (1959). Developments in the study of animal communication. *In* "Darwin's Biological Work: Some Aspects Reconsidered" (P. R. Bell, ed.). Cambridge Univ. Press, London and New York.

Marsden, H. M., and Bronson, F. H. (1964). Estrous synchrony in mice: alteration by exposure to male urine. *Science* 144, 1469.

Martin, M. M. (1970). The biochemical basis of the fungus-attine ant symbiosis. *Science* 169, 16–20.

Maschwitz, U. (1964). Gefahrenalarmstoffe und Gefahrenalarmierung bei sozialen Hymenoptera. *Z. Vergl. Physiol.* 47, 596–655.

Maschwitz, U. (1966). Alarm substances and alarm behavior in social insects. *Vitam. Horm.* 24, 267–290.

Maschwitz, U., Koob, K., and Schildknecht, H. (1970). Ein Beitrag zur Funktion der Metathoracaldrüse der Ameisen. *J. Insect Physiol.* 16, 387–403.

Matsumura, F., Coppel, H. C., and Tai, A. (1968). Isolation and identification of termite trail-following pheromone. *Nature (London)* 219, 963–964.

Mc Gurk, D. J., Frost, J., Eisenbraun, E., Vick, K., Drew, W., and Young, J. (1966). Volatile compounds in ants: Identification of 4-methyl-3-heptanone from *Pogonomyrmex* ants. *J. Insect Physiol.* 12, 1435–1441.

Meinwald, J., Meinwald, Y. C., and Mazzocchi, P. H. (1969). Sex pheromone of the queen butterfly: Chemistry. *Science* 164, 1174–1175.

Meinwald, Y. C., Meinwald, J., and Eisner, T. (1966). 1, 2 dialkyl-4(3H)-quinazolinones in the defensive secretions of a millipede (*Glomeris marginata*). *Science* 154, 390–391.

Menon, M. (1970). Hormone-pheromone relationships in the beetle, *Tenebrio molitor*. *J. Insect Physiol.* 16, 1123–1139.

Meyer, K. (1952). Über herzaktive Krotengifte (Bufogenine) 7. Mitteilung. Reisbufogenin und Artebufogenin aus Chan Su. *Helv. Chim. Acta* 35, 2444–2469.

Michael, R. P., and Keverne, E. B. (1968). Pheromones in the communication of sexual status in primates. *Nature (London)* 218, 746–749.

Michael, R. P., and Keverne, E. B. (1970). Primate sex pheromones of vaginal origin. *Nature (London)* 225, 84–85.

Michael, R. P., Keverne, E. B., and Bonsall, R. W. (1971). Pheromones: Isolation of male sex attractants from a female primate. *Science* 172, 964–966.

Moore, B. P. (1966). Isolation of the scent-trail pheromone of an Australian termite. *Nature (London)* 211, 746–747.

Moser, J. (1970). Pheromones of social insects. *In* "Control of Insect Behavior by Natural Products" (D. L. Wood, R. M. Silverstein, and M. Nakajima, eds.). Academic Press, New York.

Moure, J. S., Nogueira-Neto, P., and Kerr, W. E. (1958). Evolutionary problems among Meliponinae (Hymneoptera, Apidae). *Proc. Int. Congr. Entomol., 10th Montreal, 1956* 2, 481–493.

Mugford, R. A., and Nowell, N. W. (1970). Pheromones and their effect on aggression in mice. *Nature (London)* 226, 967–968.

Müller, D. G., Jaenicke, L., Donike, M., and Akintobi, T. (1971). Sex attractant in a brown alga: Chemical structure. *Science* 171, 815–817.

Müller-Schwarze, D. (1969). Complexity and relative specificity in a mammalian pheromone. *Nature (London)* 223, 525–526.

Müller-Velten, H. (1966). Über den angsgeruch bei der hausmaus (*Mus musculus* L.) *Z. Vergl. Physiol.* 52, 401–429.

Mykytowycz, R. (1970). The role of skin glands in mammalian communication. *In* "Communication by Chemical Signals, Advances in Chemoreception" (J. W. Johnston, Jr., D. G. Moulton, and A. Turk, eds.), Vol I. Appleton, New York.

Nicoll, R. A. (1971). Recurrent excitation of secondary olfactory neurons: A possible mechanism for signal amplification. *Science* 171, 824–826.

Pasteels, J. M. (1968). Le système glandulaire tégumentaire des Aleocharinae (Coleoptera, Staphylinidae) et son évolution chez les espèces termitophiles du genre *Termitella. Arch. Biol. Liège* 79, 381–469.

Payne, T. L. (1970). Electrophysiological investigations on response to pheromone in bark beetles. *Contrib. Boyce Thomp. Inst.* 24, 275–282.

Payne, T. L., Shorey, H. H., and Gaston, L. K. (1970). Sex pheromones of noctuid moths: Factors influencing antennal responsiveness in males of *Trichoplusia ni. J. Insect Physiol.* 16, 1043–1055.

Pfaff, D. W., and Gregory, E. (1971). Olfactory coding in olfactory bulb and median forebrain bundle of normal and castrated male rats. *J. Neurophysiol.* 34, 208–216.

Pfaff, D. W., and Pfaffmann, C. (1969). Behavioral and electrophysiological responses of male rats to female rat urine odors. *In* "Olfaction and Taste, III" (C. Pfaffmann, ed.). Rockefeller Univ. Press, New York.

Pfeiffer, W. (1962). The fright reaction of fish. *Biol. Rev.* 37, 495–511.

Pfeiffer, W. (1963). Alarm substances. *Experienta* 19, 113–123.

Pliske, T. E., and Eisner, T. (1969). Sex pheromone of the queen butterfly: Biology. *Science* 164, 1170–1172.

Ralls, K. (1971). Mammalian scent marking. *Science* 171, 443–449.

Raper, J. R. (1970). Chemical ecology among lower plants. *In* "Chemical Ecology" (E. Sondheimer and J. B. Simeone, eds.). Academic Press, New York.

Regnier, F. E., and Wilson, E. O. (1968). The alarm-defence system of the ant *Acanthomyops claviger. J. Insect Physiol.* 14, 955–970.

Regnier, F. E., and Wilson, E. O. (1971). Chemical communication and "propoganda" in slave-maker ants. *Science* 172, 267–269.

Remold, H. (1962). Über die biologische Bedeutung der Duftdrüsen bei den Landwanzen (Geocorsiae). *Z. Vergl. Physiol.* 45, 636–694.

Riddiford, L. (1967). *Trans*-2-hexenal: mating stimulant for Polyphemus moths. *Science* 158, 139–141.

Riddiford, L. M. (1970). Antennal proteins of saturniid moths—Their possible role in olfaction. *J. Insect Physiol.* 16, 653–660.

Riddiford, L. M., and Williams, C. M. (1971). Role of the corpora cardiaca in the behavior of silkmoths. I. Release of sex pheromone. *Biol. Bull.* 140, 1–7.

Rodin, J. O., Silverstein, R. M., Burkholder, W. E., and Gorman, J. E. (1969). Sex attractant of *Trogoderma inclusum. Science* 165, 904–905.

Roelofs, W. L., and Cardé, R. T. (1971). Hydrocarbon sex pheromone in tiger moth (Arctiidae). *Science* 171, 684–686.

Roelofs, W. L., and Comeau, A. (1969). Sex pheromone specificity: Taxonomic and evolutionary aspects in Lepidoptera. *Science* 165, 398–400.

Roelofs, W. L., and Comeau, A. (1971a). Sex pheromone perception: Synergists and inhibitors for the red-banded leaf roller attractant. *J. Insect Physiol.* 17, 435–448.

Roelofs, W. L., and Comeau, A. (1971b). Sex pheromone perception: Electroantennogram responses of the red-banded leaf roller moth. *J. Insect Physiol.* 17, 1969–1982.

Ropartz, P. (1966). Contribution à l'étude du déterminisme d'un effect du groupe chez les souris. *C. R. Acad. Sci. Paris* 262, 2070–2072.

Ropartz, P. (1968). Role des communications olfactives dans le compartment social des souris males. *Colloq. Int. Centre. Nat. Rech. Sci. (Paris)* **17**, 323–339.

Rossi, A. C. (1969). Chemical signals and nest-building in two species of *Colisa* (Pices, Anabantidae). *Monitore Zool. Ital. (N.S.)* **3**, 225–237.

Roth, L. M. (1945). The odoriferous glands in the Tenebrionidae. *Ann. Entomol. Soc. Amer.* **38**, 77–87.

Roth, L. M., and Dateo, G. P. (1966). A sex pheromone produced by males of the cockroach *Nauphoeta cinerea. J. Insect Physiol.* **12**, 255–265.

Roth, L. M., and Eisner, T. (1962). Chemical defenses of arthropods. *Annu. Rev. Entomol.* **7**, 107–136.

Russell, G. F., and Hills, J. I. (1971). Odor differences between enantiomeric isomers. *Science* **172**, 1043–1044.

Schildknecht, H. (1963). Abwehrstoffe der Arthropoden, ihre Isolierung und Aufklärung. *Angew. Chem.* **75**, 762–771.

Schildknecht, H. (1970). The defensive chemistry of land and water beetles. *Angew. Chem. (Int. Ed.)* **9**, 1–9.

Schildknecht, H., and Holoubek, H. (1961). Die Bombardierkäfer und ihre Explosionschemie, V. Mitteilung über Insekten-Abwehrstoffe. *Angew. Chem.* **73**, 1–7.

Schildknecht, H., and Koob, K. (1971). Myrmicacin, das erste Insekten-Herbizid. *Angew. Chem.* **83**, 110.

Schildknecht, H., and Tacheci, H. (1971). Colymbetin, a new defensive substance of the water beetle, *Colymbetes fuscus*, that lowers blood pressure—LII. *J. Insect Physiol.* **17**, 1889–1896.

Schildknecht, H., Wenneis, W. F., Weis, F. H., and Maschwitz, U. (1966). Glomerin, ein neues Arthropoden-Alkaloid. *Z. Naturforsch.* **21b**, 121–127.

Schildknecht, H., Maschwitz, U., and Wenneis, W. F. (1967). Neue Stoffe aus dem Wehrsekret der Diplopodengattung *Glomeris*. Über Arthropoden–Abwehrstoffe. XXXIV. *Naturwissenschaften* **54**, 196–197.

Schildknecht, H., Maschwitz, E., and Maschwitz, U. (1968). Die Explosionschemie der Bombardierkäfer (Coleoptera, Carabidae). III. Isolierung und Charakterisierung der Explosionskatalysatoren. *Z. Naturforsch.* **236**, 1213–1218.

Schildknecht, H., Maschwitz, E., and Maschwitz, U. (1970). Die Explosionschemie der Bombardierkäfer: Struktur und Eigenschaften der Brennkammer-enzyme. *J. Insect Physiol.* **16**, 749–789.

Schneider, D. (1963). Electrophysiological investigations of insect olfaction. *In* "Olfaction and Taste, I" (Y. Zotterman, ed.). Pergamon, Oxford.

Schneider, D. (1969). Insect olfaction: Deciphering system for chemical messages. *Science* **163**, 1031–1037.

Schneider, D. (1970). Olfactory receptors for the sexual attractant (bombykol) of the silkmoth. "The Neurosciences: Second Study Program" (F. O. Schmitt, ed.). Rockefeller Univ. Press, New York.

Schneider, I. A., and Rudinsky, J. A. (1969). Mycetangial glands and their seasonal changes in *Gnathotrichus retusus* and *G. sulcatus. Ann. Entomol. Soc. Amer.* **62**, 39–43.

Schwinck, I. (1958). A study of olfactory stimuli in the orientation of moths. *Proc. Int. Congr. Entomol., 10th Montreal, 1956* **2**, 577–582.

Seboek, T. A. (ed.) (1968). "Animal Communication." Indiana Univ. Press, Bloomington, Indiana.

Sexton, W. A. (1963). "Chemical Constitution and Biological Activity," 3rd. ed. van Nostrand Reinhold, Princeton, New Jersey.

Shorey, H. H., Morin, K. L., and Gaston, L. K. (1968). Sex pheromones of noctuid moths. XV. Timing of development of pheromone responsiveness and other indicators of reproductive age in males of eight species. *Ann. Entomol. Soc. Amer.* **61**, 857–861.

Siegel, R. W., and Cohen, L. W. (1962). The intracellular differentiation of cilia (Abstr.) *Amer. Zool.* **2**, 558.

Silverstein, R. M., Rodin, J. O., and Wood, D. L. (1966). Sex attractants in frass produced by male *Ips confusus* in ponderosa pine. *Science* **154**, 509–510.

Silverstein, R. M., Rodin, J. O., Burkholder, W. E., and Gormon, J. E. (1967). Sex attractant of the black carpet beetle. *Science* **157**, 85–86.

Sondheimer, E., and Simeone, J. B. (eds.) (1970). "Chemical Ecology." Academic Press, New York.

Starr, R. C. (1968). Cellular differentiation in *Volvox*. *Proc. Nat. Acad. Sci.* **59**, 1082–1088.

Stuart, A. M. (1970). The role of chemicals in termite communication. *In* "Communication by Chemical Signals. Advances in Chemoreception" (J. W. Johnston, Jr., D. G. Moulton, and A. Turk, eds.), Vol. I. Appleton, New York.

Tai, A., Matsumura, F., and Coppel, H. C. (1971). Synthetic analogues of the termite trail-following pheromone, structure and biological activity. *J. Insect Physiol.* **17**, 181–188.

Tokuyama, T., Daly, J., and Witkop, B. (1969). The structure of batrachotoxin, a steroidal alkaloid from the Colombian arrow poison frog, *Phyllobates aurotaenia,* and partial synthesis of bathrachotoxin, and its analogues and homologues. *J. Amer. Chem. Soc.* **91**, 3931–3938.

Tschinkel, W., Willson, C., and Bern, H. A. (1967). Sex pheromone of the mealworm beetle (*Tenebrio molitor*). *J. Exp. Zool.* **164**, 81–85.

Tumlinson, J. H., Hardee, D. D., Gueldner, R. C., Thompson, A. C., Hedin, P. A., and Minyard, J. P. (1969). Sex pheromones produced by male boll weevil: Isolation, identification, and synthesis. *Science* **166**, 1010–1012.

Tumlinson, J. H., Silverstein, R. M., Moser, J. C., Brownlee, R. G., and Ruth, J. M. (1971). Identification of the trail pheromone of a leaf-cutting ant, *Atta texana. Nature* (*London*) **234**, 348–349.

Twitty, V. C. (1955). Field experiments on the biology and genetic relationships of the California species of *Triturus*. *J. Exp. Zool.* **129**, 129–147.

Valenta, J. G., and Rigby, M. K. (1968). Discrimination of the odor of stressed rats. *Science* **161**, 599–601.

Valentine, J. M. (1931). The olfactory sense of the adult mealworm beetle *Tenebrio molitor* (Linn.). *J. Exp. Zool.* **58**, 165–227.

Vierling, J. S., and Rock, J. (1967). Variations in olfactory sensitivity during the menstrual cycle. *J. Appl. Physiol.* **22**, 311–315.

Vite, J. P. (1967). Sex attractants in frass from bark beetles. *Science* **156**, 105.

Watkins, J. R. III, Gehlbach, F. R., and Kroll, J. C. (1969). Attractant-repellant secretions of blind snakes (*Leptotyphlops dulcis*) and army ant (*Neivamyrmex nigrescens*). *Ecology* **50**, 1098–1102.

Weatherston, J. (1966). The chemistry of arthropod defensive substances. *Quart. Rev.* (*London*). **21**, 287–313.

Weatherston, J., and Percy, J. E. (1970). Arthropod defensive secretions. *In* "Chemicals Controlling Insect Behavior" (M. Beroza, *ed.*). Academic Press, New York.

Weber, N. A. (1955). Pure cultures of fungi produced by ants. *Science* **121**, 109.

190 *George M. Happ*

Whittaker, R. H. (1970). The biochemical ecology of higher plants. *In* "Chemical Ecology" (E. Sondheimer and J. B. Simeone, eds.). Academic Press, New York.

Whittaker, R. H., and Feeny, P. P. (1971). Alleochemics: Chemical interactions between species. *Science* **171**, 757–770.

Whitten, W. K. (1958). Modification of the oestrous cycle of the mouse by external stimuli associated with the male. Changes in the oestrous cycle determined by vaginal smears. *J. Endocrinol.* **17**, 307–313.

Whitten, W. K., and Bronson, F. H. (1970). The role of pheromones in mammalian reproduction. *In* "Communication by Chemical Signals. Advances in Chemoreception" (J. W. Johnston, Jr., D. G. Moulton, and A. Turk, eds.), Vol. I. Appleton, New York.

Wilson, E. O. (1958). A chemical releaser of alarm and digging behavior in the ant *Pogonomyrmex badius* (Latreille). *Psyche, Cambridge* **65**, 41–51.

Wilson, E. O. (1962). Chemical communication among workers of the fire ant *Solenopsis saevissima* (Fr. Smith). 1. The organization of mass-foraging. 2. An informational analysis of the odor trail. 3. The experimental induction of social responses. *Anim. Behav.* **10**, 134–164.

Wilson, E. O. (1968). Chemical systems. *In* "Animal Communication" (T. Sebeok, ed.). Indiana Univ. Press, Bloomington, Indiana.

Wilson, E. O. (1970). Chemical communication within animal species. *In* "Chemical Ecology" (E. Sondheimer and J. B. Simeone, eds.). Academic Press, New York.

Wilson, E. O. (1971). "The Insect Societies." Harvard Univ. Press, Cambridge, Massachusetts.

Wilson, E. O., and Bossert, W. H. (1963). Chemical communication among animals. *Rec. Progr.* **19**, 673–716.

Wilson, E. O., Bossert, W. H., and Regnier, F. E. (1969). A general method for estimating threshold concentrations of odorant molecules. *J. Insect Physiol.* **15**, 597–610.

Woodward, R. B. (1964). The structure of tetrodotoxin. *Pure Appl. Chem.* **9**, 49–74.

Wright, R. H. (1966). Why is an odour. *Nature* (*London*) **209**, 551–554.

Yamada, M. (1971). A search for odour encoding in the olfactory lobe. *J. Physiol.* **214**, 127–143.

ENDOCRINEUROLOGY

Fleur L. Strand

I. Introduction 191
II. Effects of Hormones on Neurogenesis 192
   A. Criteria for Hormonal Involvement 192
   B. Effects of Specific Hormones 193
III. Effects of Hormones on the Mature Nervous System 202
IV. Effects of ACTH and Adrenal Cortical Hormones on Peripheral
   Nerve 210
V. Conclusions 212
   References 213

I. INTRODUCTION

The study of the influence of the nervous system on the synthesis, storage, and release of hormones has been one of the most active areas of research over the last two decades. In particular, mechanisms showing the delicate control exerted by neurosecretory structures, such as the hypothalamus, over the endocrine system have been examined in meticulous detail. Attention is now turning to the obverse side of the coin—the regulatory effects of hormones on the developing and mature nervous system. Recent evidence indicates that the physiological maturation of the nervous system is deeply sensitive to hormones (Caviezel and Martini, 1971), and certainly it is to be expected that the functions of the neuron will be regulated by hormonal fluctuations, for nerve cells are demonstrably sensitive to minute changes in their environment. Whether these changes are ionic, nutritional, or humoral in nature, most of these parameters are ultimately affected by hormones; consequently hormones must play a vital role in the structural and functional development of neurons, both central and peripheral. The growth of neural circuits and their final integration into complex patterns of behavior also are influenced strikingly by hormones at different times of development. It is the study of these multitudinous and fundamental effects of hormones on the nervous system that I have chosen to call *endocrineurology*.

This paper will consider the effects of various hormones on the developing and mature central and peripheral nervous systems, drawing on evidence ranging from single cell recordings through conduction, synaptic transmission, and reflex activity to behavioral responses. For brevity, certain topics not directly pertinent to the theme of mammalian endocrineurology have been excluded; for example, embryonic transducers, neurotransmitters and the nerve growth factor, which is the subject of a separate chapter by Levi-Montalcini and Angeletti.

II. EFFECTS OF HORMONES ON NEUROGENESIS

A. Criteria for Hormonal Involvement

For a hormone to be considered of significance in the developmental processes of the nervous system, it should fulfill the following criteria:

a. Synthesis, Release, and Timing. It should be produced by the embryonic endocrine tissues in amounts large enough to be effective after dilution in the circulation and at a time when the developing neurons exhibit definite sensitivity to its actions; or it must meet criterion (b).

b. Placental Permeability. If the hormone originates from the mother, it must pass across the placenta in adequate amounts and in a metabolically active form at a time when the developing neurons are sensitive to its actions.

c. Specific Uptake. Limited regions of the nervous system should show specificity of uptake for the hormone.

d. Direct Application. Direct application of the hormone by microelectrophoresis to those areas of the nervous system that normally show specificity of uptake for that hormone, should result in the same effect seen under normal, physiological conditions of development.

e. Concentration. Alterations in the circulating levels of the hormone, whether experimentally or pathologically provoked, should change the normal pattern of neural development.

f. Antagonists. The addition of known antagonists to the hormone should prevent the expected effect of the hormone on neurogenesis.

The experimental evidence presently available does not permit any one hormone to fulfill all these criteria; consequently we cannot expect to be able to answer for some time the most fascinating enigma of all—the mechanisms by which a hormone can alter the structure, function and integrative contribution of developing neurons in such a way as to alter profoundly the behavioral patterns of the mature individual.

As the mammalian brain and certain aspects of the rest of the nervous system are not mature at birth, neurogenesis, maturation and development of synaptic connections may be considered to continue for some time after birth (for example, 30–60 days in the rat, several months in man). Thus the criteria listed above should apply also to the hormone produced by the neonatal mammal. In fact, most of the evidence we have has been derived from neonatal neurogenesis.

B. Effects of Specific Hormones

1. THYROXINE

Thyroid hormones appear to be the most important hormonal moderators of the developing brain and presumably the rest of the nervous system, if one evaluates evidence from structural, biochemical, electrophysiological, and integrative studies.

The striking effects of thyroid deficiency in humans indicate that there is a critical stage in human brain development during which lack of thyroid hormone will cause irreversible damage (Myant, 1971). The mental retardation caused by thyroid hormone deficiency is reversible only if the retardation is slight and treatment is begun early. If the cretinism is severe, even early treatment is ineffective (Hamburgh, 1969). Successful treatment of athyreotic fetuses, by administration of large amounts of thyroid hormone to the mother, has been reported by Carr *et al.* (1959). While normal titers of thyroid hormone do not cross the placenta in significant amounts, very large dosages of this hormone can reach the fetus in amounts sufficient to compensate for the lack of fetal thyroxine.

Maturation of various parts of the central nervous system, which in the human takes place *in utero*, continues after birth in the rat, making this animal a particularly suitable one for the study of hormonal effects on the developing nervous system. The embryonic rat thyroid does not mature functionally until the eighteenth day of fetal life, but the placenta is permeable to maternal thyroxine during the third trimester of the gestational period (Hamburgh *et al.*, 1971). Nevertheless, the rat fetus appears to be independent of maternal thyroxine, for newborn rats taken from hypothyroid and hyperthyroid mothers do not differ significantly in body weight or skeletal maturation from rats born to control mothers.

An extensive study of the effects of thyroid hormone deprivation on neonatal rats has been undertaken by Eayrs and his colleagues. The administration of ^{131}I at birth effectively thyroidectomizes rats and this procedure results in a decrease in the weight of the brain. While brain weight is reduced about 15%, the DNA content per gram fresh weight is increased by 10%, indicating that it is the size of the cells that is reduced, rather than the number (Eayrs and Taylor, 1951).

While the fetal rat does not appear to be responsive to changes in thyroid hormone levels, the addition or withdrawal of this hormone after birth causes significant changes in the maturation of structures within the central nervous system. When hypothyroidism is induced through the administration of phenylthiouracil, beginning on the fifteenth day of pregnancy and continuing beyond weaning, there is a delay in cerebellar maturation, a decrease in myelination, and a transitory increase in oxygen consumption. The administration of thyroxine may accelerate maturation slightly (Hamburgh *et al.*, 1964). This effect of thyroxine appears to be a direct one, for tissue culture studies on rat cerebellar neurons show the hormone to increase the rate of growth and myelination of the axons (Hamburgh and Bunge, 1964; Hamburgh, 1968). The critical period in the rat extends over the first 10–14 days of life and thereafter seems to be independent of thyroid hormone levels. Thyroid medication started after the fourteenth day is consequently unable to reverse the damage done to the developing brain deprived of this hormone during the critical period (Hamburgh, 1969).

These effects of thyroxine on the brain may be mediated partly through regulation of protein synthesis, for thyroid hormone stimulates the incorporation of amino acids into protein in the developing brain (Schneck *et al.*, 1964), while thyroidectomy depresses protein synthesis (Geel and Timiras, 1967). Various enzyme systems are affected by the hormone. Thyroidectomy soon after birth selectively decreases succinic dehydrogenase in the developing brain (Hamburgh and Flexner, 1957) but has no effect on cytochrome oxidase (see Table I for details). As

TABLE I

EFFECT OF NEONATAL THYROIDECTOMY ON THE ACTIVITY OF ENZYMES
IN THE CEREBRAL CORTEX OF DEVELOPING RATS

Enzyme	Age of rat (days)[a]	Effect	Reference
Succinic dehydrogenase	30	Decrease	Hamburgh and Flexner (1957)
Acetylcholinesterase	22	Decrease	Geel and Timiras (1967)
Glutamate decarboxylase	40	Decrease	García Argiz et al. (1967)
GABA transaminase	40	Decrease	García Argiz et al. (1967)
ATPase	40	Decrease	García Argiz et al. (1967)
Aspartate aminotransferase (latent)	40	Decrease	Pasquini et al. (1967)
Aldolase	30	None	Hamburgh and Flexner (1957)
Cytochrome oxidase	30	None	Hamburgh and Flexner (1957)
Glutamate dehydrogenase	46	None	Balázs et al. (1968)
Alanine aminotransferase	46	None	Balázs et al. (1968)
Lactate dehydrogenase	46	None	Balázs et al. (1968)

[a] Age at which brain was examined after neonatal thyroidectomy (Myant, 1971).

the synthesis of RNA is unaffected by thyroidectomy, while the incorporation of amino acids into protein is decreased, Geel and Timiras (1971) and Balázs et al. (1968) suggest that the effects of thyroid hormone on protein synthesis involve translation rather than transcription in the developing fetus.

More specifically, hypothyroidism prolongs the proliferative phase of cell division, as seen by the continuation of DNA synthesis by fetal cells in the granular zone of the cerebellum. Hamburgh suggests that thyroxine normally times the proliferative phase and so controls cell population; the hormone acts as a time clock in the developing nervous system (Hamburgh et al., 1971).

As the neonate deprived of thyroid hormone shows a decrease in the extent and complexity of neuropil and consequently a marked reduction in the number of possible synaptic contacts (Fig. 1), it is not surprising that these hypothyroid rats fail to learn, as measured by the "water escape response." This behavioral change appears to be permanent (Hamburgh et al., 1964). Subsequent treatment with thyroxine increases the density of the neuropil but does not reverse the inability of the rats to learn. The timing is off and the necessary sequential neural pathways do not develop.

Thus thyroid hormone deprivation during the critical period of brain

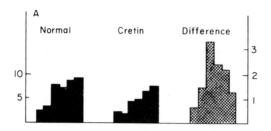

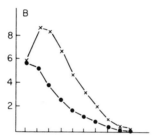

Fig. 1. Effect of neonatal thyroidectomy on the growth of cortical neuropil. A, Axonal component: density of axons (axons/mm²/10⁴) in successive cortical laminae (solid) and differences between normal and cretinous rats (dotted). B, Dendritic component: number of dendrites of successive distance from center of perikaryon. (×), normal; (●), cretinous; abscissa = 18 μm intervals. (Eayrs, 1971).

development which is probably *in utero* for the human, the first 10 days of life for the rat, is marked by a decrease in brain size due to a decrease in cellular size, a decrease in axonal myelination and a paucity of dendritic connections. This failure to develop the right synaptic connections at the proper time is the underlying cause of the subsequent inability of the hypothyroid animal to learn and the mental retardation of the human cretin. Once the critical period has passed, deprivation of thyroid hormone or its administration are ineffective; consequently, the early effects are irreversible.

The degree of brain maturation also can be correlated to the development of electrical activity in the brain and the effect of thyroid hormones upon it. Hatotani and Timiras (1967) measured the transcallosal response in normal rats, rats with neonatally induced thyroid deficiency, and in rats administered thyroxine. This response can be used as an index of cortical maturation, for excitation of the cortex can be produced by direct electrical stimulation of the callosal fibers, which are free of intermediary neuronal delay. Studies on the peak latency, duration, threshold, and amplitude of the evoked response showed retardation of the response in the hypothyroid rats, and acceleration in the hyperthyroid animals. In hypothyroid rats at 60 days of age the time course corresponds to that of the controls at 15 days, and also can be correlated to a slowing down of myelogenesis of the callosal fibers and hypoplasia of the cortical neutropil. Hyperthyroidism, on the other hand, may accelerate myelination.

Other electrical parameters have been used to determine the effect of thyroxine on the brain. The EEG is diminished (Eayrs, 1966) and the electroconvulsive threshold is increased in the adult rat (Timiras and Woodbury, 1956) but in neonatally thyroidectomized rats, the electroshock seizure threshold is considerably lower than in controls, indicating higher brain excitability. This differential effect of thyroid hormones on young and adult animals is seen not only in electroshock seizure threshold but also in brain sodium concentration, which increases in the young thyroidectomized rat (Valcana and Timiras, 1969) and decreases in the adult (Timiras and Woodbury, 1956). The nature of the effect is dependent upon the relative degree of maturation of the specific brain area affected which reinforces Hamburgh's ideas of the timing action of thyroxine in central nervous system development.

2. Sex Hormones

The importance of sex hormones for the normal development of the central nervous system is demonstrated clearly in castrated neonates, in which the patterns of subsequent sexual activity are affected profoundly by this procedure. The presence of androgens is particularly critical for the maturation of the hypothalamus from the basic female neural organization, initially common to both sexes, into the acyclic type of hypothalamus responsible for male sexual development and behavior. Males castrated surgically or by the administration of the antiandrogenic drug, cyproterone acetate, on the day after birth, show decreased copulatory activity and a greater tendency to display the female lordotic response when adult. In the female, it is the absence of androgens rather than the presence of estrogens that establishes the permanent organization of the female neural patterns, characterized by cyclic hypothalamic regulation of anterior pituitary secretion of gonadotrophins. This work is extensively reviewed by Barraclough (1967).

The sensitivity of the female hypothalamus to androgens is shown dramatically by the effects of a single injection of testosterone in neonatal life. These females become "androgenized" and display persistent vaginal cornification, are anovulatory and consequently sterile (Barraclough, 1961). Not only is female receptivity impaired but some of these androgenized females develop male-like copulatory and ejaculatory patterns (Whalen *et al.*, 1971). Unphysiologically high doses of estrogens

paradoxically inhibit female differentiation of the hypothalamus and
cause a male type of differentiation of this structure. Doerner *et al.*
(1971) suggest that some sexual malfunctions in humans may be due
to hormonal effects on brain differentiation.

Experiments involving the implantation of sex hormones in the hypothalamus or the placement of electrolytic lesions in this area suggest
that it is the preoptic anterior hypothalamic area that is responsible for
male sexual behavior while the more centrally located ventromedial nucleus controls female mating behavior (Doerner and Staudt, 1968, 1969).
These investigators have shown that cell nuclei in the ventromedial
nucleus are larger in females than in males: male castration results in
an increase in nuclear size in this region to approximately that characteristic of females. The critical period appeared to be the first 2 weeks
after birth, for castration after this time does not affect the morphology
of these cells. There is a significant linear correlation between the nuclear
sizes of the ventromedial nucleus and subsequent sexual behavior. Large
nuclear volumes are associated with an increase in female mating behavior and a decrease in male sexuality. These experiments imply that
changes in the levels of androgens during the critical period of hypothalamic organization cause permanent morphological changes that are
correlated to later sexual activity.

The centers controlling the rhythm of gonadotropin secretion are located in the preoptic suprachiasmatic region of the anterior hypothalamus (Barraclough, 1966). These cells also respond to changes in sex
hormone levels at a critical period in their development, for androgenized
female rats show changes in the type of vesicles in nerve terminals ending
on the arcuate neurons of this area. There is, however, no change in the
morphology of the cell bodies (Ratner and Adamo, 1971). Not only structural dissimilarities in the cells of the male and female hypothalamus are
found but also metabolic distinctions. These same areas show a differential sensitivity to sex hormones in terms of oxidative activity, which is
higher in prepuberal male rats than in females. Castration of the males at
birth reduces the oxidative levels to those of control females while the administration of testosterone to neonatal females increases the oxidative
levels to that of normal males (Moguilevsky *et al.*, 1969).

Some of the effects of the sex hormones on the developing nervous
system are reminiscent of the timing effects of thyroxine discussed in
Section II,B,1. The critical period of steroid sensitivity for the hypo-
thalamohypophyseal structures mediating sexual behavior has been established as being between birth and the tenth day of age in rats and
mice (Barraclough, 1967). It is logical to assume that the testes of the
neonatal rat must secrete sufficient androgen during this period to affect

permanently the organization of these structures and thus ensure male sexual activity. During this time, too, androgens must pass the blood-brain barrier in sufficient quantities to be effective. In contrast, it appears that insufficient estrogen is secreted by pre- or postnatal animals to affect the undifferentiated hypothalamus (Barraclough, 1967). This ineffectiveness may be due to timing or steroid concentration, for estrogen, administered neonatally, does have some effect on the developing brain, but a selective one. It increases the glycolipid content of the cerebrum but not the spinal cord, while cortisone affects the spinal cord and not the higher nervous structures (Casper *et al.*, 1967). Neonatal administration of estradiol also increases the sensitivity of the developing nervous system to electroshock stimulation (Vernadakis and Timiras, 1963) and markedly accelerates the appearance of the maximal electroshock seizure pattern (Heim and Timiras, 1963). This increased neural activity may be due to an increase in the sodium and chloride content of the cerebellum (Valcana *et al.*, 1967). None of these electrical or ionic changes is seen in the spinal cord and these differences in response to the steroid might well lie in the degree of maturation of these structures, for the spinal cord in rats is more mature than the cerebellum and cerebrum.

Progesterone does not appear to be of great importance in the development of the hypothalamus: its effects are confined to prolonged treatment, which does alter the estrous cycle and the fertility of the animal when it becomes adult (Saunders, 1968). Caviezel and Martini (1971) suggest that maternal progesterone may protect the hypothalamus of the female fetus from endogenous androgens, through the ability of progesterone to compete with 5α-reductase. This enzyme transforms the potent testosterone into dihydrotestosterone, an androgen metabolite which is perhaps more potent.

The crucial effects of androgens and estrogens on the developing brain are most specific and direct. This is clearly seen in experiments in which hormones are implanted into the brain (see review by Lisk, 1967) and in the meticulous investigations conducted by McEwan and his colleagues demonstrating the existence of steroid binding sites in the brain cell nuclei. This is a limited capacity mechanism for binding estrogen and testosterone in specific regions of the brain (Kato and Villee, 1967; McEwan and Pfaff, 1970). The radioactivity of the brain in male and female rats castrated neonatally and administered radioactive estradiol exceeds that of the blood by a factor of 3 or more. There is a great difference in uptake between different brain regions, with the greatest activity in the pituitary, followed by the hypothalamus, preoptic area, septum, and brain stem. Interestingly enough, there are no differences in uptake of estrogen by males and females but the retention of estradiol is significantly higher in control females than in castrated females or in normal or castrated males. The specificity of the binding sites is shown by the fact that prior administration of unlabeled estradiol reduces the uptake of labeled estradiol. (McEwan and Pfaff, 1970).

The pattern of uptake for radioactive testosterone is similar to that of estradiol, only of lesser magnitude and does not appear to change with the sex of the animal or neonatal hormonal manipulation. The antiandrogen, cyproterone, reduces testosterone uptake in both sexes, whether normal or castrated, indicating a competition for the binding sites (McEwan *et al.*, 1970a, b).

Studies as to the effects of sex hormones on neural development in other species are scanty. The data for the mouse are essentially the same as those for the rat (Edwards and Burge, 1971). Androgenized female guinea pigs, however, are more likely to show male-type sexual behavior in adulthood than androgenized female rats (Phoenix *et al.*, 1959; Whalen *et al.*, 1969). In primates, although treatment with androgens in fetal life does not prevent the development of a menstrual cycle, sexual behavior of these treated females may be male (Goy, 1970). In the adult male rhesus monkey, secretion of luteinizing hormone can be elicited by the subcutaneous implantation of 17β estradiol, a response typical of the female cycling hypothalamus (Karsch *et al.*, 1973). These investigators suggest that the sexual differentiation of the central nervous system in primates may be strikingly different from that of rodents.

The essential characteristic of sexual behavioral differentiation in the rat and the mouse is suppression of the female neurobehavioral system by androgens in the neonatal period. Nevertheless, it is unlikely that the effects of androgens during this critical period of organization of the hypothalamus are all mediated through one hormone-specific receptor or trigger mechanism, for manipulation of androgen levels also affects nonsexual behavior tests (Pfaff and Zigmond, 1971).

That the human fetus is also endowed initially with a female pattern of behavior has been misinterpreted by some writers in the social field. Millett (1970) states that as there is no psychosexual differentiation between the sexes at birth, consequently all psychosexual behavior is postnatal and learned. However, the postnatal behavior of the human male is based on profound physiological modifications of the fundamental female nervous system, as androgen secretion in the neonate directs behavioral patterns into pathways that will be characteristic ultimately of male sexual activity. The responses learned in a complex social environment then utilize, modify, and perhaps even distort the underlying neuronal patterns of behavior.

3. Adrenocorticotropin (ACTH) and the
   Adrenal Cortical Hormones

Corticosterone is produced in relatively large amounts by the adrenals of the later prenatal and newborn rat although immediately following this period of high secretory activity, a diminution of activity and lack of responsiveness to stress or ACTH occur, lasting until about the fifteenth to eighteenth day after birth (Levine, 1970). Again, it is the timing of the hormonal secretions that is the key for their subsequent effects on the developing central nervous system. Changes in adrenocortical activity in late prenatal and early postnatal life can affect neuroendocrine activity in the prepubertal and adult animal. Handling infant rats for a brief period in the first 2 days after birth permits the animals to respond to stress at 3 days of age, while the nonhandled controls are still unresponsive to stress (Levine, 1968). It is even more startling that rats born to adrenalectomized mothers are significantly more sensitive to stress and show marked differences in behavior later in life, even if reared by normal mothers (Thoman *et al.*, 1970). Apparently corticoids at this early critical stage can organize the central nervous system for secretion of ACTH in much the same way that the gonadal steroids influence central nervous control of sexual development and behavior.

It is by no means clear whether the influential hormones at this time are fetal or maternal. Adrenalectomized mothers secrete large amounts of ACTH, and, while Christianson and Jones (1957) report that ACTH cannot cross the placenta, Knobil and Briggs (1955) consider the placenta to be permeable to corticosteroids and ACTH. Male rats, exposed to prenatal or postnatal stress, show decreased male sexual activity and a greater tendency to display the female pattern of lordosis in adulthood only after prenatal stress (Ward, 1972).

One interpretation of these experiments is that the fetal or maternal ACTH, secreted as a result of the exposure of the mother to stress, modifies the fetal ratio of adrenal to gonadal steroids during the critical period of sexual development. It is also possible that the maternal ACTH may directly affect the developing hypothalamic regions involved with sexual behavior. The experiments cited do not indicate whether the site of action of the stress-evoked ACTH is on the central nervous system or directly on the fetal adrenal cortex.

A single injection of corticoids on the first day of life markedly affects brain size, especially cerebellar size. These animals have premature eye opening but visual maturation is delayed (Shapiro, 1968).

An age-dependent effect of corticoids on the excitability of the cerebrospinal axis has been reported by Vernadakis and Woodbury (1963, 1964), who found that cortisol lowered the electroshock threshold in rats between 8 and 15 days of age, but not in rats 4 to 7 days old. A possible mechanism by which corticosteroids may affect excitability at a critical period in the maturation of the central nervous system is by enhancing myelination. *In vitro* studies by deVellis *et al.* (1971) have shown that the addition of corticosteroids to glial cells induces a specific enzyme that may be involved in myelin synthesis.

It is particularly difficult to study the effects of the adrenal corticoids in the developing animal, for adrenalectomy stops the growth of the animals almost completely and administration of large quantities of cortical hormones has equally devastating results. Despite these handicaps, evidence appears to be mounting that the early programming of the central nervous system by the hormones of the adrenal cortex is vital to subsequent neuroendocrine responses to stress.

4. Growth Hormones

As this hormone apparently is unable to cross the placenta (Gitlin *et al.*, 1965), the reported increase in DNA content of the brains of newborn young of mothers administered growth hormone is of unknown significance (Zamenhof *et al.*, 1971). Growth hormone administered to rats after birth does not affect the size or weight of the brain (Zamenhof, 1942; Diamond *et al.*, 1969). The role of fetal growth hormone in the development of the central nervous system is unknown but neither hypophysectomy of rats a few days after birth nor postnatal administration of growth hormone affects brain weight or the complexity of cortical dendritical connections (Diamond *et al.*, 1969).

III. EFFECTS OF HORMONES ON THE MATURE CENTRAL NERVOUS SYSTEM

1. Thyroxine

The excitability of the adult brain is increased by thyroid hormone, whether administered exogenously or as a result of thyrotoxicosis. The frequency and the amplitude of brain activity are increased by this hormone. Thyroid deficiency, as seen in severe myxedema or following thyroidectomy, depresses brain excitability and is correlated with mental disturbances in man (Nieman, 1961). As the brain of the adult animal differs from almost all other tissues in being unresponsive to the metabolic effects of thyroid hormones (Fig. 2) these changes in electrical activity cannot be due directly to the metabolic dysfunction. While

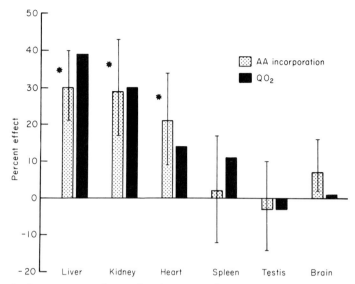

Fig. 2. Comparative effects of L-thyroxine administration on amino acid (AA) incorporation into protein. (Sokoloff and Klee, 1966.)

thyroid hormones stimulate amino acid incorporation into protein in the brain of immature animals, no such effect is seen in the adult brain. This difference in protein synthesis is attributed to a loss in the ability of adult brain mitochondria to respond to the stimulating effect of thyroxin (Klee and Sokoloff, 1965). It is likely that changes in brain excitability in the adult are due to alterations in the intracellular concentration of sodium as a result of variations in thyroid hormone titers (Timiras and Woodbury, 1956).

2. Sex Hormones

It seems quite clear from studies with radioactive steroids that these hormones enter the mammalian brain in functionally significant amounts. These investigations are reviewed by McEwan *et al.* (1970).

In adult vertebrates, with the possible exception of man (Money, 1961), sex hormones are necessary for the elicitation of complete mating behavior. The importance of the hormonal environment in regulating neural responses is elucidated by the experiments of Kawakami and Kubo (1971). Vaginal stimulation changes the rate of neuronal discharge in the brains of estrogen-primed ovariectomized rats but not in untreated spayed rats. Similar changes in excitability are seen in the spinal cord following artificial estrus. As the neuronal discharge in the brain is increased or decreased depending on specific location (the arcuate nucleus,

for instance, is facilitated while the ventromedial nucleus is inhibited);
this may be a possible mechanism by which afferent impulses from
peripheral receptors in the vagina and cervix selectively activate certain brain nuclei.

The preferential accumulation and retention of estrogen in the hypothalamus (McGuire and Lisk, 1968, 1969) result in a selective action
of estrogen on this region of the brain. The inhibitory responses of
hypothalamic neurons are enhanced by estrogens, whether recorded from
the intact diencephalon of anesthetized female rats (Lincoln and Cross,
1967: Kawakami *et al.*, 1970) or from hypothalamic islands in unanesthe-
tized, decerebrate rats (Cross and Dyer, 1970). There is a cyclic variation in the firing rate of these hypothalamic units with the highest rate
occurring during proestrus and the lowest during estrus (Fig. 3).

Progesterone is implicated in the mating behavior of mammals
(Young, 1961) but appears to require the previous priming action of
estrogen. Estrogen implants can activate the neural structures involved
in the lordosis reflex in several species: progesterone has a biphasic
effect on this behavior, first facilitating then inhibiting it (Lisk, 1969).

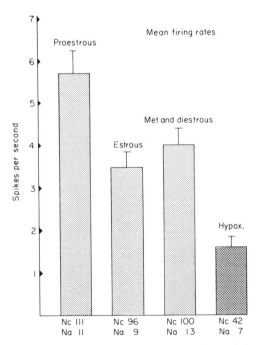

Fig. 3. Histogram of mean firing frequencies in cyclic and hypophysectomized
(hypox) female rats. Nc = number of cells; Na = number of animals. (Cross and
Dyer, 1970.)

The biphasic action of progesterone on the estrogen-primed animal is manifested also in arousal thresholds and the electroencephalogram (EEG) "after-reaction" (Kawakami and Sawyer, 1959).

A relationship between the sharp fall in progesterone levels and pre-menstrual and postpartum behavioral disturbances in humans has often been suggested but is difficult to prove (Hamburg, 1966). However, the anesthetic effect of large amounts of progesterone on the brain is unmistakable and this steroid is far more potent an anesthetic than the androgens, which in turn are more effective than estrogens (Selye, 1941). In lesser dosages, progesterone is a sedative and elevates the threshold for convulsive seizures in animals (Woolley and Timiras, 1962) and in humans (Laidlan, 1956). Again, the ratio of estrogen to pro-gesterone may be important for the progesterone effect.

The significance of these hormonal interactions appears to be in their ability to coordinate pituitary-gonad activity with the appropriate sexual behavior for fertilization through their feedback action on the limbic-hypothalamic system.

Testosterone and androstenedione are capable of maintaining mating behavior in castrated male rats (Whalen *et al.,* 1971). These steroids when administered to estrogen-primed females induce a monophasic ele-vation on the afterreaction threshold, unlike the biphasic effect of proges-terone (Sawyer, 1970). The blocking action of testosterone on ovulation in humans and experimental animals appears to be chiefly within the central nervous system rather than directly on the pituitary gland (Kawakami and Sawyer, 1967; McDonald and Gilmore, 1971).

3. ACTH and Adrenal Cortical Hormones

For many years it has been known that adrenal steroids modify the activity of the nervous system. The EEG in Addisonian patients and in adrenalectomized animals is abnormally slow but can be restored to normal by cortisone (Thorne, 1949; Bergen, 1951). Glaser *et al.* (1955) found that both cortisone and ACTH increase the number of waves of the EEG, especially in epileptic patients. These effects do not seem to be correlated with serum electrolyte changes and it has been suggested that these hormones may act directly on diencephalic structures (Streifler and Feldman, 1953). Iontophoretic application of hormones to specific brain regions in experimental animals appears to support this hypothesis (Ruf and Steiner, 1967). These subcortical effects of the hormones could then be relayed through multisynaptic pathways to influence cerebral excitability, consciousness and behavior.

Electrophysiological and behavioral studies show that cortical steroids and ACTH influence both facilitatory and inhibitory activity in the cen-

tral nervous system, from the level of single units firing in the hypothalamus to conditioned reflexes involving complex behavioral patterns. Especially interesting are the studies showing a direct effect of ACTH
on many of these functions, independent of the adrenal steroids.

Using seizure threshold as a measure of brain excitability, it is evident
that glucocorticoids exert a distinct excitatory effect on the central nervous system, in contrast to the depression of excitability resulting from
deoxycorticosterone (Woodbury, 1954). Hydrocortisone is particularly
effective in the rhinencephalon, increasing the amplitude of the theta
rhythm and causing spikes and seizures in the EEG pattern (Feldman
and Davidson, 1966).

The glucocorticoids increase conduction velocity along peripheral
axons but prolong synaptic transmission time, and these changes, together with the general increase in excitability, change the timing with
which peripheral impulses reach the higher integrative centers (Henkin,
1970). This change in timing of peripheral signals is responsible for
the decrease in sensory perception characteristic of adrenal insufficiency
in man and other animals. The concomitant increase in sensory acuity
(taste, olfaction and audition) is attributed to the increased excitability of the nervous system. These reciprocal changes in acuity and perception are restored to normal by the glucocorticoids but not by the
mineralocorticoids.

More precise information as to the specific site of hormone action
comes from studies of unit activity in various parts of the brain. Using
multiple microelectrodes, Slusher *et al.* (1966) have shown that cortisol
rapidly changes the firing pattern of units in the hypothalamus and
midbrain. Iontophoretic application of dexamethasone, a synthetic gluco-
corticoid more soluble than the natural hormone, inhibits hypothalamic
neurons (Fig. 4), which are then activated by subsequent local applica-

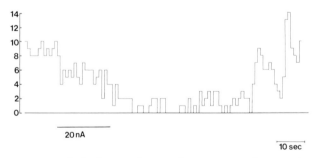

Fig. 4. Frequency of discharge of a single neuron in the area hypothalamica
anterior, plotted against time. Inhibition by microelectrophoretically applied dexamethasone phosphate (20 nA). ——duration of microelectrophoresis. (Steiner, 1970).

tion of ACTH (Steiner *et al.*, 1969). This depressant action of dexa-methasone is also seen at multiunit levels in the hypothalamus (Sawyer *et al.*, 1968). However, Feldman and Dafny (1966) found an increase in firing rate of anterior hypothalamic cells following intravenous cortisol injection. An initial positive feedback action of ACTH on the neurons controlling its secretion is postulated by Sawyer (1970) in a comprehensive review article on the applications of hormones to electrophysiology.

If the sensitive negative feedback control of ACTH secretion is due to increasing levels of corticosterone in the blood, then this steroid must be able to enter the brain in amounts proportional to its plasma concentration. McEwan *et al.* (1969) have shown that radioactive corticosterone is taken up by all parts of the brain of normal and adrenalectomized rats but that superimposed upon this general pattern of uptake is the tendency of limbic structures, the hippocampus and septum, to retain and concentrate labeled corticosterone (Fig. 5). McEwan and Weiss (1970) suggest that corticosterone retention by the hippocampus may

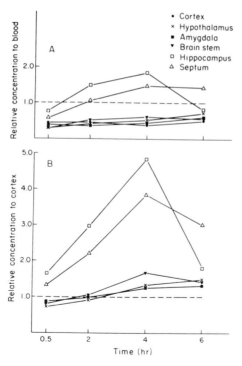

Fig. 5. Time course of radioactivity injected as (1,2-^{3}H) corticosterone in 6 brain regions expressed as a ratio to the concentration in blood (A), and cerebral cortex (B). (McEwan *et al.*, 1969.)

enhance the inhibitory action of this structure on ACTH release from the pituitary.

ACTH and the adrenocortical steroids have wide-ranging effects on higher nervous mechanisms and behavioral reactions. ACTH exerts an inhibitory action on the central nervous system and in high doses can completely block conditioned reflexes in the rabbit (Fig. 6). This inhibitory effect is seen also in rats, where the administration of ACTH leads to a delay of the extinction of conditioned avoidance responses in normal and adrenalectomized animals (Murphy and Miller, 1955; Miller and Ogawa, 1962; De Wied, 1966; and Bohus, 1968). On the other hand, these responses are facilitated by cortisol (Bohus, 1968). As ACTH is effective in adrenalectomized animals, the opposite effects of these hormones on behavior cannot be due to suppression of ACTH secretion by adrenal corticoids but rather to a direct action of each of these hormones on the central nervous system.

The use of polypeptide fragments of ACTH that are devoid of corticotropic activity confirms the extra-adrenal action of ACTH on behavior. The peptide fragment ACTH (4–10) appears to be the most potent of the several fragments tried (De Wied, 1969). Implantation of ACTH (1–10) in various brain regions indicates that it exerts its effects within the central nervous system (van Wimersma Greidanus and De Wied, 1971). This observation is corroborated by studies on single unit activity in freely moving rats, which show ACTH and adrenal steroids to have opposite and independent effects (Pfaff *et al.*, 1971). While the specific site of ACTH action is not known, the mesencephalic reticular formation has been implicated by Feldman *et al.*, (1961) and the thalamus by De Wied (1969).

4. Other Hormones

a. Melanocyte Stimulating Hormone (MSH). The effects of MSH on the central nervous system are very similar to those evoked by ACTH, which is not surprising in view of the fact that they share the same active polypeptide sequence 4–10. MSH increases polysynaptic action potentials in the spinal cord of the cat (Krivoy and Guillemin, 1961) and inhibits the rate of extinction of the avoidance response (De Wied, 1966). The release of both hormones can be triggered by changes in environmental conditions such as cold stress (Kastin *et al.*, 1967), but it is not known what factors control the rate and extent of breakdown of the polypetide hormones to the short and active fragments.

b. Insulin. Insulin-induced hypoglycemia causes changes in the EEG, including a progressive decline in α wave frequency. These changes

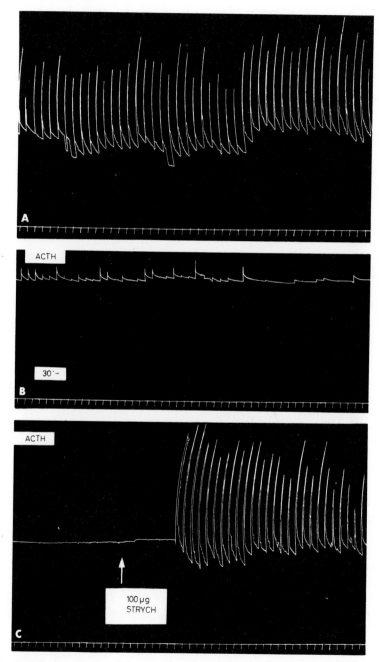

Fig. 6. Polysynaptic spinal flexor reflex activity before and after ACTH administration and recovery after strychnine injection. (Korányi and Endröczi, 1967.)

are reversed by glucose administration (Berger, 1937; Hoagland *et al.*, 1937). Insulin increases brain excitability (Timiras *et al.*, 1955) and has been utilized in the "shock-therapy" of mental diseases. As both hypoglycemia and hyperglycemia are accompanied by increased brain excitability it is unlikely that changes in carbohydrate metabolism are directly responsible.

c. Parathormone. Low serum calcium levels following hypoparathroidism result in increased susceptibility to seizures and paroxysmal abnormalities in the EEG (Woodbury and Davenport, 1949). High serum calcium levels are almost always associated with neurological disorders (Karpati and Frame, 1964).

d. Melatonin. Melatonin administration depresses the EEG in rabbits (Supnieski *et al.*, 1961) and in cats (Marczynski *et al.*, 1964) but does not appear to affect behavior in rats. In man and other animals melatonin has a mildly sedative effect but does not induce any marked behavioral alterations.

Wurtman *et al.* (1963) have shown that melatonin is the active principle of the pineal that inhibits gonadal function in the rat. The site of inhibition is unknown but is most probably the ovary rather than the central neuroendocrine mechanisms.

IV. EFFECTS OF ACTH AND ADRENAL CORTICAL HORMONES ON PERIPHERAL NERVE

In a comprehensive review of the relation between the adrenal cortex and the central nervous system, Woodbury (1958) also discusses alterations in the peripheral nervous system as a result of hypo- and hyperfunction of the adrenal cortex. Unlike brain excitability, the excitability of peripheral nerve is decreased following adrenalectomy. Cortisone and ACTH decrease the excitability of somatic motor nerves, according to Chauchard (1952). The electrical excitability of skeletal muscle is increased in the absence of the adrenal steroids. Wright and Lester (1959) found that adrenalectomy increases the velocity and lowers the threshold of isolated sciatic nerve. These results are difficult to interpret in terms of metabolic changes, for *in vitro* effects differ markedly from *in vivo* experiments. This is not surprising for many hormones exert their influence indirectly through alterations in salt and water metabolism, etc.

In intact animals, Walker (1955) reported that the repetitive nature of action potentials (AP's) characteristic of the adrenalectomized rat could be prevented by deoxycorticosterone. Other electrophysiological changes following adrenalectomy are an increase in the latent period, duration, and amplitude of AP's (Van Hof-Van Duin, 1958; Strand *et al.*, 1962).

In my laboratory, it has been found that both adrenalectomy and cold stress increase the amplitude of muscle action potentials during repetitive stimulation in the rat (Fig. 7). The preparation used is the intact sciatic nerve–gastrocnemius muscle *in situ,* with the rat under pentobarbital anesthesia. The nerve is stimulated with supermaximal stimuli (5/sec; duration 0.1 msec) for 30 minutes and AP's recorded from a coaxial needle electrode inserted in the muscle.

As both stress and adrenalectomy increase the titers of circulating ACTH (Hodges and Vernikos, 1959) the results suggest that the influence of ACTH on these electrophysiological parameters is direct and not mediated through adrenal cortical steroids. To test this hypothesis, ACTH (0.25 mU and 5 mU) was administered to rats 15 minutes before nerve stimulation commenced and the resulting changes in AP amplitude compared with normal and cold stressed rats. It can be seen from Fig. 8 that increasing amounts of ACTH increase AP amplitude and decrease the rate of fall during fatigue. Presumably the endogenous release of ACTH during cold stress (4 hours at 3°C) is less than the amounts of administered ACTH. Preliminary experiments on hypophysectomized rats show a drop in AP amplitude and increased fatigue; these changes

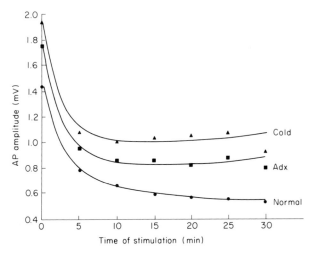

Fig. 7. Change in amplitude of muscle action potentials (AP) with stimulation in normal, adrenalectomized (adx) and cold-stressed rats. Further details in text.

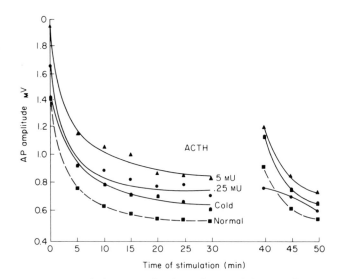

Fig. 8. Action potential (AP) amplitude and rate of fatigue during stimulation following ACTH administration or cold stress.

are reversed by the administration of 10 mU ACTH. As the adrenals in these animals are atrophied, the restorative action of ACTH must be extra-adrenal (Strand *et al.*, 1973). The polypeptides MSH and ACTH 4–10 appear to have similar effects on muscle AP amplitude and fatigue (Cayer and Strand, 1971, 1973).

Investigations are underway to determine whether this effect of ACTH and related polypeptides is on the nerve, neuromuscular junction, or muscle (Strand and Stoboy, 1969; Strand, 1969).

V. CONCLUSIONS

Hormonal control over the developing nervous system appears to have two significant parameters: timing and regional specificity. Hormones influence the immature nervous system of mammals by directing the development of the neurons into specific morphological, biochemical and synaptic pathways. The timing of these hormonal controls is exquisite: the organization of the rat hypothalamus is fixed in the first 10 days after birth. Lack of thyroid hormone during this critical period results in a paucity of the synaptic contacts essential for learning, and subsequent thyroxine administration stimulates only a disorganized growth of neuronal processes. Similarly, a lack of androgens during the first few days of life prevents the establishment of a normal substrate

for male sexual behavior in adulthood. The complexity of this timing
scarcely can be appreciated without the realization that different parts
of the nervous system mature at different rates and consequently lose
their sensitivity to hormonal directon at various periods in neurogenesis.

Gonadal and adrenocortical steroids bind specifically to cell nuclei
localized in regions of the hypothalamus, pituitary and limbic systems.
These cells respond with structural, enzymatic and electrophysiological
changes to increased or decreased titers of these hormones or to their
direct application to steroid-sensitive cells in the brain. ACTH has been
shown also to have a direct, extra-adrenal effect on the central and
peripheral nervous systems.

As hormones affect the excitability of the central and peripheral nervous systems, different regions may be facilitated or inhibited. The balance between these influences, together with alterations in synaptic transmission threshold and time, may provide a coding mechanism whereby
hormones prime the nervous system in preparation for selective triggering of behavioral responses.

While it is beyond the scope of this paper to discuss the mechanisms
by which hormones may evoke their actions, biochemical and electrophysiological studies indicate that they may act at the level of the cell
membrane (Sawyer, 1968; Weiss and Kidman, 1969). Certainly the evidence that shows specificity of binding sites for various steroids in different parts of the nervous system as well as on target organs, indicates
an action at the level of cellular or intracellular membranes. The ability
of hormones to interact with cell membranes would indicate that their
effects on the nervous system are intimately involved with consequent
changes in bioelectric potentials across these membranes, as well as to
enzyme systems activated through cyclic AMP. We can contemplate
readily then that neurons may be facilitated or inhibited as a result
of changes in hormonal environment; what is far more difficult to explain
on this basis is how minute amounts of specific hormones at critical
periods in the development of the nervous system can organize neural
pathways. Steroid binding within the nucleus, probably to DNA, may
be the responsible agent for the development of the finely coordinated
and delicately balanced patterns of behavior characteristic of the adult
of the species.

REFERENCES

Balázs, R., Kovács, S., Teichgraeber, P., Cocks, W. A., and Eayrs, J. T. (1968).
J. Neurochem. **15**, 1335–1349.

Barraclough, C. A. (1961). *Endocrinology* **68**, 62–67.
Barraclough, C. A. (1966). *Recent Progr. Hormone Res.* **22**, 503–539.
Barraclough, C. A. (1967). *In* "Neuroendocrinology" (L. Martini and W. F. Ganong, eds.), pp. 62–100. Academic Press, New York.
Bergen, J. R. (1951). *Amer. J. Physiol.* **164**, 16–22.
Berger, H. (1937). *Nervenkrankh.* **106**, 165–187.
Bohus, B. (1968). *Neuroendocrinology* **3**, 355–365.
Carr, E. A., Jr., Beierwaltes, W. H., Raman, G., Dodson, V. N., Tanton, J., Betts, J. S., and Stambaugh, R. A. (1959). *J. Clin. Endocrinol.* **19**, 1–18.
Casper, R., Vernadakis, A., and Timiras, P. S. (1967). *Brain Res.* **5**, 524–526.
Caviezel, F., and Martini, L. (1971). *Advan. Exp. Med. Biol.* **13**, 215–226.
Cayer, A., and Strand, F. L. (1971). *Fed. Proc.* **30**, 312.
Cayer, A., and Strand, F. L. (1973). *Fed. Proc.* **32**, 296.
Chauchard, P. (1952). *Rev. Sci.* **90**, 120–134.
Christianson, M., and Jones, C. I. (1957). *J. Endocrinol.* **15**, 17–42.
Cross, B. A., and Dyer, R. G. (1970). *In* "The Hypothalamus" (L. Martini, M. Motta, and F. Fraschini, eds.), pp. 115–122. Academic Press, New York.
de Vellis, J., Inglish, D., and Galey, F. (1971). *In* "Cellular Aspects of Neural Growth and Differentiation" (D. Pease, ed.), pp. 23–32. Univ. of California Press, Berkeley, California.
De Wied, D. (1966). *Proc. Soc. Exp. Biol. Med.* **122**, 28–32.
De Wied, D. (1969). *In* "Neuroendocrinology" (W. F. Ganong, ed.), pp. 97–141. Oxford Univ. Press, London and New York.
Diamond, M. C., Johnson, R. E., Ingham, C., and Stone, B. (1969). *Exp. Neurol.* **23**, 51–57.
Doerner, G., and Staudt, J. (1968). *Neuroendocrinology* **3**, 136–140.
Doerner, G., and Staudt, J. (1969). *Neuroendocrinology* **4**, 278–281.
Doerner, G., Doecke, F., and G. Hinz. (1971). *Neuroendocrinology* **7**, 146–155.
Eayrs, J. T. (1966). *Sci. Basis Med.* 317–339.
Eayrs, J. T. (1971). *In* "Hormones and Development" (M. Hamburgh and E. J. W. Barrington, eds.), pp. 345–355. Appleton, New York.
Eayrs, J. T., and Taylor, S. H. (1951). *J. Anat.* **85**, 350–358.
Edwards, D. A., and Burge, K. G. (1971). *Hormones Behav.* **2**, 49–58.
Feldman, S., and Dafney, N. (1966). *Israel J. Med. Sci.* **2**, 621–623.
Feldman, S., and Davidson, J. M. (1966). *J. Neurol. Sci.* **3**, 462–472.
Feldman, S., Todt, J. C., and Porter, R. W. (1961). *Neurology* **11**, 109–115.
García Argiz, C. A., Pasquini, J. M., Kaplun, B., and Gomez, C. J. (1967). *Brain Res.* **6**, 635–646.
Geel, S., and Timiras, P. (1967). *Brain Res.* **4**, 135–142.
Geel, S., and Timiras, P. (1971). *In* "Hormones and Development" (M. Hamburgh and E. J. W. Barrington, eds.), pp. 391–401. Appleton, New York.
Gessner, P. K., McIsaac, W. M., and Page, I. H. (1961). *Nature (London)* **190**, 179–180.
Gitlin, D., Kumate, J., and Morales, C. (1965). *J. Clin. Endocr.* **25**, 1599–1608.
Glaser, G. H., Kornfeld, D. S., and Knight, R. P. (1955). *Arch. Neurol. Psychiat.* **73**, 338–344.
Goy, R. W. (1970). *Phil. Trans. Roy. Soc. London B* **259**, 149–162.
Hamburg, D. A. (1966). *In* "Endocrines and the Central Nervous System" (R. Levine, ed.), pp. 251–265. Williams and Wilkins, Baltimore, Maryland.
Hamburgh, M. (1968). *Gen. Comp. Endocrinol.* **10**, 198–213.

Hamburgh, M. (1969). *In* "Current Topics in Developmental Biology" pp. 109–148. Academic Press, New York.

Hamburgh, M., and Bunge, R. P. (1964). *Life Sci.* 3, 1423–1430.

Hamburgh, M., and Flexner, L. B. (1957). *J. Neurochem.* 1, 279–288.

Hamburgh, M., Lynn, E., and Weiss, E. P. (1964). *Anat. Rec.* 150, 147–159.

Hamburgh, M., Medoza, L. A., Burkart, J. F., and Weil, F. (1971). *In* "Cellular Aspects of Neural Growth and Differentiation" (D. C. Pease, ed.), UCLA Forum Med. Sci. No. 14, pp. 321–329. Univ. of California Press, Los Angeles, California.

Hatotani, N., and Timiras, P. (1967). *Neuroendocrinology* 2, 147–156.

Heim, L. M., and Timiras, P. S. (1963). *Endocrinology* 72, 598–606.

Henkin, R. I. (1970). *Progr. Brain Res.* 32, 289–293.

Hoagland, H., Rubin, M. A., and Cameron, D. E. (1937). *Amer. J. Physiol.* 120, 559–570.

Hodges, J. R., and Vernikos, J. (1959). *Acta Endocrinol.* 30, 188–196.

Karpati, G., and Frame, B. (1964). *Arch. Neurol.* 10, 387–397.

Karsch, F. J., Dierschke, D. J., and Knobil, E. (1973). *Science* 179, 484–486.

Kastin, A. J., Arimura, A., Viosca, S., Barrett, L., and Schally, A. V. (1967). *Neuroendocrinology* 2, 200–208.

Kato, J., and Villee, C. A. (1967). *Endocrinology* 80, 567–575.

Kawakami, M., and Kubo, K. (1971). *Neuroendocrinology* 7, 65–89.

Kawakami, M., and Sawyer, C. H. (1959). *Endocrinology* 65, 631–643.

Kawakami, M., and Sawyer, C. H. (1967). *Endocrinology* 80, 857–871.

Kawakami, M., Terasaw, E., and Ibuki, T. (1970). *Neuroendocrinology* 6, 30–48.

Klee, C. B., and Sokoloff, L. (1965). *Fed. Proc.* 24, 399.

Knobil, E., and Briggs, F. N. (1955). *Endocrinology* 57, 147–152.

Korányi, L., and Endröczi, E. (1967). *Neuroendocrinology* 2, 65–75.

Kostowski, W., Rewerski, W., and Piechocki, T. (1970). *Neuroendocrinology* 6, 311–318.

Krivoy, W. A., and Guillemin, R. (1961). *Endocrinology* 69, 170–175.

Laidlan, J. (1956). *Lancet* 271, 1235–1237.

Levine, S. (1968). *Develop. Psychobiol.* 1, 67–70.

Levine, S. (1970). *Progr. Brain Res.* 32, 79–85.

Lincoln, D. W., and Cross, B. A. (1967). *J. Endocrinol.* 37, 191–203.

Lisk, R. D. (1967). *In* "Neuroendocrinology" (L. Martini and W. F. Ganong, eds.), Vol. II, pp. 197–259. Academic Press, New York.

Lisk, R. D. (1969). *Neuroendocrinology* 5, 149–160.

Logothetis, J., Harner, R., Morrell, F., and Torres, F. (1959). *Neurology* 9, 352–360.

Marczynski, T. J., Yamaguchi, N., Ling, G. M., and Grodzinska, L. (1964). *Experientia* 20, 435–437.

McDonald, P. G., and Gilmore, D. P. (1971). *Neuroendocrinology* 7, 46–53.

McEwan, B. S., and Pfaff, D. W. (1970). *Brain Res.* 21, 1–16.

McEwan, B. S., and Weiss, J. M. (1970). *Progr. Brain Res.* 32, 200–211.

McEwan, B. S., Weiss, J. M., and Schwartz, L. S. (1969). *Brain Res.* 16, 227–241.

McEwan, B. S., Zigmond, R. E., Azmitia, E. C., and Weiss, J. M. (1970). *In* "Biochemistry of Brain and Behavior" (R. E. Bowman and S. P. Datta, eds.), pp. 123–167. Plenum Press, New York.

McEwan, B. S., Pfaff, D. W., and Zigmond, R. E. (1970a). *Brain Res.* 21, 17–28.

McEwan, B. S., Pfaff, D. W., and Zigmond, R. E. (1970b). *Brain Res.* 21, 29–38.

McGuire, J. L., and Lisk, R. D. (1968). *Fed. Proc.* 27, 270.

McGuire, J. L., and Lisk, R. D. (1969). Neuroendocrinology 4, 289–295.

Miller, R. E., and Ogawa, N. (1962). J. Comp. Physiol. Psychol. 55, 211–213.

Millett, K. (1970). In "Sexual Politics," P. 28 ff. Doubleday, Garden City, New York.

Moguilevsky, J. A., Libertun, C., Schiaffini, O., and Scacchi, P. (1969). Neuro-endocrinology 4, 264–269.

Money, J. (1961). In "Sex and Internal Secretions" (W. C. Young, ed.), 3rd ed., Vol. II, pp. 1383–1400. Williams and Wilkins, Baltimore, Maryland.

Murphy, J. V., and Miller, R. E. (1955). J. Comp. Physiol. Psychol. 55, 211–213.

Myant, N. B. (1971). Advan. Exp. Med. Biol. 13, 227–237.

Nieman, E. A. (1961). J. Neurol. Neurosurg. Psychiat. 24, 50–57.

Pasquini, J. M., Kaplun, B., García Argiz, C. A., and Gomez, C. J. (1967). Brain Res. 6, 621–634.

Pfaff, D., Silva, M. T. A., and Weiss, J. (1971). Science 172, 394–395.

Pfaff, D. W., and Zigmond, R. E. (1971). Neuroendocrinology 7, 129–145.

Phoenix, C. H., Goy, R. W., Gerall, A. A., and Young, W. C. (1959). Endocrinology 65, 369–382.

Ratner, A., and Adamo, N. J. (1971). Neuroendocrinology 8, 26–35.

Ruf, K., and Steiner, F. A. (1967). Science 156, 667–669.

Saunders, F. J. (1968). Physiol. Rev. 48, 601–643.

Sawyer, C. H. (1970). In "The Hypothalamus" (L. Martini, M. Motta, and F. Fraschini, eds.), pp. 83–101. Academic Press, New York.

Sawyer, C. H., Kawakami, M., Meyerson, B., Whitmoyer, D. I., and Lilley, J. J. (1968). Brain Res. 10, 213–226.

Schneck, L., Ford, D. H., and Rhines, R. (1964). Acta Neurol. Scand. 40, 285–290.

Selye, H. (1941). J. Pharmacol. Exp. Therap. 73, 217–141.

Shapiro, S. (1968). Gen. Comp. Endocrinol. 10, 214–228.

Slusher, M. A., Hyde, J. E., and Lauffer, M. (1966). J. Neurophysiol. 29, 157–169.

Sokoloff, L., and Klee, C. B. (1966). In "Endocrines and the Central Nervous System" (R. Levine, ed.), pp. 371–386. Williams and Wilkins, Baltimore, Maryland.

Steiner, F. A. (1970). Progr. Brain Res. 32, 102–107.

Steiner, F. A., Ruf, K., and Akert, K. (1969). Brain Res. 12, 74–85.

Strand, F. L. (1969). Fed. Proc. 28, 438.

Strand, F. L., and Stoboy, H. (1969). Pflueger's Arch. 312, 97.

Strand, F. L., Friedebold, G., and Stoboy, H. (1962). Acta Physiol. Pharmacol. Neerl. 11, 213–234.

Strand, F. L., Stoboy, H., and Cayer, A. (1973). Neuroendocrinology (in press).

Streifler, M., and Feldman, S. (1953). Confinia Neurol. 13, 16–27.

Supnieski, J., Misztal, S., and Marczynski, T. J. (1961). Dissertationes Pharm. (Warsaw) 13, 205–217.

Thoman, E. B., Sproul, M., Seelet, B., and Levine, S. (1970). J. Endocrinol. 46, 297–303.

Thorn, G. S. (1949). In Adrenal Cortex Conf., 1st p. 189. Josiah Macy, Jr. Foundation. New York.

Timiras, P. S., and Woodbury, D. M. (1956). Endocrinology 58, 181–192.

Timiras, P. S., Baker, D. H., and Woodbury, D. M. (1955). J. Pharmacol. Exp. Therap. 113, 50.

Valcana, T., and Timiras, P. S. (1969). J. Neurochem. 16, 935–943.

Valcana, T., Vernadakis, A., and Timiras, P. S. (1967). Neuroendocrinology 2, 326–329.

Van Hof-Van Duin, J. (1958). *Acta Physiol. Pharmacol. Neerl.* **7**, 289–360.

Vernadakis, A., and Timiras, P. S. (1963). *Nature (London)* **197**, 906.

Vernadakis, A., and Woodbury, D. M. (1963). *J. Pharmacol. Exp. Therap.* **139**, 110–113.

Vernadakis, A., and Woodbury, D. M. (1964). *J. Pharmacol. Exp. Therap.* **144**, 316–320.

Walker, S. M. (1955). *Amer. J. Physiol.* **182**, 393–395.

Ward, I. (1972). *Science* **175**, 82–84.

Weiss, B., and Kidman, A. D. (1969). *Advan. Biochem. Psychopharmacol.* **1**, 131–164.

Whalen, R. E., Edwards, D. A., Luttge, W. G., and Robertson, R. T. (1969). *Physiol. Behav.* **4**, 33–39.

Whalen, R. E., Luttge, W. G., and Gorzalka, B. B. (1971). *Hormones Behav.* **2**, 83–90.

Wimersma Greidanus, Tj. B., van, and de Wied, D. (1971). *Neuroendocrinology* **7**, 291–301.

Woodbury, D. M. (1954). *Rec. Progr. Hormone Res.* **40**, 65–107.

Woodbury, D. M. (1958). *Pharmacol. Rev.* **10**, 275–357.

Woodbury, D. M., and Davenport, V. D. (1949). *Amer. J. Physiol.* **157**, 234–240.

Woolley, D. E., and Timiras, P. S. (1962). *Endocrinology* **70**, 196–209.

Wright, E. B., and Lester, E. J. (1959). *Amer. J. Physiol.* **196**, 1057–1062.

Wurtman, R. J., Axelrod, J., and Chu, E. W. (1963). *Science* **141**, 277–278.

Young, W. C. (1961). *In* "Sex and Internal Secretions" (W. C. Young, ed.), Vol. II, pp. 1173–1239. Williams and Wilkins, Baltimore, Maryland.

Zamenhof, S. (1942). *Physiol. Zool.* **15**, 281–292.

Zamenhof, S., Van Marthens, E., and Burszstyn, H. (1971). *In* "Hormones in Development" (M. Hamburgh and E. J. W. Barrington, eds.), pp. 101–119. Appleton, New York.

AGING AS A PROCESS

D. Bellamy

I. Definitions 219
   A. Life Tables 221
   B. The Nature of Aging 223
   C. Environment and Life Span 224
II. Homeostasis and Aging 226
   A. Physiological Aging 226
   B. Aging and the Endocrine System 233
   C. Hormones and the Aging Process 244
   D. Homeostasis and Mortality 253
III. Aging at the Tissue Level 256
   A. Extracellular Components 256
   B. Cellular Aging 257
IV. Evolution and Aging 269
V. General Conclusions 271
   References 271

I. DEFINITIONS

Three landmarks in the life of organisms are the time of sexual matura-tion, the time when growth ceases, and the time of death. This raises the question: "Is the period of growth and development related to the duration of life?" Put more specifically: Is aging a continuation of the same general process constituted by the phenomena of growth and de-velopment? This is an interesting starting point in any discussion of the nature of aging because development and aging are viewed as pro-

cesses which relate to the pattern of the total life history. It also makes
it possible to ask subsidiary questions concerning the likelihood of control of aging through the interplay of humoral factors which are known
to be important in the regulation of growth and development.

That there is a close interplay between aging and development is
evidenced by the difficulties in defining the term aging. To many biologists the term is used to describe any time-dependent change with respect to life history. However, it is necessary to place some restriction
on the definition in order to arrive at a category of changes which
is useful in formulating problems open to experiment. For example,
aging may be distinguished from the time-dependent changes of development; developmental events aid survival until the individual is a repro-
ductively competent adult, whereas aging processes lead to a failure
to adapt to the environment and may ultimately result in death. Points
of difficulty arise because many developmental events are the obvious
precursors of aging phenomena and some changes which begin before
or shortly after birth continue unabated throughout life. Despite this,
many workers would restrict the term aging and eliminate from consideration any changes which do not render the individual more likely
to die in a given time interval as it grows older. A good example of
a postreproductive decline in function, which does *not* show up as an
increase in mortality, is the regression of the human female reproductive
system. This particular difficulty of definition has been recognized and
another category, "senescence," has been introduced to include only
those events which contribute to the decreased resistance to death. However, it is generally accepted that of the many aging phenomena, very
few could be proved not to influence mortality. Also, the force of mortality depends very much on the environment. For example, it is of
no obvious disadvantage for civilized man not to be able to run a 5-min-
ute mile at the age of 50. Nevertheless, the well-documented decline
in human athletic performance from the second or third decade could
be included as an early aging process in a less sophisticated society,
where predators had the ability to run faster than human prey.

Perhaps our difficulties arise because it is implicit in most current
definitions of aging that aging phenomena have unique causes. It is
by no means clear that this is so and the view may be advanced that
differentiation and aging result from the operation of the same basic
cellular processes. Differences between early and late events are that
early events lead to perfection of function and late events result in
the deterioration of function.

For the purposes of the present discussion the term senescence will
not be used and aging is defined as a loss of adaptability resulting

from a decline in tissue and functional reserves. The definition implies that a major feature of aging is that the homeostatic systems of the body become less efficient in combatting fluctuations in the external world. On this basis, aging begins long before there is a marked increase in mortality. In the human, a steady decline in physiological performance is first noticeable during the third decade of life; the corresponding stage in the laboratory rat (life span 800 days) occurs at about 200 days; in the fruit fly *Drosophila* (life span 27 days), aging of some systems begins at about 10 days. Thus, it appears that in this context, aging phenomena are noticeable for well over half the life span.

A. Life Tables

When a population of animals is maintained under controlled laboratory conditions, all members have the same environmental history, yet they do not die at the same instant in time. If the number of individuals remaining alive at set intervals out of a given number recorded at birth is plotted against time, instead of a rectangular graph, a curve is obtained. There is a wide spread in the age at death and this is taken to mean that there is variability in resistance to death in the population. The actual shape of the curve, called a life table, gives no information about the rate of aging in individuals. It is merely an accurate age-frequency distribution at the moment of death; the shape of the curve is not fundamentally related to the rate or kind of aging of a given population. Ruling out accidents, the spread in the timing of death indicates that either the population is not uniform in genotype, or that if it is genetically homogeneous, events have occurred at random since birth which have resulted in different rates of aging. Life tables are useful in making comparisons between aging in two or more populations which differ with respect to environment, genetics, or the experience of random internal events which may influence the rate of aging. They show the times of decreasing "fitness" in terms of death rate and it is possible by comparing life tables to detect the differential effects of a changed environment upon the life expectancy at different ages, implying that the fitness of individuals varies with age.

Animals tend to have a life span which is characteristic of the species, implying that aging is under genetic control. It is this genetic element of longevity which accounts for the 1000-fold variation in the mean life span between species in captivity. With regard to human populations it has been estimated that longevity is about 90% heritable (Szilard, 1959). However, many geneticists would say that the environmental or nongenetic component is much greater than 10%. Data on the heritabil-

ity of the human life span comes from studies on identical twins, which
have a much smaller mean difference in ages at death compared with
fraternal twins (Kallmann, 1961). Further, having all four grandparents
surviving to 80 years of age gives a statistical chance of an individual
living 4 years beyond the average life expectancy of the present population in the U.S.A. On the other hand, it has been estimated that city life
reduces average life expectancy by 5 years compared with rural living.

Life tables are simply constructed by counting the numbers of individuals of different ages in a population. This tends to give a static
picture, whereas the population is made up of at least five basic numeri-
cal changes. An input of individuals occurs through births and immigra-
tion; an output occurs through deaths (natural and by predation) and
emigration; also, with the passage of time, the total number of individuals changes due largely to environmental factors influencing the
steady state.

Mortality in natural populations is simple to measure when reproduction and migrations can be ignored and the population examined repeatedly as it decreases in size. However, this is only possible with
animals which have an annual life cycle, a short reproduction period
and a small range of movement. More often than not, the death rate
in a population is difficult to measure and one must resort to estimating
the accumulation of animal remains which are resistant to decay, such
as bones in vertebrates, the shells of mollusks, and the cuticle of insects.

It is also difficult to assess the age of many wild type species. Evidence
of age may be obtained from annual events which leave a permanent
alteration in the hard parts, such as rings or bands of different con-
sistency and effects of continuous wear in bone, teeth, and shells. On
the other hand, physiological periodicity leaves its mark in some organs,
particularly the ovary, where the number of empty ovarioles in the
ovary of insects or the number of corpora lutea in the ovary of mammals
may sometimes give a good assessment of chronological age. Few of
these methods are sufficiently reliable to be used on individuals and
the sparse physiological and biochemical data that are available for
noncaptive animals are statistical mean values referred to average size
as the only practical temporal quantity.

It appears that a differential mortality rate is a sex-linked feature
of many animals. In humans, the male death rate is higher than that
of the female over the entire life-span. At birth, the ratio of males to
females is 106:100; there is evidence that a differential death rate of
males is a feature of interuterine life with estimated primary sex ratios
of between 110–160 males to every 100 females. In some birds, reptiles,
and some insects, the female has the shorter life span. However, a con-

sistent feature appears to be that the shorter-lived sex is the one which has the XY chromosome combination. The higher death rate in XY individuals is possibly due to the action of recessive alleles on the X chromosome for which there are no counteracting alleles in the Y chromosome.

Because of all the difficulties in obtaining reliable data from natural nonhuman populations, most life tables have been constructed using laboratory species or domesticated populations. It is well-known that these tables frequently show a life expectancy which is much higher than for the same species in the wild. For example, the mean life expectancy of wild mice which are not subjected to predation is only a few months (Berry and Jakobsen, 1971), whereas in the laboratory the same strain lives for 2–3 years. It may be that part of this increase in life span in captivity is due to the experimenter selecting genotypes for good adaptation to laboratory conditions, but a more likely explanation is that the major causes of death in wild populations are physical and biotic insults of a fluctuating environment. It is the natural life span in the wild that must be under selection pressure. Furthermore, in asking the question which processes render the individual more likely to die in a given time interval as it grows older, we are likely to obtain different answers from laboratory strains compared with wild types.

B. The Nature of Aging

Aging of machines occurs because frictional forces result in the wearing-out of moving parts and also because there is an inevitable deterioration of key components through chemical changes defined variously as rusting and perishing (Fels, 1969). With regard to the former category, there is no evidence that components of living tissue age in proportion to normal usage, although excessive loading of joints in human subjects may result in subsequent malfunctions of the joints in old age. In this respect it is a characteristic of life that the working capacity of organs increases with the demands made upon them. The two relevant phenomena of work-hypertrophy and compensatory-hypertrophy are, to a great degree, independent of age. Regeneration, on the other hand, which has no counterpart in the nonliving world, does decline with age (Needham, 1949). Intermolecular changes in the category of chemical aging do contribute to biological aging, particularly with regard to extracellular structures. This is most evident in the aging of protein molecules and will be described later.

The fact that organisms age and die implies that imperfections appear at the chemical level which cannot be corrected either during the individual life span or in evolutionary time. Any imperfection of function which results in a progressive deterioration in the organism could be classed as an aging effect and from this viewpoint alone, it is likely that there are many causes and consequently many processes of aging. A common cause of aging in the various organs is also unlikely on the grounds that they differ so much in both structure and function. In addition, the available experimental evidence supports the idea of organ aging being mainly self-contained and not dominated by systemic factors. This multifactorial theory of aging is opposed to the less likely unitary hypothesis which states that there is only one process which is responsible for the general loss of adaptability in all organs.

When we come to examine the reasons for the increased force of mortality in old organisms, there are two possible levels of function, corresponding to the organ and the cell, at which aging may be considered as a fundamental process. At the level of the organ, there appears to be little evidence that malfunctions, either through disproportionate growth or the failure of repair mechanisms, occur on a scale sufficient to increase the chances of the organism dying. On this plane, however, one is clearly dealing with the deterioration of complex interacting multicellular systems, comprising sense organs, nervous system, endocrine system and effectors. Taken together these fail to produce the correct degree of response in relation to alterations in the external and internal environment. Lack of ability to respond appropriately to a disturbed environment does not appear to be due to the failure of any single component. For example, the endocrine system probably fails to meet demand because of both a fall in the rate of secretion of hormones and the inability of the target tissue to make the appropriate response to the hormones. Often, structural changes in several components of the effector organs are alone sufficient to limit mechanical aspects of an endocrine response.

Death in the context of organ function involves a series of events which occur over a small fraction of the life span, possibly amounting to only a few hours. That is to say, organisms die as the result of internal fluctuations in body chemistry which can no longer be contained through regulation; a small shift in metabolism which could be counteracted in youth becomes amplified to the point of preventing a vital function. Thus, death in old organisms may result from minor changes in the environment.

C. Environment and Life Span

Advancing age brings with it a general decline in physical fitness with a reduction in muscular strength and a prolongation of the period of recovery from exertion. There is also a greater susceptibility to disease and consequences of accidental damage, either of which may result in death. These general features of aging are found at all levels of organization and it is instructive in this connection to examine the features of aging in primitive multicellular animals.

Rotifers exhibit a life cycle in which cell division is completed prior to hatching. Growth occurs by an increase in cell size and most organs appear to have a constant cell number between individuals. Reproduction occurs usually after the attainment of maximum cell size with a rapid and copious egg-laying cycle. Gradually, the reproductive capacity declines and aging is characterized by slow movement, a decline in feeding activity, together with marked morphological changes in the somatic organs.

This life cycle results in the manifestation of a characteristic life span, which, nevertheless, is modified by the environment. The mean life span is inversely related to temperature and is markedly affected by the pH of the external medium. The latter effect is most noticeable in the post-reproductive life which increases as the pH increases. Changing the salt content also affects mean life span—in particular, high external calcium concentrations tend to reduce longevity. There are also indications that life span depends on nutritional status (Lansing, 1942).

There can be little doubt that aging in rotifers has effects on the life span and reproductive capacity of offspring. In this respect, parental age is the determining factor. Eggs taken from old, mature, individuals give rise to parthenogenic clones (orthoclones) which die out earlier and have a shorter mean life span than those obtained from eggs of younger animals. (Lansing, 1947). This implies that the reproductive system of the parents deteriorates in old age. Both egg size and egg shape show a greater variability with increasing parental age and it has been argued that this is the basis for the observed variability in fecundity of offspring; offspring from large eggs have a very high fecundity; and those coming from small eggs have a low fecundity (Jennings and Lynch, 1928).

The shortening of mean life span leading to extinction of offspring from old parents is not the result of a simple genetic change in the offspring, because individuals developing from eggs taken early in the life of a clone doomed to a short life span with subsequent early extinction (because it was derived from a senile parent) show the life span and pattern of development of early eggs produced by the original stock. Lansing's contention is that the early extinction and shortened mean life span are due to the transmission of an aging factor through the population which has a cumulative deleterious action on life span; but the picture is far from clear.

Nevertheless, these phenomena observed in rotifers stress two common findings in other animals: the influence of environment on longevity and the increased phenotypic variation with age.

Another noteworthy feature of these experiments is that the rate of sexual maturation bears an inverse relationship to longevity. As longevity decreases in old orthoclones, the age of sexual maturation (measured by the beginning of egg laying) becomes progressively lower; while, as longevity increases in "young" orthoclones, the age of sexual maturation becomes increasingly greater.

There is evidence that similar relationships exist in other animals. Brody (1945), accumulated data for a large number of species of mammals and birds which indicate that the duration of life is directly proportional to the length of time taken to reach mature size and inversely proportional to the speed of maturation. This also appears to hold for phenotypic differences in growth rate and longevity within species (Markovsky and Perlmutter 1973). Such findings support the view that there is a definite chronological relationship between growth, development, and aging. This view is strengthened by the demonstration that prolongation of the period of growth and development results in prolongation of the life span of a wide range of organisms. The relevant work involves semistarvation in protozoans (Rudzinska, 1951), rats (McCay, 1942; Carlson and Hoelzel, 1946; Templeton and Ershoff, 1949), and mice (Tannenbaum, 1947); and changes in endocrine status in mice (Bellamy, 1969), cats (Hamilton *et al.*, 1969), and humans (Hamilton and Mestler, 1969).

II. HOMEOSTASIS AND AGING

A. Physiological Aging

As aging is manifest in a progressive loss of vigor and the lowering of resistance to the many and varied onslaughts of the environment, it is ultimately a failure in homeostasis which results in death through old age. Failure in the homeostatic systems of the body with age may be visualized at three levels: receptors tend to show a higher stimulus threshold, the central nervous system handles information at a slower rate, and effectors become less efficient in making the appropriate response.

Anthropomorphic data clearly indicate that the elderly are subject to change (Dalderup *et al.*, 1966). However, if the composition of the blood is taken as an indicator of the overall state of homeostasis, there appears to be no impairment of functional capacity. For example, the acid–base balance of the blood in resting humans varies little between the ages of 25 to 85 and there are no significant changes in the carbon dioxide tension, total carbon dioxide content, and bicarbonate concentration. In agreement with this, the pH only changes from 7.400 to 7.368 between the second and eighth decades of life (Shock and Yiengst, 1950). Organic constituents of blood show a similar stability (Horvath, 1946; Praetorious, 1951; Rogers, 1951). The general constancy of the ionic composition of the body is also borne out by studies on individual tissues. No significant age changes are observed in total water, fat, potassium, and the sodium content of liver, while the intracellular water, nitrogen, potassium, and phosphorus in muscle decrease only by between 5% and 8% (Yiengst *et al.*, 1959). There is no change in muscle lactic acid and creatine (Horvath, 1946), and analysis of human aortas taken over an age range from 2 to 69 years reveals no significant alterations in either total nitrogen or sulfur; aortic elastin and creatine also remain relatively constant (Kanabrocki *et al.*, 1963; Myers and Lang, 1946). A constancy of composition is also found to be a feature of the ribonucleic acid in ventricular muscle of the rat from 100 to 1200 days of age (Wulff *et al.*, 1963). In keeping with this, the mean DNA per nucleus and the mean volume of nuclei in rat liver do not change over an age range of 12–27 months (Falzone *et al.*, 1959). In general there is little evidence for large changes in the functional integrity of subcellular components (Griswold and Pace, 1956; Gold *et al.*, 1968).

The failure to detect age changes indicative of a large-scale deterioration in tissue function impressed the early workers using histological methods. Only moderate signs of aging are found in the histochemistry of liver and there are no obvious large-scale functional deficiencies at this level in the brain. Kidney tissue is marked by having the largest age-alterations in structure, but this appears to be a consequence of a large decrease in the relative cell mass without a marked change in composition of the remaining cells (Lowry *et al.*, 1946).

Despite the evidence in favor of an overall stability of tissue structure, age changes in composition are frequently observed. This applies to a small number of blood constituents (Ackermann and Kheim, 1964; Das, 1964). Often, age changes are not uniform between tissues; although there is no variation in the riboflavin content of human brain, heart, and skeletal muscle over 7 decades (Schaus and Kirk, 1956), aortic tissue is characterized by a 60% fall in the riboflavin content over the same period (Schaus *et al.*, 1955). Also, in contrast to ventricular tissue, the folia tissue of the cerebellum loses 30% of its ribonucleic acid throughout the life span (Wulff *et al.*, 1963).

Some of the largest alterations in tissue composition have been found in the lipid fraction. Cholesterol in the blood of healthy women increases 2-fold in concentration from the age of 20 to 60 (Swanson et al., 1955). There are also substantial increases in the concentration of tissue elastin, collagen, mucopolysaccharides, and calcium (Lansing et al., 1950; Hall et al., 1955; Kirk and Dyrbye, 1956; Schaub, 1963; Yu and Blumenthal, 1963; Dalderup et al., 1967). However, these changes, although a general feature of aging in many tissues, are not a universal phenomenon (Streicher, 1958; Bashey et al., 1967; Sobel and Hewlett, 1967; Sobel et al., 1968).

A common feature of aging in many types of postmitotic cells is the progressive accumulation of "age pigment." This is a yellow-brown intracellular material with the properties of a highly oxidized lipid. It emits a bright yellow-green fluorescence when excited by ultraviolet light. In the brain of old rodents the quantities of age pigment range from 0.6% in the granular cells of the cerebellum to 17% in the Purkinje cells (Reichel, 1968).

Taking the available evidence on the gross composition of tissues, it appears that, on the whole, age changes are not very marked in those intracellular components, such as water, nitrogen, and inorganic ions, which are concerned with fundamental cellular organization. On the other hand, there are marked changes in the chemistry of the extracellular compartment, notably in the ground substance. These changes are probably a reflection of the general tendency for there to be an age-dependent shift in the balance of cell populations. For example, the eosinophil count in rats decreases progressively with age, there being a 50% drop between the ages of 300 and 800 days. Despite this loss, there is little change in other blood cells (Everitt and Webb, 1958). In aging cattle, it has been observed that a decrease in blood lymphocytes is accompanied by a rise in the proportion of blood neutrophils and a fall in the total leukocyte count (Riegle and Nellor, 1966).

Changes in cells are most marked at the histological level. An almost universal feature is that of cell loss (age involution). This is a well-established characteristic of the thymus where the process results in a progressive general decrease in cellular mass which is accompanied by shifts in the ratio of lymphocytes to macrophages and fibroblasts (Bellamy and Alkufaishi, 1972). Age-involution is also a feature of skeletal muscle, and some rats show a 30% loss of thigh muscle between the first and second year of life (Yiengst et al., 1959). Taking total body potassium as a measure of cell mass, the number of cells per unit body weight shows a steady decline in humans from the late teens to the age of 80 (Allen et al., 1960). In another study (Shock et al., 1963), it was found that the extracellular space did not change although the total body water diminished significantly with increasing age. Intracellular water, calculated as the difference between total body water and extracellular space, showed a significant age regression. This is interpreted as a reflection of the loss of functioning cells with increasing age. From evidence of this nature it is generally held that involution is one of the most consistent features of aging in the higher vertebrates.

The kind of study that has been discussed so far, although pointing to a number of age-changes in metabolism has not thrown light on the reasons for the increased chances of mortality as time passes. There is no consistent trend that can be definitely interpreted as a deterioration in function, and apart from tissue involution there seems little change in functional organization as measured by the mean composition of tissue samples. This points to the fact that aging is apparent for the most part at a physiological level. Although physiological parameters that depend on a complete interaction of several organs, such as ion balance and blood pressure hardly change with age when measured in the resting state, there is an increasing tendency for departures from these norms to take place. This is manifest in the cardiac arrhythmia of the aging male rat (Jones *et al.*, 1967). In this context the systolic blood pressure of CBA mice under controlled conditions remains within the range 126 ± 12 mm Hg throughout the life span of $2\frac{1}{2}$ years (Henry *et al.*, 1965), but the normal distribution curve for rat blood pressure becomes biphasic with age (Berg and Harmiston, 1955). When the rats pass the age of 600 days, about 30% of males and 20% of females become hypertensive with marked renal disease. The latter condition is probably the main cause of the hypertension.

Aging is more clearly seen when physiological mechanisms are activated to respond to increased internal or external demands. A decline in the ability to cope with increased functional demands due to endogenous processes can be seen in relation to the reproductive capacity of rodents. In polytocous rodents a decline in overall fertility occurs well before death and this is associated with a failure to conceive again immediately after parturition. There is also a failure of maternal behavior and lactation. Although reproduction must come to an end when the ovaries no longer contain oocytes, it appears that animals stop producing litters a considerable time before they exhaust their complement of germ cells. The available evidence indicates that ovulation and conceptions proceed normally well into old age but the embryos fail to develop normally and are usually aborted in the middle of pregnancy.

A decline in the ability to cope with increased functional demands due to endogenous processes can be clearly seen in the reproductive capacity of rodents. After 14 months the litter size of hamsters is greatly reduced and the length of gestation increases (Solderwall et al., 1960). Similarly, data for the mouse suggest that parental age influences the length of the period of fertilization plus gestation, the litter size, and the stillborn mortality rate (Roman and Strong, 1961). Resorption during the late phase of gestation is thought to be a contributory factor in the reduction of litter size in senescent females. It is also possible that the basic cellular cycles of reproduction change with age. In this connection it is known that the cell cycles of mitosis in esophageal epithelium of mice gradually increase in length (Thrasher, 1971).

A failure of physiological adaptation is most clearly seen under laboratory conditions in response to specific stresses. Old $C_{57}B_2$ female mice have a high mortality rate following exposure to a temperature of 6°–7°C which is not seen in young, mature females. The failure of integrative function which is responsible for the death of older animals appears early in the experimental period. Old mice which survive the first 2 days of cold exposure have a good chance of surviving for the next 12 days in the cold (Grad and Kral, 1956). Mortality of old mice is reduced when they are slowly adapted to the lower temperature, but the adaptive process is less effective than in the young.

A number of the individual physiological responses of rodents to cold, such as the increased oxygen uptake, rise in blood sugar, loss of body weight, and changes in blood cell count, also become less marked with age (Grad and Kral, 1956). Other studies on humans suggest that the mechanisms for minimizing heat loss deteriorate in old subjects (Krag and Kountz, 1950).

The response of the human body to heat stress has also been examined for possible age effects (Shattuck and Hilferty, 1932; Friedfeld, 1949). In general, there is an increased degree of discomfort at high temperatures and the physiological responses in the old become more variable (Krag and Kountz, 1952). One relevant feature is that men over the age of 40 have a greatly increased vasodilation in response to heat (Hellon and Lind, 1958). Since in a hot environment heat loss is almost entirely by evaporation, the greater skin blood flow in old subjects would not be expected to have a marked effect on their heat regulation. Age differences in rectal and skin temperatures are not significant in heat-exposed subjects, therefore the view may be taken that the faster skin circulation of older men is unnecessary for heat regulation and places a needless burden on the general circulation.

Responses to temperature are intimately linked with the basal metabolic rate. There is a significant fall in basal heat production and carbon dioxide elimination as humans age (Shock *et al.*, 1963). There is also a larger day-to-day variation in several aspects of respiratory function (Shock and Yiengst, 1955) which may be of significance with regard to the efficiency of adaptation.

Starvation, another common stress encountered by animals in the wild, may be examined conveniently in the laboratory. However, there have been few investigations of starvation in relation to age. In rats no significant age differences have been observed in either the rate of weight loss in long-term starvation or the survival time after food deprivation (Jakubczak, 1967).

With regard to experimental hypoxia there is a marked loss of adaptation in old animals (Fluckiger and Verzár, 1955; Sulkin and Sulkin, 1967).

Data for humans do not support the concept that older individuals have less chance of surviving traumatic injury (Collins, 1950). On the other hand, this conclusion relates only to traumatic abdominal injuries. Regeneration of the liver of rats after partial hepatectomy occurs to an equal extent in the young and old but old animals take more time to form the requisite number of new cells (Bucher and Glinos, 1950). Following unilateral nephrectomy senescent rats have the same capacity for compensatory growth as young ones (McKay *et al.*, 1924). However, as for liver regeneration, the kidney of old animals responds slowly (Reiter *et al.*, 1964). There is no evidence that new cells produced in response to unilateral nephrectomy by old rats are biochemically different from those in young rats (Barrows *et al.*, 1962). Despite this finding, the chronic hypertrophy of kidneys in rats given a pathogenic diet is associated in old animals with a decreased capacity of mitochondrial oxidative phosphorylation (Fedorčákova *et al.*, 1968).

Responses to drugs, anesthetics and toxic agents have also been used to measure age changes in adaptation (MacNider, 1946). Increased age is often a factor which is responsible for a loss of drug responsiveness. For example, patients with arthritis are more responsive to steroid therapy in the third decade than in later years. This difference is not related to the duration of the disease. Hexabarbital anesthesia is prolonged in old rats compared with young ones. Alongside this age change there is an increased mortality attributable to the drug (Streicher and Garbus, 1955). The endocrine status and the immunocompetence are both thought to be important aspects of this rise in toxicity with age, but neurophysiological and neuropathological factors may well be involved. In this connection, the toxic action of uranium salts is manifest locally by specific degenerative changes in the kidney tubule which are more marked in old animals. These histological changes in the epithelial cells go in parallel with the loss of functional capacity of the kidney (Mac-Nider, 1917).

The increased sensitivity of old animals to infection suggests a failure of the immune response. Here it has been suggested that the loss of ability to produce specific antibodies to thyroglobulin in rats is due to a loss of immunocompetent cells with age (Schumacher and Premachandra, 1968). There is also an age-related increase in susceptibility of mice to attack by foreign immunocompetent cells manifest, for example, in the graft-versus-host reaction (Troup and Walford, 1969).

Age-dependent failures of adaptive renal responses have been the subject of much detailed investigation. Clearance studies with diodrast and aminohippuric acid indicate that renal plasma flow is diminished in old subjects (Miller *et al.*, 1951). There is also a fall in the uptake of aminohippuric acid by isolated tubules. Similarly the capacity to reabsorb glucose from the glomerular filtrate shows a linear decline with age (Miller *et al.*, 1952) which amounts to about 10% from the third to the ninth decade. Also, the changes in filtration rate parallel the changes in reabsorptive capacity which may be of significance in relation to the age-dependent depression of glucose tolerance in humans (Schneeberg and Finestone, 1952) and the failure to induce a reduction in the alkali reserve after treatment of old dogs with anesthetics (MacNider, 1943). However, there are many physiological functions that deteriorate with age, any one of which may be equated with a loss of vigor (Conrad, 1960). For example, a considerable body of statistical evidence attests that on average, the ability to exercise and carry out physical work declines with age (Ruger and Stoessiger, 1927; Burke *et al.*, 1953). Between the ages of 30 and 70 there is also a 2-fold increase in the time of recovery from exercise. This may be related to changes in several factors such as oxygen uptake, pulmonary ventilation, diffusing capacity of lungs, work pulse, and blood lactate.

An important aspect of physical fitness and many adaptive responses is the ability to make rapid neurophysiological adjustments. Thus there is a consistent decline in the capacity to see at low light intensities and an increased limitation upon the extent of dark adaptation (McFarland and Fisher, 1955). There are also age-dependent losses in the ability to move rapidly in response to environmental changes (Birren, 1955; Birren and Botwinick, 1955; Birren and Kay, 1958; Pierson and Montoye, 1958; Hodgkins, 1962).

In long-term responses involving locomotory activities the impairment in ventilation and gas exchange appears to underlie the age deterioration (Norris *et al.*, 1955). However, in short-term reflex delays in the aged it is suggested that the increased reaction time is due to decreased excitatory influences from higher levels (Magladery *et al.*, 1958). Apart from the well-established age involution of nervous tissue and the appearance of "age pigment" in neurones of the central nervous system (see Section II A) there are few indications of the cause of the failure of nervous coordination in old individuals. The excitation threshold for electrical stimulation of nerves is higher in old animals than in young ones. Also there is a decrease in the activity of choline acetylase and choline esterase (Frolkis, 1966), the key enzymes in transmission of the nerve impulse. It may well be that the decline in the number of neurones and motor end-plates with age also contributes to the loss of flexibility of nervous responses (Gutmann *et al.*, 1968). However, the most likely cause of the decline in vigor in relation to long-term environmental changes lies in the organization of the endocrine system. This possibility will now be considered in detail.

B. Aging and the Endocrine System

Hormones are integrated into the pattern of development in two ways. They act as controllers, in that they release or trigger the developmental potential inherent in certain tissues. Actions of this type are sometimes irreversible, but usually the target tissue returns to its former state when the concentration of hormone falls to the initial level. Second, hormones invoke responses that offset undesirable changes in the external environment. From the latter point of view, the aging of endocrine systems takes place in two stages. Initial development of endocrine responses make the animal more adaptable and independent of its surroundings; later changes decrease the capacity for adaptation to the environment.

Several biological aspects of development are of importance to the endocrinologist dealing with the aging of endocrine systems. These are related to the source of the initial stimulus that leads to the modification of an endocrine response. The stimulus may be internal or external.

The most obvious modifications of the endocrine system that fall into the first category are those linked with the attainment of sexual maturity. This applies particularly in the female, where there is a precise chemical control of the sequence of events in the reproductive cycle. In this context, the corresponding alterations in the endocrine system are brought about by intrinsic mechanisms, although the various responses may be synchronized by external factors.

The basis for alterations of the second type, where the stimulus lies outside the organism, underlies the development of many lower vertebrates. The chemical nature of the stimulus is most clearly seen in the environmental factors encountered by euryhaline fish, where a spawning migration from freshwater to the sea brings about profound changes in the endocrine system (Chester Jones and Bellamy, 1964). Some of the endocrine changes appear to accelerate the deterioration of those organs not directly involved with reproduction; other changes, such as those connected with the regulation of salt balance, seem to be reversible. However, even though the immediate stimulus which activates the endocrine system comes from the environment, the impulse to undertake the migration may be intrinsic.

The above features of the developing individual determine the changing patterns of homeostasis with age, and, in order to discuss the aging of mammalian endocrine systems, four aspects related to intrinsic stimuli will be considered as follows: (1) changes in endocrine-controlled norms; (2) changes in the responsiveness of target tissues to hormones; (3) changes in the function of endocrine organs; (4) changes in the physiological and biochemical action of hormones.

1. Changes in Norms

At each stage of life, the chemical composition of the mammalian body is maintained within narrowly defined limits, through the action of endocrine systems. However, new intracellular norms for enzymes and metabolites are established as the animal moves from one stage of development to another. The most rapid changes of this type are found in the early postnatal period (Gartner and Arias, 1963; Withrow et al., 1964). Few measurements have been made in the older age groups with regard to intracellular metabolites, but age variations have been noted in the concentration of several plasma constituents (Vranic and Pokrajac, 1961; Ackermann and Kheim, 1964; Das, 1964; Gsell et al., 1964). Bearing in mind the fall in metabolic mass with age it is likely that a decreased rate of metabolism would tend to increase the concentration of most plasma metabolites which have a direct dietary origin. The fact that there is no general rise in the concentration of these substances with age suggests that adaptive changes in the endocrine system occur which offset this tendency, but we are largely ignorant of the details.

Every endocrine response, even though the end point is the restoration or maintenance of a constant concentration of some chemical in the body fluids, is always associated with changes in the expenditure of energy. Thus, by its nature, a hormone brings about an alteration in metabolism and the temporary or permanent attainment of new norms. For example, an increased rate of synthesis of the hormone and also, particularly in long-term hormone action, a stimulation of the formation of a component of the effector system in the target tissue may accompany hormone secretion. In this way there may be requirements for an increased supply of energy-yielding substances and precursors of macromolecules. These changes are important in responses that result in enzyme synthesis. Therefore many aspects have to be considered, and although there may be no age-dependent alterations in, say, the concentration of a particular metabolite, several new norms may be established associated with a tendency for the steady-state equilibrium of this metabolite to shift due to the aging of other components of the system.

In aging mammals it is difficult to relate the observed changes in plasma metabolites to a variation in the endocrine organ, rather than to alterations in the enzyme activity of the target tissue. If there was a fall in the efficiency of either the appropriate endocrine organ or the enzymic process that previously maintained a particular metabolite at a constant level, the concentration in the body fluids would tend to change with age. However, in this complex control system, a change in one of the three potential variables, availability of metabolite, activity of endocrine gland and rate of enzyme synthesis, could well lead to alterations in the other two.

2. Response of Target Tissues

It is well known that all organs of the body undergo chemical and morphological changes with age. In discussing the aging of tissues that are targets for hormones, only those features will be considered that have been experimentally linked with alterations in response to exogenous hormones.

Little work has been done on the early development of hormone sensitivity. However, it is known that certain embryonic tissues show the normal qualitative response that is characteristic of the mature animal. For example, there is evidence that fetal liver and thymus are affected by adrenal steroids in the same way as in postnatal life (Jost, 1951; Angervall and Lundin, 1964). Other embryonic and early postembryonic tissues are also sensitive to hormones. An adequate hormonal status is essential for several normal morphogenic events to occur, such as sex differentiation, and for ontogenesis and maturation of some physiological and behavioral functions. Hormones may also impose new permanent and irreversible orientations on crucial structures. Unusual actions may be elicited (Burrows, 1945) which can sometimes only be invoked during a short, well-defined period (Moore, 1945). Often, effects are obtained which result in the abnormal differentiation of target tissues, as in the case of androgens acting on the secondary sexual organs in the female fetus. Presumably these abnormal reactions are connected with the incomplete differentiation of the target tissue. An abnormal response to androgens is not confined to vertebrate sexual tissues. It is also found in the embryonic lymphoid tissues in birds (Meyer *et al.*, 1959), and may result from common mechanisms of differentiation within the vertebrates.

From work on the subsequent changes in the postembryonic response to hormones, it appears that target organs for hormones become sensitive to the appropriate hormones at stages of early development that coincide with the differentiation of the target tissues. A gradual rise in the degree of hormone sensitivity is implied from some experiments but few investigators have been concerned with this aspect in detail. In the later stages of life it has been found that tissues become less sensitive to some, but not all, hormones.

The complicated nature of these changes is illustrated by the endocrine control of renal function. In the human kidney, electrolyte control is not precise until 2–3 weeks after birth. This situation is thought to be connected with the slow development of the loop of Henle (Hubble, 1957). At the other end of the time scale, it is known that the kidney of old animals is not so versatile in its response to hormones as in early adult life. Experiments with the rat have shown that the antidiuretic response to exogenous vasopressin depends on the age of the animal. More hormone is required to produce a given response in very young, and old animals compared with animals of intermediate age. This will be considered in detail later. With respect to other polypeptide hormones, it has been shown that there is an age-dependent decrease in sensitivity to insulin (Shock, 1952; Silverstone *et al.*, 1957; Giarnieri and Lumia, 1961; Hrůza, 1970), growth hormone (Moon *et al.*, 1956; Everitt, 1959; Jelinkova, 1964; Root and Oski, 1969; Hrůza, 1970), and ACTH (Shaw, 1952; Riegle *et al.*, 1968). A loss of sensitivity is not general for all hormones. For example, the characteristics of lipid mobilization and deposition in humans treated with noradrenaline suggest that age has little effect on the endocrine response (Eisdorfer *et al.*, 1965). Also despite the drop in sensitivity of tissues to ACTH the assessment of the hypothalamic–pituitary–adrenal axis indicates no decrease in functional integrity (Friedman *et al.*, 1969).

Some hormones are more effective in old animals. Thus exogenous thyroid hormone may bring about a loss of body weight, which for a given dose is greater in old rats than in young ones (Bodansky and Duff, 1936). Also, the effect of thyroxine or thyrotropic hormone on the basal oxygen consumption of rats is more marked in old animals (Belasco and Murlin, 1941). In apparent contradiction to these findings, thyroidectomy has less effect on the metabolism of old rats compared with young ones (Nagorny and Golubitzkaia, 1947; Hrůza, 1970). Possibly, in addition to differences in tissue sensitivity to thyroid hormones, there are changes in the role of the thyroid gland with age.

Only a small number of results are available on the changes in sensitivity of tissues to steroid hormones. From the work of Desaulles (1958) it seems that there is a diminution in the responsiveness of the kidney to aldosterone and cortisol. As far as the metabolic action of corticosteroids is concerned, cortisol is more effective in inhibiting growth in 5-week-old rats than in 3-week-old animals (Palmer, 1966). This difference may be connected with changes in the endocrine control of growth with age (Walker *et al.*, 1950). Further, it is not possible to obtain a loss of weight with cortisol treatment in 3-week-old rats, whereas 5-week-old animals show this response at quite low doses of steroid (Palmer, 1966).

During the early postnatal period there are large changes in the response of enzyme synthesis to glucocorticoids. The basic reaction in the process of enzyme induction, the formation of RNA polymerase, does not respond to the corticosteroids until 3 weeks after birth (Barnabei *et al.*, 1966). Also, the extent of enzyme induction after cortisol treatment depends on the age of the animal. Young rats show a large increase in glutamic-pyruvic transaminase but the percentage change in enzymic activity gradually diminishes with age (Harding *et al.*, 1961). With tryptophan pyrrolase, another enzyme that is induced by cortisol, the kinetics of enzyme synthesis vary with age (Correll *et al.*, 1965). With regard to thyroid control of glycerophosphate dehydrogenase in rat liver, all animals in all age groups are fully capable of responding to the presence and absence of thyroxine in the same manner (Bulos *et al.*, 1971), suggesting that age effects on enzyme induction are not general. Despite this evidence, an age-dependent modification in the ability of tissues to synthesize enzymes in response to a variety of environmental stimuli has been proposed as a common biochemical expression of aging (Adelman, 1971). Recent work supports this view in that impairment of enzyme induction occurs widely and may take the form of a decreased level of enzyme, a drop in magnitude of response and a delayed response (Rahman and Peraino 1973).

In the assessment of a declining sensitivity to injected hormones the view may be taken that this reflects a lower steady-state concentration resulting from an increased rate of metabolism or clearance. Another viewpoint is to consider the possibility of a loss or modification of hormone receptors. A third approach requires an examination of the biochemistry of the target cells in relation to possible variations in the enzymatic potential of target tissue to respond to the hormone.

Little work has been done with regard to all three methods of investigation. There appears to be a fall in the rate of metabolism of steroid hormones with age (Romanoff *et al.*, 1963), possibly related to a decrease in metabolic mass. Incidental to this, it might be expected that a given dose of hormone would be more effective in an older animal; this is not always so.

With regard to hormone uptake many studies have involved the use of steroids, where methods are available for the isolation and characterization of these compounds and their metabolites. However, few workers have been concerned with age variations. In the human red cell, which is probably the simplest model system for the study of transport processes, there is a rise in the uptake of triiodothyronine with age, which reaches a plateau in early adult life (Soltz, Horonick, and Chow, 1963). Here, as in other studies concerned with changes in the uptake of hormones from the plasma, it is difficult to assess the contribution due to changes in the uptake mechanism as opposed to alterations in the proportion of total hormone available to the cell because the capacity of plasma proteins to bind the hormone and render it inactive may alter with age (Gala and Westphal, 1965). In this context, recent findings indicate that the renal enzyme system which inactivates parathyroid hormone becomes less active late in life (Fujita *et al.*, 1971). Similarly it has been postulated that there is a diminished rate of excretion and/or inactivation of thyroxine in older rats (Grad, 1969).

With reference to possible age variations in receptor structure, we are still largely ignorant of the chemical nature of the site of hormone action. For the steroids, it is known that cell proteins are important quantitatively, in the adsorption of hormones and metabolites (Bellamy, 1963), but no information is available on the age-related variations in protein structure and their influence on binding affinities.

It is thought that cell proteins reflect the process of aging in their tertiary structure (Bjorksten, 1963). Such changes in the configuration of proteins might affect the binding of steroid hormones to cells as well as the biochemical properties of the steroid–protein complex (Bellamy and Leonard, 1966). But in skeletal muscle, which is a target tissue for steroids and undergoes considerable changes in function with age, the age-related alterations in protein structure, as determined by indirect enzymatic methods, are not very marked (Kohn, 1963).

As to the biochemical potential of target tissues, it appears that the enzyme pattern in some tissues is not affected very much by age, but this is not general. Muscle undergoes marked changes during the life span. There is a decline in the work performance (Burke *et al.*, 1953), which occurs at a time when there is a large fall in relative cell mass.

At the biochemical level, there is a diminution in the capacity for aerobic energy production (Kiessling, 1962) and a rise in the activity of glycolytic enzymes (Stave, 1964). It has also been suggested that an observed drop of 80% in the Mg-activated ATPase in the muscle of old rats is connected with the age-dependent loss of contractile efficiency (Rockstein and Brandt, 1962). All of these biochemical changes could well alter the response of muscle to hormones.

It has already been pointed out that the aging kidney becomes less sensitive to antidiuretic hormone (Dicker and Nunn, 1958). The kidney of old animals is also less responsive to stimuli that bring about compensatory growth (Reiter *et al.*, 1964), suggesting that there is a general decline in biochemical efficiency. This is reflected in the slow rate of turnover of nucleic acids after unilateral nephrectomy in old rats. An age-dependent deterioration in renal transport processes found in intact animals (Lewis and Alving, 1938) is also found in *in vitro* preparations of kidney (Beauchene *et al.*, 1965). The phenomenon coincides with a fall in the concentration of the ATPase thought to be a component of the active transport systems. It is clear that the loss of hormone sensitivity is only one aspect of the aging process in the kidney but it may well be linked with specific changes in the enzymes basic to ion transport. Other experiments on the lipolytic response of rats to adrenaline and theophylline lend support to the idea that the loss of responsiveness to catechol amines is not due to faulty receptors (Jelinkova *et al.*, 1970). On the other hand, mitochondria in the liver of old rats show aberrant growth responses to thyroxine treatment of the intact animal indicating that there is a loss of the capacity to make an integrated response (Shamoto, 1968). In summary, the generalization may be made that age variations in metabolic pattern of target tissues are important factors leading to a decrease in tissue sensitivity to hormones. However, these changes at the molecular level are not specifically confined to effectors of the endocrine system.

3. Changes in Endocrine Organs

In the following discussion, a change in the function of an endocrine organ is defined relative to the rate of secretion of the hormone. Changes in secretion occur either through an aging process intrinsic to the gland, or through an alteration in the rate of metabolism or excretion of the hormone. In the latter two situations it is necessary to postulate a feedback mechanism that operates to maintain a fixed concentration of plasma hormone. A fall in the rate of hormone metabolism is counteracted by a decrease in the secretion rate. This appears to apply to the decline in the secretion of aldosterone and thyroxine in aging humans

(Gregerman *et al.*, 1962; Tait *et al.*, 1966). Here the change in the function of the endocrine organ may be compounded of several factors such as the quantity and concentration of metabolic enzymes, hepatic blood flow and the uptake of hormone by tissues other than the target organ.

Because of technical difficulties surrounding the methods of measurement, little is known of the variations in the concentration of circulating hormones throughout the life span in most animals. From a biological point of view, a fall in the concentration of plasma growth hormone and sex steroids would be expected with age, but little unequivocal work has been done on this aspect.

For the sex steroids it may be inferred from the age-related fall in the excretion of urinary steroids (Mason and Engstrom, 1950; Pincus *et al.*, 1954; Brown and Mathew, 1962) that there is either a fall in the concentration of plasma hormones or a decrease in their synthesis and metabolism. The decrease in 17-ketosteroid excretion is due largely to a fall in androsterone. Urinary 17-ketogenic steroids also decline with age. Both changes are reversed by thyroid hormones (Trout *et al.*, 1970). There is also a decline in the excretion of 17-hydroxycorticosteroids in human male urine which occurs at about middle age and correlates with a 50% fall in the excretion of pregnanediol (Romanoff *et al.*, 1971). As there are no age differences in the concentration of plasma corticoids (Grad *et al.*, 1967), it may be concluded that there is a lowered rate of secretion by the adrenal cortex. Based on these changes in urinary steroids it has been proposed with the 17-ketosteroid to 17-hydroxycorticosteroid ratio of urine, which on average declines steadily with chronological age, is a useful measure of the physiological age of the human adrenal cortex (Abbo, 1966). Recent work has demonstrated changes in metabolic clearance and blood production rates of estrogens which supports the idea of a postmenopausal decline in ovarian estrogen production (Longcope, 1971).

There is no change in the concentration of plasma-bound iodine with age (Rapport and Curtis, 1950), and from this it is inferred that the concentration of thyroid hormones is not affected in old animals. For corticosteroids there is sound chemical evidence for age changes in the concentration of circulating hormones in rats (Grad and Khalid, 1968). This applies during the early phase of growth (Gala and Westphal, 1965). In humans there does not appear to be a marked difference comparing the concentration of plasma corticosteroids in young adults with that in people over 65 years of age (Tyler *et al.*, 1955; Samuels, 1956). On the other hand, there is a phase shift in the circadian rhythm (Serio *et al.*, 1970).

A complication in this work is that neither the total plasma hormone nor its secretion rate is a guide to the concentration of biologically active hormone when a large proportion of the hormone interacts with the plasma proteins. The specific binding proteins in the plasma may vary independently of endocrine secretion, so that a variable proportion of the total hormone is in immediate diffusion equilibrium with the cells (Bellamy *et al.*, 1962). In this connection, it is established that large changes in the plasma protein pattern occurs during the life span of a number of mammals (Rafsky *et al.*, 1949; Oberman *et al.*, 1956; Halliday and Kekwick, 1957; Das and Bhattacharya, 1961; Ringle and Dellenback, 1963).

However, the main problem concerns the inadequate analytic methods that are available to measure small concentrations of hormones in peripheral plasma. An indirect approach is to study the capacity of preparations of endocrine organs to synthesize hormones *in vitro*. By this means it becomes easier to measure the larger amounts of steroids produced in the experiment. This method has been used to study age differences in the rate of synthesis of androgens (Axelrod, 1965), corticosteroids (Kemeny *et al.*, 1964) and growth hormone (Meites *et al.*, 1962).

A third approach to the problem of hormone availability may be made by measuring the amount of hormone stored in the endocrine gland. It is difficult for workers in this field to agree on a common interpretation of the changes in the quantities of stored hormones. For example, a decrease in the concentration of hormone in the endocrine gland may be regarded by some to indicate a fall in the rate of synthesis, while others would see the same phenomenon resulting from an increased rate of secretion with a concomitant high rate of synthesis.

With respect to the anterior pituitary, it appears that there are no age-related changes in the amount of stored growth hormone, ACTH, and TSH (Blumenthal, 1954). In the case of growth hormone, it has been shown that a constant quantity is maintained in the gland throughout life without an impairment of the synthetic capacity of the tissue (Meites *et al.*, 1962). The major factor that limits the amount of circulating growth hormone appears to be the concentration of releasing substance passing from the hypothalamus. Old rats have none of this releaser (Pecile *et al.*, 1965). In contrast to growth hormone, the amounts of pituitary gonadotropins, FSH and LH in the pituitary increase considerably with age (Albert *et al.*, 1956).

Comparative work on the insulin content of the pancreas shows the dangers of generalizing from the situation with respect to stored hormones in a single species. In aging cattle there is a decrease in the insulin concentration (Fisher and Scott, 1934), in rats an increase

(Griffiths, 1941), and in humans no change with age (Scott and Fisher, 1938). It is not known how these age differences are related to secretory activity.

Two factors that bear on changes in endocrine activity are connected with variations in the structure of endocrine cells and the amount of endocrine tissue relative to body volume. Like all other tissues in the body, endocrine glands show marked histological changes with age but none of the alterations are sufficiently unique to show that endocrine cells have a different pattern of aging compared with other tissue types (Pokrajac, et al., 1959; Charipper et al., 1961; Bourne and Jayne, 1961; Steward and Brandes, 1961; Batali et al., 1961; Jayne, 1963). Most glands generally develop an increased amount of connective tissue with age. This appears as a thickening of the capsule, a rise in the density of collagen fibers and the replacement of secretory cells with connective tissue elements. In the endocrine cells, fragmentation of mitochondria and nuclear damage usually appear with age and there may be a decreased rate of mitosis.

In some glands specific histological changes are related to the characteristic structure or function of the tissue. For example, in the thyroid, age brings about a loss of follicles and a drop in the amount of stainable colloid. In the adrenal cortex, there is an alteration in the lipid composition and pattern of zonation. Vasodilation also occurs, with an increasing frequency of hemorrhage. With regard to aldosterone secretion, a reduction in the granularity of the juxtaglomerular cells and a decreased width of the zona glomerulosa points to a morphological basis for the fall in the aldosterone secretion rate (Dunihue, 1965). However, as with other specialized tissues, in no case have histological changes in endocrine organs been linked directly with a specific endogenous variation in functional activity.

The proportion of endocrine tissue in the body varies with age. In the case of the thyroid, there is a decrease in the weight of the gland relative to the body (Robertson, 1916). The change is most prominent during the early stages of postnatal life. For example, in the dog, during the first 8 weeks of life, there is an almost 50% fall in the relative weight of the thyroid (Haensly et al., 1964). The relative weight of the adrenals and parathyroids also declines with age (Haensly and Getty, 1965; Baca and Chiodi, 1965; Raisz et al., 1965). Similar, but opposite changes in the weight of the gonads also occur. These alterations in the relative weight of endocrine organs might have a bearing on the effective concentration of circulating hormones because of corresponding variations in the relative volume for distribution of the hormone. A large fall in relative cell mass of an endocrine organ also indicates a corresponding rise in the rate of hormone synthesis per cell.

Little is known of age-related changes in the sensitivity of endocrine organs to the primary stimulus that results in hormone secretion. The problems are similar to those encountered in the assessment of the mechanisms responsible for changes in target tissue sensitivity. It is established that homeostatic adjustments to the ingestion of water, sodium, and potassium initially become more effective with age (Talbot and Richie, 1958), but more fundamental information is needed on the relationship between stimulus and response. For example, from the point of view of changes in the endocrine organs, it would be of interest to know the effect of age on both the secretory capacity and the time of response of the zona glomerulosa of the adrenal cortex to various concentrations of plasma sodium.

4. Changes in Function

Most of the evidence that points to changes in the nature of hormone function with age has come from experiments on the action of exogenous corticosteroids and neurohypophysial hormones. The underlying principle is that age-dependent changes in the characteristics of the response are related to intrinsic changes in the homeostatic mechanisms (Krecek et al., 1958). This applies particularly to the action of neurohypophysial hormones and adrenocorticosteroids on electrolyte excretion in rats. In the early stages of life, vasopressin increases renal water loss in very young animals but inhibits water excretion in old animals (Krecek et al., 1958). Similarly, cortisol increases the excretion of sodium and water in young rats but reduces water loss in older animals. The age difference here is most likely related to the fact that the younger in contrast to the older animals had not reached an age at which they were normally weaned. It is in this period that a difference is noted in the metabolic effects of adrenal hormones. For example, after adrenalectomy, a 12-hour period of starvation proved fatal to the majority of animals in a group of very young rats (20 days old) but old and young adult rats (60 days old up to 14–18 months) survived longer than 24 hours. (Vranic and Pokrajac, 1961). Again this difference is probably linked with the endocrine control of energy stores and related to a change in feeding behavior after weaning. Differences in hormone action are not confined to this early period. Aldosterone has little or no action on the excretion of potassium in 5-week-old, weaned rats, but produces a marked inhibition of potassium output in rats 1 year old or more (Desaulles, 1958).

It is not known how these different effects of hormones are related to the development of homeostatic systems. During the first weeks of life after birth there are many changes in the rat's physiology. In addition, there is also the possibility that some of the alterations in responses to hormones are connected with age-linked variations in hormone sensitivity.

C. Hormones and the Aging Process

There are indications from parabiosis in rats that humoral factors are important in controlling some aspects of aging (Lunsford et al., 1963; Hruza, 1971). The view has also been taken that the neurosecretory system in insects is either directly or indirectly involved in the rate of aging (Rockstein et al., 1971). Some of the age-related changes in tissues that are target organs for hormones seem to be reversed when old tissue is transplanted into younger animals (Franks and Chesterman, 1964), but experiments involving the transplantation of grafts, particularly of target tissues (Daniel et al., 1971) and the injection of hormones, indicates that hormones do not exert full control over the aging process (Horton, 1967). Endocrine glands are involved in the manifestation of some signs of old age (Bellamy, 1967), but hormones can only modify the rate of aging in certain tissues, particularly when animals are treated during the period of growth and development (see Section 3).

1. HORMONES AND TISSUE INVOLUTION

Although there have been no comprehensive comparative studies within the animal kingdom, it appears that as organisms grow, there is a progressive decrease in the amount of water relative to the weight of the body. It appears that by far the greatest part of the decrease is due to a fall in extracellular fluid, rather than a dehydration of the cells (Needham, 1931). This explanation seems reasonable in view of the decrease in sodium and chloride during human development, as these ions are predominantly extracellular (Schol, 1939; Stearns, 1939). More detailed work has been done with laboratory animals. In the chick, between the sixth and eighteenth days of life, while the dry body weight increases over 2-fold, the chloride content drops by 25% (Murray, 1926). Such a fall in the proportion of extracellular fluid is likely to be a phenomenon common to vertebrate embryos (Iob and Swanson, 1934).

The above conclusions apply to several tissues. The water content of individual fibers of rat skeletal muscle remains constant from 15–336 days of age, whereas the proportion of extra-fiber water decreases (Hines and Knowlton, 1939). A similar change was found in skeletal muscle of rats between 1 and 6 months or more of age. This study also showed an increase in size of the individual muscle fibers and a relative decrease in the size of the spaces between the fibers (Yannet and Darrow, 1938). Similar but less marked changes in the extracellular fluid occur in liver and brain of cats, with no change in the fluid content per cell; the older animals showed a 12% decrease in extracellular fluid (Yannet and Darrow, 1938). Cardiac muscle of dogs falls into the same category in that the proportion of fibers increases from 739 gm/kg in puppies to 798 gm/kg for adult dogs (Hastings *et al.*, 1939).

Detailed investigations on the development of the guinea pig cerebral cortex shows that extracellular fluid, calculated from chloride content, rises slightly at an early period when the nerve cells are becoming more spaced out, and then falls by about 50% by the time that the animal is adult (Flexner and Flexner, 1949, 1950a, b; Peters and Flexner, 1950). There is no doubt that the state of the endocrine system is a controlling factor in the maintenance of the balance between cells and extracellular fluid; growth hormone for example, favors the formation of cells as opposed to connective tissue.

All the above evidence points to growth and development being characterized by a rise in the proportion of cells to extracellular fluid. When growth ceases, some tissues maintain the balance between cells and extracellular space, whereas others show a decline in cellularity and a rise in the proportion of noncellular material—a phenomenon described as "age involution" (Korenchevsky, 1942). Skeletal and cardiac muscle of rats both show an increase in water, sodium, and chloride, and a decrease in acid soluble phosphorus and potassium, comparing adult (603 days) with senile (988 days) animals (Lowry *et al.*, 1942). Taking a histochemical view, this means that the extracellular fluid has almost doubled, while no essential change has occurred with regard to fiber water content; the lipid concentration also increased and all of these changes are the opposite of those seen with growth. Rat kidney is also characterized by an increase in extracellular fluid between 603 and 988 days, suggesting a decrease in the proportionate cell mass (Lowry *et al.*, 1942).

On the other hand, there is very little change with age after maturity in heart, liver, and brain. Heart, for example, shows a slight increase in the amount of extracellular fluid and an increase in phospholipid, and in both heart and brain a little intracellular potassium is replaced by sodium. Total solids in human brain show evidence of a consistent decline with senescence amounting to 8–15% (Strobel, 1939).

To summarize, just as growth in the old literature has been described as a partial "desiccation" of protoplasm, so may aging be described as the antithesis of growth, in that it represents a state of increasing "hydration." Both terms reflect the corresponding rise and fall in the proportion of cells.

Age involution is only one of four types of atrophy. These types have recently been compared and contrasted in skeletal muscle. Muscular atrophy may occur by loss of nervous connections, disuse, changes in the concentration of sex steroids and corticosteroids and through old age (Gutmann, 1970). It appears that senile involution does not have a distinct cause, but results partly from a reduction in axonal flow and transmitter release from the central nervous system, together with contributions from disuse—due to behavioral aging—and gonad hypofunction.

The thymus has long been a favorite organ for speculation concerning the nature of age involution. The gland grows very rapidly during the embryonic state in most mammalian types and is the first organ to show the histological characteristics of lymphoid tissue. The maximum size, relative to the body, is reached in the embryo or during the neonatal period. Subsequently, although the thymus continues to grow, it does not keep pace with the rest of the body (Fenger, 1915; Robertson, 1916; Boyd, 1932). At about the time corresponding to the onset of sexual maturity, a gradual decline in weight occurs and this continues into old age.

Taking into account its growth rate, the thymus appears to be of particular importance during the early stages of development. Thus, Fenger (1915), from a cursory examination of the chemical composition of the thymus in cattle, concluded that the organ fulfilled its role during the early embryonic period. It is now known that the thymus plays an important part in the development of the immunological response.

Thymus involution begins at about the time of onset of sexual maturity when there is also a fall in the growth rate of several organs of the body. Thus, there is no sound reason to connect involution with sex steroids rather than with the largely unidentified factors that control the development of other tissues. Apart from this, there is experimental evidence demonstrating that sex hormones do not normally influence the rate of thymus involution (Bellamy, 1966).

Castration results in a rise in the weight of the thymus and this effect has been taken to mean that sex hormones play an important role in involution (Andersen, 1932). In mice, the increase in thymus weight after castration takes place rapidly within a few days of the operation. Subsequently involution, as measured by the rate of weight loss, is not affected. Exogenous androgens and estrogens also bring about thymus involution (Inay and Thompson, 1938). The quantities of sex hormones necessary to induce involution of the thymus probably represent plasma concentrations far in excess of those normally found in the animal, as they also reduce the size of the gonads and therefore inhibit the release of pituitary gonadotropins (Greep and Chester-Jones, 1950). Thus, it may be said that endogenous sex steroids are not likely to reach a high enough concentration during the life cycle to control involution of the thymus.

It is well known that the size of the thymus bears an inverse relationship to the concentration of circulating steroids of the cortisol type (Bellamy *et al.*, 1966). But, although involution is initiated by the injection of corticosteroids, the histological changes after cortisone or ACTH treatment show that corticosteroid-induced involution is different from that occurring with advancing age (Weaver, 1955), mainly in regard to the deposition of collagen and appearance of adipose tissue, which are special features of old thymus (Bellamy and Alkufaishi, 1972). Also, the effect of corticosteroids is rapid and reversible (Bellamy *et al.*, 1966) so that, in order to bring about the slow, progressive fall in thymus weight seen with age, there would have to be a corresponding slow continuous rise in the concentration of biologically active plasma corticosteroids. There is no evidence for either a rise in the concentration of circulating corticosteroids or a fall in the concentration of specific corticosteroid-binding protein (Gala and Westphal, 1965). Other evidence against the involvement of the adrenal cortex in the control of involution comes from experiments on the effect of adrenalectomy. As with castration, there is a rapid rise in the weight of the thymus, but the operation does not affect the rate of loss of weight with age (Bellamy, 1967).

The thyroid gland, as well as having an indirect effect on thymus size through the adrenal cortex (White and Dougherty, 1947), has been implicated directly in the process of involution in that a decrease in thymus size was observed after thyroidectomy (Marine *et al.*, 1924). It was pointed out earlier, that there is no detailed knowledge of the concentration of thyroid hormones throughout the life span. The experiments often give contradictory results and it is not known whether there is a progressive decline in the activity of thyroid hormones with age, coinciding with the fall in thymus weight. From measurements of plasma-bound iodine, it is inferred that the concentration of thyroid hormones varies little with age so that, in the main, it is unlikely that involution is a consequence of a decline in thyroid activity.

In conclusion, at any age, the size of the thymus is dependent on a number of endocrine glands. But from the available evidence there is no reason to suppose that the age-related changes in thymus weight and histology are due to alterations in endocrine activity, rather than to some intrinsic aging process.

Similarly, although muscle mass is reduced after both castration (Korenchevsky *et al.*, 1953) and treatment with glucocorticoids (Faludi *et al.*, 1964; Tonelli *et al.*, 1965), there is no evidence that changes

2*D. Bellamy*

in endocrine activity of either gonads or adrenal cortex are responsible for age involution of muscle. In this connection, there is scope for a detailed investigation of the various forms of atrophy. On the face of it, atrophy brought about by starvation, tenotomy, and denervation have much in common with age involution, in that the water content of the fibers remains relatively constant, the total amount of tissue decreases markedly, and there is little change in extracellular mass (Hines and Knowlton, 1939). With regard to thymus involution brought about by either age or cortisol injection, it appears that there is an inbalance between the division and death of lymphocytes and the resultant drop in the density of lymphocytes results in an increased density of fibroblasts and macrophages (Bellamy and Alkufaishi, 1972), which is a common feature of both age and cortisol-induced involution.

2. Aging and Hormone Deficiency

If it is anticipated that a hormone deficiency is responsible for an aging phenomenon, the obvious test is to treat the animal with the hormone in an attempt to restore the youthful characters. From studies of this nature with sex steroids, it appears that age differences in biochemistry, physiology, and behaviour are not related solely to hormone deficiencies (Caldwell and Watson, 1954; Pauker *et al.*, 1958; Ackermann and Kheim, 1964; Rigby *et al.*, 1964). Recent work comparing eunuchs with intact men (Hamilton and Mestler, 1969) and the effects of castration on life span of domestic cats (Hamilton *et al.*, 1969) have indicated that endogenous sex steroids have a life-shortening action, particularly if they are present during the growth period.

With reference to the anterior pituitary, although the maximum body weight attained during growth of rodents is increased by exogenous growth hormone (Everitt, 1959), the normal termination of growth is not due to a lack of this hormone in the plasma (Moon *et al.*, 1956). Also, it is well known that several features of old age are similar to those encountered in hypothyroidism (Starr, 1955), but there is no evidence for a general decline in the concentration of thyroid hormones with age and most symptoms of old age cannot be alleviated by thyroxine.

In connection with the neurohypophysis, it is the view of Friedman (Friedman *et al.*, 1965) that the characteristic age-related changes in the water and electrolyte content of cells are similar to those in diabetes insipidus and point to an impairment of neurohypophysial function. Support for this idea comes from experiments in which the pattern of salt and water distribution in old rats was partially restored to that characteristic of earlier stages of development by treatment with neurohypophysial hormones (Friedman *et al.*, 1963). Further, the combined treatment of old rats with posterior pituitary powder and cortisol resulted in a marked prolongation of life (Friedman *et al.*, 1965). In the old rats used in the above experiments, the neurohypophysis was depleted of neurosecretory material and failed to respond adequately to an osmotic stimulus (Friedman *et al.*, 1956). An impairment of pituitary function is also inferred from other experiments (Dunihue, 1965; Rodeck, 1966), but the histological evidence is contradictory (Morrison and Staroscik, 1966). It is also difficult to tell whether or not these phenomena are due to the onset of an age-dependent syndrome centered on the hypothalamus and pituitary, and in this respect they may well be strain-dependent. From similar work, it is thought that changes in water metabolism in old rats are due mainly to an intrinsic regression in renal function and are not mediated solely through an alteration in the neurohypophysial system (Rodeck *et al.*, 1960).

3. HORMONES AND THE PHARMACOLOGICAL CONTROL OF AGING

It has already been pointed out that dietary restriction is associated with a prolongation of life. From these studies, factors that determine life span have been linked with the period of rapid growth; the rate of aging is thought of as being inversely related to the rate of development. There are other ideas that senescence is due to a decline in growth rate (Lansing, 1948; McCay, 1952). Each view has led to an experimental approach to the control of longevity, through a modification of endocrine balance.

With respect to the latter view, there is no evidence that senescence is due to a lack of pituitary growth hormone. When growth hormone was given daily in the long-term treatment of middle-aged rats, although there was a significant augmentation of body weight gain and water consumption, no effects were observed on several other metabolic and physiological variables, and there was no change in life span. Further, the hormone failed to prevent the final decline in body weight as rats approached death (Everitt, 1959).

As to the other view, although there is clearly a parallelism between the time taken to reach maturity and longevity in rats, no direct connection has been established between development and aging. Growth in rats also follows a special pattern, in that it continues for most of the life span (Lesser *et al.*, 1970). However, it is established that enriched diets consumed throughout life accelerate growth and shorten life span while prolonged undernutrition, which retards growth, may also increase longevity providing that it is not associated with a lowered resistance to environmental factors. Recently, work has been carried

out to investigate the effects of high temperature on longevity in male
rats which shows that the associated voluntary restriction of food, together with an ability to maintain a low body temperature when young,
tend to lengthen life span. Voluntary restriction of food intake at high
temperatures increases life span more than high temperature shortens
it (Kibler and Johnson, 1966).

Apart from experiments on dietary restriction, few other methods have
been used to inhibit growth and study the relationship between growth
rate and aging. Hypophysectomy retards the growth rate of a number
of animals, but also results in several hormone deficiencies. The achieve-
ment of the correct balance of replacement hormones in order to investigate the action of a single hormone presents great problems. It is true
that hypophysectomized animals age at a slower rate than normal, as
judged by the rate of development of endocrine organs dependent on
the pituitary for maturation. It is also notable that hypophysectomy
will retard the aging process in connective tissue (Verzar and Spichtin,
1966), and will prevent age changes in renal function that may be
regarded as detrimental (Everitt and Duvall, 1965). However, the
life span of hypophysectomized animals is shorter than normal (Verzar
and Spichtin, 1966).

Thyroidectomy is known to retard the growth rate and the rate of
general development in mammals but a hypothyroid state does not increase longevity. The condition results in the unbalanced development
of various organs and in several respects the hypothyroid animal resem-
bles one that has aged prematurely (Korenchevsky, 1961). In contrast
recent research into the relationship between castration and the duration
of life indicates that castrated human males live longer than normal
subjects, particularly if the operation was carried out before maturation
(Hamilton, 1965; Hamilton and Mestler, 1969; Hamilton *et al.*, 1969).
Treatment with androgens and anabolic steroids although it may produce
some specific and temporary antiaging effects, does not retard the overall
process of aging (Korenchevsky, 1961; Dontas *et al.*, 1967).

With respect to other hormones that affect the growth process, little
attention has been paid to the adrenocorticosteroids. In this connection,
it is established that the treatment of rodents with glucocorticoids inhibits growth. From early work, this was thought of as a toxic action
(Ershoff, 1951). The acceptance of the ideas of Selye, in which increased
adrenocortical activity was cited as a factor leading to a shortened life
span (Tanner, 1959), strengthened the view that corticosteroids were
detrimental to the life process. However, it is known that in some circumstances rats treated with corticosteroids can be maintained for a
considerable time despite a marked inhibition of growth (Winter *et*

al., 1950). Also, some of the effects of the cortisol type of steroid may lead to a rise in biochemical efficiency. This applies particularly to the rat, where cortisol appears to increase the oxidative capacity of skeletal muscle (Palmer, 1966) and leads to an improvement in work performance (Ingle *et al.*, 1952).

The metabolic characteristics of cortisol-treated animals suggest that undernutrition might contribute to the slow rate of growth, but the effect of corticosteroids on growth differs in some respects from that of undernutrition (Bernick and Ershoff, 1963). Some workers have shown that glucocorticoid-treated rats eat less food than normal (Van Putten *et al.*, 1963; Tonelli *et al.*, 1965). Others have not confirmed this (Grossie and Turner, 1965) and it may well be that the fall in food intake is secondary to a fundamental action on the growth process (Bellamy, 1964). In this connection, it has been observed that steroid-treated rats eat more food than animals maintained at the same body weight by food restriction (Winter *et al.*, 1950). Indeed, some of the biochemical characteristics of cortisol-treated animals are similar to those found after hypophysectomy. Also, in humans where suitable analytic methods are available, it has been shown that cortisol treatment brings about a fall in the concentration of plasma growth hormone (Frantz and Rabkin, 1964; Hartog *et al.*, 1964).

Although undernutrition increases the life span of rodents, it also makes them abnormally sensitive to environmental factors that contribute to mortality (Widdowson and Kennedy, 1962). This may be partly due to the restriction of vitamins and other dietary substances required for tissue maintenance. From this point of view, cortisol treatment, resulting in growth inhibition without a proportional decrease in a balanced diet, might be more advantageous in slowing down the rate of development and increasing longevity. Unfortunately, cortisol, like undernutrition, interferes with the defense mechanisms of the body (Cameron, 1953). Here a contributory factor may be atrophy of the adrenal cortex with a consequent imbalance of corticosteroid hormones. Accordingly, some of the synthetic cortisol derivatives, such as prednisolone, with a high potency in growth inhibition and a small effect on the adrenal (D'Arcy and Howard, 1962) might be more useful for long-term treatment.

Early experiments with prednisolone increased the life-span of a short-lived strain of mice to that characteristic of most inbred strains (Bellamy, 1969). At the same time, several biochemical features of the aging animal were retarded. Later work with a variety of rat strains indicated that the effect of prednisolone on life span varied considerably according to the time and duration of treatment, the genotype and nutritional status. A common feature was that no effect was obtained if treatment was given only after the time of maximum growth. From this work, there does not seem to be a direct action of corticosteroids on events that happen after maturation.

It is also evident from comparative studies in other animals that corticosteroids have a more general effect in reducing mortality. Of 38 drugs tested on *Drosophila*, cortisol and cortisone were the most effective, giving up to a 40% increase in life span (Hochschild, 1971). Cortisol is also specifically effective in prolonging the post mitotic life span of primary amnion cells *in vitro* (Yuan *et al.*, 1967). With regard to these experiments, the effectiveness of corticosteroids may be related to their action in stabilizing intracellular membranes, particularly those of the lysosomes (Hochschild, 1971). However, without firm knowledge of the cause of death, which in mammals may vary between inbred strains of the same species, it is fruitless to speculate on the exact nature of the effects. There are considerable differences in life span between various strains of the laboratory mouse (Storer, 1967) and some of the results on the pharmacological prolongation of life span (Bender *et al.*, 1970) may be achieved by allowing animals to attain a normal rate of aging (Gordon *et al.*, 1967).

4. Conclusions

The endocrine organs constitute one of Korenchevsky's five physiological "pillars," the stability of which largely determines the normal course of aging. Although the importance of age changes in endocrine organs has been appreciated from the beginning of gerontological research, we are far from an understanding of the causes of aging in any hormonally controlled system.

The major difficulty is that little experimental work has been carried out and, in this respect, the situation in the field of endocrinology is no different from that in other branches of gerontology. It is surprising that few workers have been concerned with age variations in the concentration of circulating hormones. Also, little information is available on the fundamental chemical changes in both endocrine organs and target tissues. It appears that the bases for age-related changes in homeostatic mechanisms are sometimes due as much to modifications in the function of target tissues as to changes in the activity of endocrine glands.

Apart from the obvious need for the collection and correlation of basic facts on aging tissues, it may be concluded from the present discussion that several well-defined experimental approaches offer scope for future studies on the aging of endocrine control. The basis for these investigations is the change in the chemical composition of the body fluids that appears with age. Once a particular chemical difference is established, in the plasma or intracellular environment, it is possible to seek an explanation by examining the biochemistry of the target tissue on the one hand, and the factors that control the rate of hormone secretion on the other. Clearly, a complete explanation of the aging phenomenon requires the cooperation of the endocrinologist and the biochemist. One of the most fruitful fields for future collaboration concerns the changes in the response of target organs. For the steroids, large, well-defined age differences are found in the biochemical response to exogenous hormones. Also, reliable methods are available to determine the concentrations of hormones and metabolites and to investigate the metabolic fate and mode of excretion of the hormones.

If an excess or deficiency of a hormone is linked circumstantially with an aging phenomenon, an attempt may be made to counteract the change by experimental means. Apart from the work of Friedman and his co-workers, little success has been achieved with gland extirpation and hormone therapy, and changes in the biologically effective concentration of hormones do not appear to be general causative factors in the development of many signs of old age.

D. Homeostasis and Mortality

With regard to the failure of homeostasis during aging, it has been suggested that, in general, there is a diminution of neural influences upon effector organs while the sensitivity to hormones increases (Frolkis, 1966). There is much evidence in support of the failure of nervous coordination, but evidence relating to hormone action is, in the main, in favor of a decreased sensitivity, although the degree of hormonal stimulation may depend on the previous experience of the animal. In many glands, there is a fall in the rate of hormone turnover.

There can be little doubt that a marked age deterioration at the physiological level is a feature of many organs. Despite this, the individual cells of kidney, brain, and muscle do not appear to undergo a drastic loss of functional capacity; it is probably true to say that for these organs, the deficiency in old age lies in the loss of cells, together with a shift in the balance between cell types from that which was attained at maturity. Although there are changes in cell mass, the ability of old animals to maintain day-to-day resting stability in an unchanging environment is largely unimpaired. Lack of adaptability in a disturbed environment is not due to the failure of any single component. It has been pointed out in the previous section that the endocrine system probably fails to respond on several counts, bound up with a decreased capacity to secrete sufficient hormone and the failure of the hormone to evoke the correct degree of response in the target organ.

The major problem in the field of gerontology is to relate age changes in function to the increasing probability of death. Where a particular reflex action or rapid neuromuscular response is of importance to an organism in its search for food and shelter, the reasons for the increased chances of mortality may be more or less defined. For example, the sensitivity of rat aorta to catecholamines is reduced in old animals. This is associated with a coarsening of the muscle fibers of the tunica media and an increased thickness and disarrangement of the elastic lamella. At a purely chemical level, the old tissue appears to be more efficient in that it contains a higher concentration of protein and DNA. However, it is concluded that the structural changes alone are sufficient to account for the observed decrease in hormonal sensitivity through a limitation on the mechanical response. The gradual deterioration of effector organs containing muscle fibers and elastic tissue is probably an important factor in the poor adaptive responses of old organisms, where the adaptation requires a rapid neurophysiological response at a much higher intensity than normal. Nevertheless, animals still die over a narrow age-range when maintained under constant laboratory conditions where no sudden large demands are made upon the body.

It must be admitted that we are largely ignorant of the immediate cause of death in old age. This problem is distinct from the underlying molecular events that produce age changes at a higher level of organization. From this standpoint, death is a consequence of events of a more complex order and is open to study without reference to the long-term progressive changes in function that lead up to it. It is true that the slow linear trend by which some physiological characters change with age is theoretically sufficient to account for the exponential increase in mortality observed at the population level (see following Section). But death is an individual process and involves events which take place over a small fraction of the life span, possibly amounting to only a few hours. Death, in this context, is the result of an internal fluctuation which cannot be contained through homeostatic regulation; a small shift in metabolism which could be counteracted in youth becomes amplified to the point of preventing a vital function. Thus, death in old organisms may result from minor changes in the environment.

Few experiments have been designed specifically to test the possibility that there is an increased oscillation of physiological control mechanisms in old animals. With regard to future work, it is desirable to concentrate on the fluctuations in homeostatic control. In the past, too much empha-

sis has been placed on the maintenance of mean values which are regarded as constant norms. Data for many individual animals are required
to establish these statistical norms and the short-term fluctuations within
any one individual are largely ignored. In order to understand the increased tendency for homeostasis to fail during aging, we must first
determine just how stable the "constant" internal environment is at all
ages, and investigate the reasons for changes in norms and "overshoot
phenomena" which in the old organism may ultimately result in death.

VARIABILITY AND AGE

It is frequently observed that, as a population of animals ages, the
variability of physiological and biochemical parameters increases. There
are three possible sources of variability. First, an increased error of
measurement. Published data usually give no indication of experimental
error and this can only be ruled out as a source of increased variability
where there is no change in the mean value with age. The second source
of variation is the drift in norms within individuals, as expressed by
increased genetic and phenotypic variation. In this sense, there are certainly shifts in population norms with age. Third, there is a possibility
of increased random variation within individuals due to a greater overshoot and undershoot among the various homeostatic systems at both
a physiological and a biochemical level.

The only way to distinguish between an age-dependent variability
in norms and a variability in individual homeostasis is to carry out
longitudinal studies, but as yet few such studies have been made. There
are many indications of increased population variability with age and
it is unlikely that they are all related simply to measurement errors.

One of the areas of possible functional age variation is that of blood
homeostasis, since the composition of the blood would be expected to
mirror any general deterioration in regulatory systems. There is certainly
a great variability in the blood constituents among humans, but an exhaustive analysis of 21 norms in 478 individuals and 15 in 284 subjects
has led to the conclusion that the observed variability is independent
of age (Das, 1969). This does not rule out age-linked loss of accurate
homeostasis because homeostatic oscillations could be masked in natural
populations by virtue of great intrinsic genetic and phenotypic variation.

Within inbred laboratory stock, variability does increase with age.
One of the first indications of this came from work on aging in rotifers,
where it was observed that homozygous populations derived from a
single egg gradually became more variable with regard to reproductive
function. It is difficult to avoid the conclusion that this is due in large

part to some kind of random phenotypic drift in individuals. Inbred rodent populations also show increased variability at both the physiological and biochemical level. However, from the point of view of the underlying mechanism, an analysis of variability in structure may offer the opportunity of defining aging at the level of individual cells. Thus, in human subjects within the age group 65–69 years, there is an increased coefficient of variation for the numbers of mitochondria per human hepatic cell, and the circumference and total size of these organelles (Tauchi and Sato, 1968). Further, the inability to maintain the norm for mitochondrial structure is manifest in the injurious effects of thyroxine which appear with increasing frequency in the cells of senile rats (Shamoto, 1968). There are also similar age differences in the response to chronic hypoxia in the fine structure of cardiac muscle and cells of the autonomic ganglia (Sulkin and Sulkin, 1968).

Increased variability within a population as a concomitant of age is important because it may increase the chances of death of individuals showing extreme variations. This is particularly so if the variation is associated with a decline in the mean resistance to death. A useful model system has been examined where the resistance to death was measured by the volume of blood that had to be taken from a rat in order to kill it. It was observed that a steady decline in the resistance to death from experimental hemorrhage of only 12% per decade, coupled with an increased variability in resistance, resulted in an exponential rise in the chances of death occurring at a particular blood loss (Simms, 1942); the decline in resistance to death from hemorrhage matches the mortality curve for normal rats.

III. AGING AT THE TISSUE LEVEL

A. Extracellular Components

The macromolecules that make up the bulk of the organic fraction of connective tissue appear to undergo a series of transformations akin to the perishing of rubber whereby, through cross linking, crystalization, or molecular reordering, their mechanical properties change. These chemical alterations occur after the molecules have been laid down outside the cells and are responsible for the wrinkling of skin and the loss of elasticity of connective tissue elements which are two of the most telling characteristics of aging in human subjects.

Most investigations into the aging of extracellular materials have been carried out using easily accessible tissue such as skin, tendon, arterial walls, ligaments, and lungs. All of these tissues contain cellular components but usually these are of minor importance in terms of bulk and have been ignored in most generalized studies which have been concerned with aging of connective tissue.

At the highest level of chemical organization, there is much evidence for degenerative changes in connective tissue. In old skin, for example, hemorrhages frequently occur (senile purpura) which are due to the failure of the matrix of extracellular fibers surrounding the capillaries to cushion them against mechanical damage.

Experimental work on the aging of connective tissue has centered on collagen and elastin, two well-defined protein components which comprise the bulk of the organic fraction. There is a large body of evidence indicating that these components change both quantitatively and qualitatively with corresponding alterations in their physical and mechanical properties (Verzár, 1957).

Deterioration in the collagen/elastin stroma of key organs such as lungs, heart, and kidneys has been proposed as a key event in the aging process, where degenerative changes interfere with various effector-aspects of homeostasis by reducing the mechanical efficiency. The continuous accumulation of collagen is also thought to impede the supply of nutrients to the cells. Although the evidence on molecular aging of collagen and related compounds has firm chemical and histological backing, there is no clear-cut experiment showing that these changes actually result in death. However, it is reasonable to assume that accumulation of collagen and the loss of flexibility of elastin contribute to the mechanical failure of some effectors, notably skeletal muscle.

B. Cellular Aging

Working on the assumption that the ultimate control of the direction and force of development resides in the cells, it is logical to investigate cell function for the underlying causes of aging. It is also sensible to set up models of the aging process at the cellular level, although it is no exaggeration to say that most models at the present time pose questions of fact for which we must find the answers before the model may be tested.

For the purposes of discussion, there are basically two types of cells in the body. Adult mammalian tissues consist of cells that may be classed as "mitotic"; they are capable of multiplication and replacement. A few highly specialized tissues, such as the central nervous system, together with cardiac and striated muscle, are composed of cells which have lost the power to divide. These "postmitotic" cells make up the bulk of living tissue.

1. MITOTIC CELLS

For many years, it was held that dividing cells were potentially immortal. However, recent experimental work has clearly shown that populations of vertebrate somatic cells cultured *in vitro* cannot be maintained indefinitely (Hayflick, 1965). Cultures of human diploid embryonic fibroblasts undergo approximately 50 doublings then die out, often showing well-defined terminal histological changes with marked chromosome abnormalities. Similar conclusions on the finite life of cells come from experiments in which normal cell populations of skin, bone marrow, and mammary tissue have been serially transplanted as their hosts age. There is as yet no clear indication of the mechanism involved in the eventual dying out of these cell populations. Diploid fibroblasts from adults grown *in vitro* show less than half the doubling potential of embryonic cells, suggesting that the cells change progressively from birth. This limitation of the life span of cultured cells may not be directly related to the doubling potential. Instead, it may be a function of the total metabolic time of the cell strain *in vitro*. There is also some evidence that the postmitotic life span of human amnion cells *in vitro* may be extended by modifying the environment, particularly by increasing the concentration of steroid hormones (Yuan *et al.*, 1967). There are several possible explanations for the gradual deterioration of cell cultures. There may be a progressive loss of some vital self-duplicating cellular entity—death occurring when the concentration of this entity falls below a critical level. The loss of such an entity could occur either through the cumulative effects of random losses or because the entity was itself dividing at a rate somewhat slower than that of the cells. On the other hand, the cells may be progressively changing by the operation of an epigenetic processes akin to differentiation.

Workers in this field have suggested that the evidence for a finite lifetime of diploid cells *in vitro* may be the cellular manifestation of aging *in vivo* where the main feature is a loss of biochemical information. In this connection, there is an age-related lag in the response of cell division *in vivo*, associated with both regeneration and the diurnal rhythms in cell division (Falzone *et al.*, 1959) and there is evidence for degenerative changes taking place in dividing cells *in vivo* (Tonna, 1973). Against this, cell doublings *in vivo* greatly exceed the maximum achieved for cell strains *in vitro* (Cameron, 1973).

Normal cell populations which duplicate *in vitro* have been described as cell "strains" to distinguish them from cell "lines" which may arise during the culture of normal cells (Hayflick and Moorhead, 1961). Cell lines are capable of indefinite proliferation *in vitro*, but have many abnormal properties. In particular, they may behave like cancer cells on inoculation into living organisms. The existence of cell lines suggests that mechanisms exist for perfect control of cell metabolism. Indeed, one would expect this from the fact that a continuous series of perfect cell division lineages has been maintained through evolutionary time to the germ cells of all present-day organisms. It is also a strong possibility that cell proliferation in other tissues, such as the intestinal epithelium, is a perfect process.

As an alternative to perfect duplication and self-maintenance, it is possible that mechanisms exist to counteract imperfections which inevitably occur. Elimination of imperfections may take place through the presence of error-correcting codes. Cell lines may appear in tissue culture because of the unmasking of a self-correcting code which is normally only expressed in germ cells. With regard to errors occurring in mitotic cells, it is possible that these could be brought about by any of the well-defined mechanisms such as nondisjunction, translocation, inversion, deletion, and repetition. In general, these processes are likely to result in malfunction of the nucleus, either through loss of chromosome material or by changes in relative gene activity; it may be predicted that malfunctions would be greatest in the most rapidly dividing cells.

The fact that there are cells in many organisms which are capable of division raises the possibility that a process of selection could occur within the body to favor the multiplication of cell variants which, during the lifetime of the individual, might prove harmful to the organism. This idea forms the basis for the "clonal selection" theory of aging. A particular development of this theory states that, through random mutation or other kinds of nuclear malfunction, cells which generate antibody-forming lymphocytes could change their properties and give rise to a lymphocyte population which would not be able to distinguish its own cells as being "self." This could result in the progressive loss of cells from a wide range of organs by a phenomenon termed "autoimmunity" (Walford, 1967). It is believed that autoimmunity is the chief process which leads to a gradual deterioration of animals (Walford, 1969). Evidence has been obtained for a decline in isoantibodies and for autoimmune reactions developing in old mammals, and the view has also been expressed that adsorption of certain antigen-antibody or hapten-antibody complexes, could result in aging-phenomena (Ram, 1967). Where an increased titre of autoantibodies has been determined in humans, the incidence in males has always been lower than that in females (Goodman *et al.*, 1963; Doniach and Roitt, 1964; Irvine *et al.*, 1965; Herbeuval *et al.*, 1967). If autoimmunity was a major factor in determining longevity, the incidence in shorter-lived males should be higher than in the female. Also, the situation in invertebrates, where the classical aspect of immunity has no application, raises questions as to the general validity of the autoimmune theory. Immunosuppressive drugs exert only a marginal beneficial effect on mammalian longevity, but the consequences of such treatment with regard to the destruction of normal immunological reactions make it difficult to make a valid test of the autoimmunity theory by this means.

Aging in Protozoa. Although cultures of microorganisms cannot be regarded as analogous to metazoan cell cultures, they have been examined from time to time in an attempt to throw light on cellular aging (Sonneborn and Schneller, 1960a, b, c). Early work predicted that because protozoans were pure germ plasma, they must inevitably be immortal. However, it was soon found that individual ciliates underwent a series of temporal changes akin to sexual maturation, followed by senescence; they aged and died. Also, in the laboratory, a particular protozoan clone will eventually die out after passing through this life cycle a certain number of times. Now clones may arise by the processes of conjugation and autogamy which entail, respectively, the introduction of individuals of a different clone and a kind of internal fertilization. The nuclear apparatus is replaced during the both processes. In conjugation, male nuclei are reciprocally exchanged between organisms and a fertilized nucleus is formed in each male; in autogamy, haploid male and female micronuclei from the same individual unite to form a fertilization nucleus. Both kinds of reorganized cells become the starting points for new clones. Progeny of autogamy (exautogamants) produced early or late in the senescent period of an individual develop in several possible ways: to give fully viable clones, which begin new life cycles; to give clones which are either nonviable or which divide at markedly reduced rates and soon die out; to give clones with intermediate characteristics. Despite this variability of response, it is clear that the probability of obtaining a new viable clone from a senescent clone is universely proportional to the age of the parental organisms. In extreme old age, new clones either cannot be formed at all, or if formed, always die off after a few generations have passed.

The earliest evidence for aging in *Paramecium aurelia* is the appearance of abnormal numbers of micronuclei. In young cells, the norm is 2; in old cells, the range is from 1 to 4. Old cells also have abnormalities in the micronuclear spindles and chromosomes. Other cell components, such as cilia and the gullet, also experience characteristic age changes. These abnormalities increase in number as the clone ages. It has been found that old clones accumulate many detrimental and even lethal mutations which are expressed as both dominant and recessive characters. These genetic changes in the macronucleus appear rapidly when conjugation is not allowed to occur; lethal changes are noted after only 80 fissions and by about 200 fissions, all of the progeny are nonviable. Under circumstances where a culture has the opportunity of periodic nuclear reorganization, it may be maintained for over 1000 fissions and "clonal" old age is avoided.

There is much evidence pointing to the macronucleus as the central organelle which controls the activities of the micronuclei—via the intervening cytoplasm. Late in the life span, various structural abnormalities appear in the macronucleus which are not inconsistent with a central hypothesis that aging involves progressive changes in the function of macronuclear genes. There is also ample data indicating that aging in ciliates is the result of differential gene expression. The two questions which are central to the phenomenon of aging are: What is the mechanism by which cells with a common mitotic lineage, cultivated in a constant environment, develop an orderly sequence of diverse phenotypes? Also, how does nuclear reorganization halt the progress of aging, transforming a degenerating cell into a totipotent individual? The answers to these questions are also pertinent to aging in higher organisms which resemble protozoans in the progressive manifestation of two inevitable signs of aging: phenotypic instability and loss of reproductive function.

Despite the obvious age deterioration in the nucleus, which is a common feature in a wide range of organisms, it is by no means certain that the key to aging phenomena lies primarily in this organelle. There is no reason why we should not turn our attention to the cytoplasm or even the cellular environment for causes of aging, because extranuclear processes influence nuclear function. It may be also wise not to overlook the fact that highly differentiated cells can be derepressed and revert to an unspecialized form so that the aging process may begin all over again. Neglected material in this context may be found in the invertebrates (dedifferentiation of planarians), vertebrates (conversion of iris epithelial cells to lens cells), and plants (conversion of leaf epithelial cells into meristematic tissue). "Rejuvenation" in this context appears to be initiated by virtue of the cells finding themselves in a new environment.

2. Postmitotic Cells

Because postmitotic cells cannot be replaced if there is a failure in function, changes in nondividing cells with the certainty of accumulated faults are likely to be an important cause of aging. This aspect was highlighted by early work on flatworms (Sonneborn, 1930), where the separation of anterior tissues from the posterior portion of the animals showed, through the subsequent process of regeneration, that senescence was a feature only of the anterior tissues which had a preponderance of postmitotic cells. There are three views as to the way in which post-mitotic cells could deteriorate with time.

a. Accumulation of Inert Metabolites. One of the most commonly observed age deteriorations accepted widely by histologists is the progressive accumulation of lipofuscin granules in certain cells of the body. This "age pigment," as it has been termed, appears in the light microscope as brown granules in nerve cells, seminal vesicles, adrenal cortex, and interstitial cells of the gonads, and is particularly prominent in cardiac muscle (Reichel, 1968). The accumulation of lipofuscin in the myocardial cells results in a large displacement of muscle volume, and one would anticipate that this would result in a loss of efficiency of the contractile elements. However, there is a clear absence of correlation with cardiac disease or heart failure.

Age pigment apparently occurs in increasing amounts in a variety of nondividing cells in long-lived animals, but in general is not so marked in the cells of those with a short life span. Nevertheless, the view has been expressed that the presence of lipofuscin is the only constant cellular alteration that can be correlated with age in human subjects and laboratory mammals (Whiteford and Getty, 1966).

Age pigment has been isolated and attempts made to characterize it in chemical terms. It has a blue-green fluorescence and is partially soluble in aqueous and lipid solvents. Chemical analysis indicates that it is predominantly composed of lipid and protein. The lipid factor is similar to the constituent lipid of the tissue of origin—some of the components appear to be oxidized products of unsaturated fatty acid residues which are probably responsible for the color, the fluorescence, and the general resistance to enzymic degradation. The protein component is relatively rich in glycine and valine.

Another accumulation, termed "an amyloid deposit," has been observed in old tissues from a range of animal types. There is no firm evidence as to its role in the aging process, although one group of workers has suggested its appearance indicates the development of auto-immune reactions.

Recent refinements of the "accumulated waste" theory hold that there are interactions between macromolecules and metabolites or highly reactive intermediates of metabolism, such as free radicals, which result in the linking together of proteins into large inert aggregations (Bjorksten, 1968; Harman and Piette, 1968). It is postulated that many of these links are formed by nonenzymatic mechanisms, particularly through oxidative reactions. This theory awaits the firm chemical identification of cross-linked polymers and the establishment of a rise in concentration with age. Taking the viewpoint that aging may be due in part to deleterious side effects of free radicals normally produced in metabolism, mice have been treated with compounds which are capable of reacting rapidly with free radicals. Results show that treatment with antioxidants from weaning does increase mean life span in some, but not all, strains (maximum 26% increase) (Harman, 1961).

b. Somatic Mutations. One of the most widely held theories on aging holds that spontaneous chemical changes occur in the DNA of somatic cells which are analogous to mutations in germ cells, in that they result in deletion of proteins or give rise to proteins with abnormal amino acid sequences. Somatic mutations would be expected to alter the function of those organs composed largely of postmitotic cells as the tissues gradually become occupied by malfunctioning cells, and of course, mutations could be held responsible for any kind of deterioration in cells, giving this theory a dominant position in gerontology.

Alternatively, mutations could result in rapid cell death with organ function deteriorating through the loss of its parts. Major problems surround the identification of mutations. Chromosome abnormalities appear in histological sections of cells in regenerating liver and the frequency of the aberrations increases with age. These aberrations have been equated with mutations. However, there are serious difficulties in accepting the theory of somatic mutations. The frequency of aberrations does not always fit observed differences in life span. Also, the life-shortening action of radiation, which has been taken to involve somatic mutations, may be explained in other ways (Atlan *et al.*, 1969). For example, γ-irradiation of *Drosophila* imagos at specific times between 1–20 days of age results in death at a constant time postexposure. Age at irradiation influences survival time only when the flies are irradiated 30–90 days after eclosion. These results are taken to mean that the life-shortening effects of irradiation result from a radiation-induced sickness which, at least for 1- to 20-day-old flies, is unrelated to aging. The same conclusions apply to irradiated mice where, with some strains, the shortened life span is associated with an abnormal endocrine syndrome.

It is established that ionizing radiations may cause specific mutations in the skin pigment cells of mice at the same rate as in spermatogonia. On the other hand, irradiations of experimental animals at doses that would be expected to result in somatic mutations, does not affect the rate of aging. This applies particularly to reproductive capacity, aging of collagen, and the timing of the incidence of cancer in mice.

For human subjects, evidence against the mutation theory is somewhat indirect and comes mainly from the reports of the United States Atomic Bomb Casualty Commission. These reports cover more than two decades of longitudinal studies on the survivors of the atomic bomb explosions at Hiroshima and Nagasaki. The relevant evidence arises from a comparison of a group who were within 2 kilometers of the explosion of one of the bombs and had survived acute radiation symptoms, indicating that they had been subjected to about one-third of the acute lethal radiation dose, with a matched nonexposed group. There is no evidence in the reports of any intergroup differences in cardiovascular function, sensory acuity, and the cosmetic aspects of aging. Immunological tests gave no indication that the irradiated group behaved in any respect as if they were older than their chronological age.

In summary, while it is highly probable that somatic mutations occur with age, the available evidence is inadequate to establish that this process is responsible for a general deterioration in function. Even in Protozoa, where both dominant and lethal mutations are produced on a fairly large scale in very old individuals of "old clones," there is some considerable doubt as to whether this is a cause or an effect of aging.

c. Gene Suppression. Another theory which has been proposed also uses the central theme of inactivation of genes. The theory rests on the assumption that there are age-related losses in the ability of cells to synthesize RNA and protein and further postulates that these losses in synthetic ability are due to the irreversible binding of repressors to corresponding structural or operator genes (Von Hahn, 1966).

d. Incorrect Translation. Failing a primary change in the chemical state of DNA through mutation or the irreversible binding of repressors to genes, abnormal protein patterns could arise due to a loss of competence of the cytoplasmic apparatus to correctly translate the genetic code into the polypeptide specified by the nucleus. Such an error would concern the specificity of information handling enzymes and lead to the synthesis of malfunctioning enzymes by virtue of the substitution of abnormal amino acids (Orgel, 1963). Following treatment of microorganisms and invertebrates with amino acid analogs which give rise to abnormal proteins through amino acid displacement, the organisms show premature senescence and a reduced life span (Harrison and Holliday, 1967; Holliday, 1969). These results suggest the probable outcome of disturbing the process of translation but, as yet, there is no direct evidence for amino acid substitution occurring on a large scale in aging organisms.

The outcome of either mutation or faulty translation would most likely be a steady decline in the biological activity of tissue proteins and a change in their physical properties. As will be made clear in the next section, this is not a general feature of aging at the molecular level. Evidence is being obtained that may be taken as indicative of changes in molecular configuration of some enzymes in old animals, but this kind of change may be explained by a partial denaturation occurring because enzymes synthesized by old animals are released into a suboptimum cellular environment which alters their tertiary structure. The crucial test of the incorrect translation theory awaits the isolation of proteins from old animals which have abnormal amino acid sequences.

e. Unselected Differentiation. Once a mitotic cell has differentiated into a postmitotic cell, it has entered upon an irreversible developmental pathway and there is a certain degree of probability that this pathway will result in cell death. An extreme form of this theory states that there is an aging pathway in all postmitotic cells which is controlled by a special group of "aging genes" and that in this sense, all specialized cells are dying cells (Bullough, 1971). The origins of this idea stem from a consideration of the proliferation of epidermal cells which, in the process of dying, are converted into the important protective covering of the body.

A "program or epigenetic theory" of aging is a reasonable alternative to the mutation theory as an explanation for a shift in the direction of metabolism; changes in the relative activities of various genes would produce progressive alterations in the concentrations of intracellular enzymes. At no period in life is there a perfect steady state expressed as a stationary body weight and body composition, and the corresponding shift in enzyme pattern need not involve special aging genes and would be no different in principle from those events responsible for differentiation. That is to say, a pattern of genetic information develops in postmitotic cells with time, which is influenced by events in the extranuclear phase. New protein patterns would differ from those of development, in that the changes would be quantitative rather than qualitative. Aging, by this mechanism, differs from development because the process leads to a loss of efficiency. This occurs, either because the forces of natural selection have favored death following rapidly after reproductive maturation, or as a result of organisms existing in a special protected environment where natural selection cannot operate. In the latter situation there can be no selection against random disproportionate alterations in enzyme activity which, on the time scale of aging, may cause inefficiency at the organ level and possibly random death of cells.

Up to a certain point in time, it may be postulated that these epigenetic changes are potentially reversible. In other words, there has been no change in the chemistry of DNA, only a progressive opening and closing of information channels leading to other parts of the cell. From this viewpoint, the regeneration of protozoa after autogamy may well be the key to an understanding of the nature of aging, in that "regeneration" could occur by a process which was, in effect, the uncoupling of DNA from cytoplasmic events and its reestablishment in a "youthful" environment, so allowing the specific developmental program to unfold once more. Evidence that aging is linked with development comes mainly from experiments showing that slowing-down development increases life span (see Section IIC).

There is no shortage of evidence that changes in enzyme pattern occur in postmitotic tissues, although only a few of these changes are obviously connected with a failure in tissue function and many of them do not take a downward trend. On the whole, the age-dependent variations are not very marked. Few experiments have been designed to follow age changes in enzyme activity throughout the life span. Data from laboratory rodents suggest that the period of most rapid change occurs during the first few months of life and by the end of the first year of life, changes in activities occur very slowly.

Although few comprehensive comparative studies have been accomplished, it would appear that a high proportion of cells in the body maintain enzyme concentrations in old age which are close to those found in the mature adult. As is the case for physiological and morphological processes, it appears that biochemical processes involving enzyme synthesis exhibit aging mostly in terms of a less effective response to a given stimulus.

Conclusions as to the stability of enzyme concentration patterns rest, for the most part, in the lack of statistical evidence for change. Many times, a large difference in mean values has been found in two-point experiments, but the deviation from the mean is often a large fraction of this difference. Some of the differences in means may become significant if more individual measurements were made (n is usually not more than 10 and often less than 10). Another factor is that by sampling at more than two points in time, consistent trends might be established. Third, by analyzing for many enzymes simultaneously instead of the usual one or two, patterns might be established.

As an indication of the biochemical constancy at the cellular level, the basal oxygen uptake of humans declines with age, but when expressed per unit of intracellular water, does not show any regression (Shock *et al.*, 1963). From this it is concluded that there is no general impairment in the respiratory metabolism of old cells. The basal whole-body oxygen consumption of adult rats actually increases with age between 10% and 13% (Ring *et al.*, 1964). This discrepancy between rats and human subjects has not been resolved, but it is likely that the decrease in muscle mass of rats is counteracted by a higher rate of respiration of cells in other tissues. On the other hand, no age changes have been detected in aerobic metabolism of tissue preparations from rat liver (Barrows *et al.*, 1958). Similar experiments with kidney indicate only a small age decrement in aerobic metabolism, possibly due to a fall in the number of mitochondria per cell (Barrows *et al.*, 1960). The maintenance of aerobic metabolism at the cellular level is borne out by results obtained for guinea pig tissues. Further, the specific dynamic action of a standard protein meal expressed as excess oxygen uptake is essentially the same in old (mean age 77 year) and young (mean age 24 year) men, although older subjects show a slower response (Tuttle *et al.*, 1952).

Little is known about anaerobic metabolism. Indirect evidence is suggestive of a decrease in the demand of the aging brain for anaerobically produced energy (Enzmann and Pincus, 1934; Samson *et al.*, 1958). Some glycolytic enzymes in rat kidney have been observed to decrease with age by between 15% and 20%. However, these changes are not associated with an impairment of glycolysis measured *in vitro* (Zorzoli and Li, 1967).

Aging rat liver and brain are remarkable for their metabolic stability. Liver does not undergo age involution (Yiengst *et al.*, 1959; Kurnick and Kernen, 1962), and there is no general alteration in protein turnover (Fletcher and Sanadi, 1961; Barrows and Roeder, 1961; Beauchene *et al.*, 1967) although catalase in both liver and kidney shows an age related decrease in both synthesis and degradation (Haining and Legan, 1973). It also appears that the enzyme pattern of liver and brain is largely independent of age (Schmukler and Barrows, 1967; Hollander and Barrows, 1968; Gold *et al.*, 1968). Despite this, there is a decrease in nitrogen content of liver attributable mainly to losses of mitochondria and other membrane structures and a decrease in the RNA content per cell (Detwiler and Draper, 1962). In humans, there is a rise in the volume-ratio of nucleus to cytoplasm (Tauchi and Sato, 1962). The actual quantity of RNA per nucleus is increased and there is an alteration in the base composition. These latter changes may be connected with the marked increase in gross RNA turnover which is observed in nuclei of a number of tissues in old rats (Wulff *et al.*, 1964). Despite an alteration in RNA metabolism, cytophotometric measurements of Feulgen-stained nuclei in neurones of the cerebellum and liver cells indicate that there is no loss of DNA from individual cells during aging (Enesco, 1967). A histochemical study of the aging rat submandibular gland, revealed a decrease in the activity of several enzymes and a fall in RNA concentration (Bogart, 1967). Work involving the separation of isoenzymes in nematodes also suggest that the enzyme pattern may well vary with age (Erlanger and Gershon, 1970). Detailed biochemical work on the aging of RNA metabolism appears to be moving in favor of template inactivation (Britton *et al.*, 1973).

Most work on the biochemistry of aging has been carried out using the laboratory rat and it could be argued that the biochemical features of aging in the laboratory rat are a consequence of inbreeding. This objection has been overcome with regard to some of the phenomena described above, in that they are also found in the aging wild rat (Barrows *et al.*, 1962), although there are quantitative differences indicative of a slower rate of aging in the wild strain when both strains are housed in the laboratory (Chvapil and Roth, 1964).

Taken together, the biochemical work suggests that each tissue is characterized by a particular pattern of aging at the enzymatic level but to support this idea it is desirable that comprehensive studies be made on each tissue. So far, only one such analysis has been carried out on human blood vessels which indicates that there is a considerable variation in both the extent and pattern of change in enzyme activity (Kirk, 1969). Some enzymes increase in activity (maximum, 3.5-fold rise), others hardly change, and some at first increase and then decrease in activity. They may be divided into four groups according to the time when the activity reaches a maximum, but there are no obvious similarities in function of enzymes within any group (Bellamy, 1970). Enzymes in group 1 show a steady decline in activity from the first decade amounting to about −5% per decade. Those in group 2 have a peak activity in the second decade (mean 50% increase) then show a decline, which between the third and eighth decade proceeds on average at a rate similar to that for enzymes of group 1. Enzymes in group 3 reach their maximum activity in the third decade (mean, 2-fold increase) then decline at a rate similar to those in group 1 and 2. Group 4 contains enzymes that reach their peak between the third and ninth decades. The mean rate of increase to the peak is about the same for all groups.

Taking all enzymes together, there is no evidence to support the idea that there is a general deterioration in metabolism with age. By the ninth decade about 50% of the enzymes retain more than 90% of the activity characteristics of the first decade; only one enzyme retains less than 50% of its activity in the first decade.

Another study, although not so many enzymes were measured, has been carried out on mouse prostate. In this organ, there is no detectable change in total DNA content as the animals age, indicating that cell loss did not occur on a large scale. Enzyme activities, expressed per unit of DNA, fell into two basic patterns: starting at age 9 months, some enzymes showed a steady decline; others showed a rise, followed by a fall. At the age of 30 months, only 5 enzymes out of 11 showed an activity per unit DNA which was less than that at 9 months (Mainwaring, 1967).

Available evidence, therefore, although stressing biochemical stability, suggests that there is some kind of drift in enzyme activity which is not always toward a drop in activity. It is within the bounds of probability that genotypic or phenotypic influences would, given time, eventually lead to ratios of key enzymes in some cells which were incompatible with life. The expression of this phenomenon at the organismic level might be expected to take the form of first an increase in variability followed by a decline in variability in the relevant enzymes at the time of peak mortality. The shift in mean plus the lowered variability would be the result of selection against these phenotypes with greatly disturbed enzyme patterns, leaving a small fraction of the population to survive into extreme old age with an enzyme pattern closer to that of youth.

It is not an essential feature of this model that all cells in a given tissue change in the same way at the same time. There are variations in the environment of cells dependent upon local differences in the rate of blood flow and the presence of adjacent extracellular structures and other cell types. Thus, basic chemical differences probably exist for cells to take up an individual phenotypic aging pattern. Indeed, there is clear histological evidence for age-dependent random variations in cells. For example, hepatic parenchymal cells, scattered either individually or in small groups, show variations in the morphology of cell organelles and functional differences are apparent through histochemical tests. Changes in enzyme pattern, once set in motion, may lead to an early death of the cells and may prove the explanation for the apparent random nature of cell death in postmitotic tissues.

IV. EVOLUTION AND AGING

In many organisms, death may be regarded as a programed event, in that it is an essential part of the life cycle. For example, annual plants develop in such a way that food is transferred from parent to seeds during the terminal phase of the reproductive cycle, resulting in the rapid death of the parent plant. In many animals, both invertebrate and vertebrate, similar developmental processes have evolved. In the salmon, for example, during the up-river spawning migration, the irreversible utilization of postmitotic tissues, mainly skeletal muscle, to provide amino acids for the formation of eggs and sperm, culminates in rapid death after the gametes are shed, and it is difficult to escape the conclusion that the reproductive process is the major cause of death. In these organisms, death is an obvious outcome of the natural selection of an efficient reproductive process. On the other hand, for most organisms, aging and death appear to be merely the outcome of the extension of a sequence of chemical events which was selected to establish an efficient mechanism for maturation and reproduction in a habitat where it was highly probable that death would occur at, or soon after, peak reproductive efficiency had been attained.

Malfunctions which occur beyond the reproductive phase cannot be corrected in evolutionary terms because events characteristic of this period, which are advantageous in prolonging life, cannot be perpetuated in the gametes through the forces of natural selection. From this aspect, cell loss begins in some organs before the animal is adult. Indeed, it is an essential feature of differentiation that entire organ systems be dismantled to make way for more advantageous structures. During postnatal development of vertebrates, postmitotic cells begin to disappear from key organs well before the onset of sexual maturity, and the rate of loss continues steadily throughout life. In humans cell loss on a large scale appears to begin in the third decade (Novak, 1973). The existence of this phenomenon may be explained in that organisms can tolerate a certain degree of imperfections in function which are in keeping with the maintenance of the correct degree of reproductive efficiency in relation to selection pressure. The random appearance of imperfections seriously interferes with the life process only when the organism is allowed to live beyond the natural life span selected through evolution. This applies to animals in captivity and modern man.

According to Needham (1959), the individual must age because there is a need for the reproductive replacement of a certain number of individuals, but it is difficult to see how death rate alone serves a positive selection purpose. On an evolutionary plane, reproduction tends to create novelty in the form of new combinations of genes as well as tending to preserve novelty in the form of mutant genes. A fixed life span may be advantageous in conferring genetic flexibility, although the mechanism by which it evolved is obscure.

V. GENERAL CONCLUSIONS

The readily observable deterioration in physiological adaptability after maturation, with the concomitant increasing force of mortality that comes with age, allows us to define aging but it does not tell whether the phenomenon results from accumulated disease, repeated trauma, progressive cellular damage, or if it is inherent in the common chemical plan upon which living organisms are constructed. Only in the latter situation is there something inherent in the organism itself which would strictly justify the use of the term "process" to describe an intrinsic sequence of events controlled largely by genotype and environment.

Although it is not possible to pinpoint a direct humoral control of aging, life span is certainly bound up with the speed of development which, in turn, is linked closely with maximum body size, absolute brain weight, and rate of energy production. The three latter aspects have all been separately implicated at one time or another in the control of senescence. As they are all closely interconnected, to take any of these views is really to state the problem of aging in terms of a epigenetic theory expressed through the complex DNA code which allows genetic continuity in the populations. The genetic program increases and maintains Darwinian fitness for as long as it promotes relative survival of the species and it gradually reverts to a state of disorder as individuals live beyond the mean life span, which is a characteristic of a particular evolved population in equilibrium with other populations of an ecosystem. This is not to say that aging is a process, although it is likely that aging phenomena in general occur through changes in the function of the cell nucleus.

REFERENCES

Abbo, F. E. (1966). *J. Geront.* **21**, 112.
Ackermann, P. G., and Kheim, T. (1964). *J. Geront.* **19**, 207.
Adelman, R. C. (1971). *Exp. Geront.* **6**, 75.
Albert, A., Randall, R. V., Smith, R. A., and Johnson, C. E. (1956). "Hormones and Ageing." Academic Press, New York.
Allen, T. H., Anderson, E. C., and Langerham, W. H. (1960). *J. Geront.* **15**, 348.
Andersen, D. H. (1932). *Physiol. Rev.* **12**, 1.
Angervall, L., and Lundin, P. M. (1964). *Endocrinology* **74**, 986.
Atlan, H., Miquel, J., and Binnard, R. (1969). *J. Geront.* **24**, 1.
Axelrod, L. R. (1965). *Biochim. Biophys. Acta* **97**, 551.
Baca, Z. V., and Chiodi, H. (1965). *Endocrinology* **76**, 1208.
Barnabei, O., Romano, B., Bitonto, G. di., Thomasi, V., and Sereni, F. (1966). *Arch. Biochem. Biophys.* **113**, 468.

Barrows, C. H., and Roeder, L. M. (1961). *J. Geront.* **16**, 321.

Barrows, C. H., Yiengst, M. J., and Shock, N. W. (1958). *J. Geront.* **13**, 351.

Barrows, C. H., Falzone, J. A., and Shock, N. W. (1960). *J. Geront.* **15**, 130.

Barrows, C. H., Roeder, L. M., and Falzone, J. A. (1962). *J. Geront.* **17**, 144.

Barrows, C. H., Roeder, L. M., and Olewine, D. A. (1962). *J. Geront.* **17**, 148.

Bashey, R. I., Torii, S., and Angrist, A. (1967). *J. Geront.* **22**, 203.

Batali, M., Rogers, J. B., and Blumenthal, H. T. (1961). *J. Geront.* **16**, 230.

Beauchene, R. E., Fanestil, D. D., and Barrows, C. H. (1965). *J. Geront.* **20**, 306.

Beauchene, R. E., Roeder, L. M., and Barrows, C. H. (1967). *J. Geront.* **22**, 318.

Belasco, I. J., and Murlin, J. R. (1941). *Endocrinology* **28**, 145.

Bellamy, D. (1963). *Biochem. J.* **87**, 334.

Bellamy, D. (1964). *J. Endocrinol.* **31**, 83.

Bellamy, D. (1966). *Proc. Int. Congr. Geront., 7th* p. 147. Hegermann, Vienna.

Bellamy, D. (1967). *Symp. Soc. Exp. Biol.* **21**, 427.

Bellamy, D. (1969). *Exp. Geront.* **3**, 327.

Bellamy, D. (1970). *Mem. Soc. Endocrinol.* **18**, 303.

Bellamy, D., and Alkufaishi, H. (1972). *Age Ageing* **1**, 88.

Bellamy, D., and Leonard, R. A. (1966). *Biochem. J.* **98**, 581.

Bellamy, D., Phillips, J. G., Chester-Jones, I., and Leonard, R. A. (1962). *Biochem. J.* **85**, 537.

Bellamy, D., Janssens, P. A., and Leonard, R. A. (1966). *J. Endocrinol.* **35**, 19.

Bender, A. D., Kormendy, C. G., and Powell, R. (1970). *Exp. Geront.* **5**, 97.

Berg, B. N., and Harmiston, C. R. (1955). *J. Geront.* **10**, 416.

Bernick, S., and Ershoff, B. A. (1963). *Endocrinology* **72**, 231.

Berry, R. J., and Jakobsen, M. E. (1971). *Exp. Geront.* **6**, 187.

Birren, J. E. (1955). *J. Geront.* **10**, 437.

Birren, J. E., and Botwinick, J. (1955). *J. Geront.* **10**, 429.

Birren, J. E., and Kay, H. (1958). *J. Geront.* **13**, 374.

Bjorksten, J. (1963). *Gerontologia* **8**, 179.

Bjorksten, J. (1968). *J. Amer. Geriat. Soc.* **16**, 408.

Blumenthal, H. T. (1954). *Arch. Pathol.* **57**, 481.

Bodansky, M., and Duff, V. B. (1936). *Endocrinology* **20**, 541.

Bogart, B. I. (1967). *J. Geront.* **22**, 372.

Bourne, G. H., and Jayne, E. P. (1961). "Structural Aspects of Ageing." Pitman, London.

Boyd, E. (1932). *Amer. J. Dis. Child.* **43**, 1162.

Britton, V. J., Sherman, F. G., and Florini, J. R. (1973). *J. Geront.* **27**, 188.

Brody, S. (1945). "Bioenergetics and Growth." Van Nostrand Reinhold, Princeton, New Jersey.

Brown, J. B., and Mathew, G. D. (1962). *Recent Progr. Horm. Res.* **18**, 337.

Bucher, N. L. R., and Glinos, A. D. (1950). *Cancer Res.* **10**, 324.

Bullough, W. S. (1971). *Nature (London)* **229**, 608.

Bulos, B., Sacktor, B., Grossman, I. W., and Altman, N. (1971). *J. Geront.* **26**, 13.

Burke, W. E., Tuttle, W. W., Thompson, C. W., Janney, C. D., and Weber, R. J. (1953). *J. Appl. Physiol.* **5**, 628.

Burrows, H. (1945). "Biological Actions of Sex Hormones." Cambridge Univ. Press, London and New York.

Caldwell, B. M., and Watson, R. (1954). *J. Geront.* **9**, 168.

Cameron, G. R. (1953). "Suprarenal Cortex." Butterworths, London and Washington, D. C.

Cameron, I. L. (1973). *J. Geront.* **27**, 157.

Carlson, A. J., and Hoelzel, F. J. (1946). *J. Nutr.* **31**, 363.

Charriper, H. A., Pearlstein, A., and Bourne, G. H. (1961). "Structural Aspects of Ageing." Pitman, London.

Chester-Jones, I., and Bellamy, D. (1964). *Symp. Soc. Exp. Biol.* **18**, 195.

Chvapil, M., and Roth, Z. (1964). *J. Geront.* **19**, 414.

Collins, D. D. (1950). *J. Geront.* **5**, 241.

Conrad, R. A. (1960). *J. Geront.* **15**, 358.

Correll, W. W., Turner, M. D., and Haining, J. L. (1965). *J. Geront.* **20**, 507.

Dalderup, L. M., Opdam-Stockmann, H., and Rechsteiner-de Vos, H. (1966). *J. Geront.* **21**, 22.

Dalderup, L. M., Keller, G. H. M., and Stroo, M. M. (1967). *Gerontologia* **13**, 86.

Daniel, C. W., Young, L. J. T., Medina, D., and De Ome, K. B. (1971). *Exp. Geront.* **6**, 95.

D'Arcy, P. F., and Howard, E. M. (1962). *J. Pharm. Pharmacol.* **14**, 294.

Das, B. C. (1964). *Gerontologia* **9**, 179.

Das, B. C. (1969). *Gerontologia* **15**, 275.

Das, B. C., and Bhattacharya, S. K. (1961). *Can. J. Biochem. Physiol.* **39**, 569.

Desaulles, P. A. (1958). *Ciba Foundation Colloq. Ageing* **4**, 180.

Detwiler, T. C., and Draper, H. H. (1962). *J. Geront.* **17**, 138.

Dicker, S. E., and Nunn, J. (1958). *J. Physiol.* **141**, 332.

Dontas, A. S., Papanicolaou, N. T., Papanayiotou, P., and Malmos, B. K. (1967) *J. Geront.* **22**, 268.

Doniach, D., and Roitt, I. M. (1964). *Semin. Hematol.* **1**, 313.

Dunihue, F. W. (1965). *Endocrinology* **77**, 948.

Eisdorfer, C., Powell, A. H., Silverman, G., and Bogdonoff, M. D. (1965). *J. Geront.* **20**, 511.

Enesco, A. E. (1967). *J. Geront.* **22**, 445.

Enzmann, E. V., and Pincus, G. (1934). *J. Gen. Physiol.* **18**, 163.

Erlanger, M., and Gershon, D. (1970). *Exp. Geront.* **5**, 340.

Ershoff, B. H. (1951). *Proc. Soc. Exp. Biol. Med.* **78**, 836.

Everitt, A. V. (1959). *J. Geront.* **14**, 415.

Everitt, A. V., and Duvall, L. K. (1965). *Nature (London)* **205**, 1016.

Everitt, A. V., and Webb, C. (1958). *J. Geront.* **13**, 255.

Faludi, G., Mills, L. C., and Chayes, Z. W. (1964). *Acta Endocrinol. Copenhagen* **45**, 68.

Falzone, J. A., Barrows, C. H., and Shock, N. W. (1959). *J. Geront.* **14**, 2.

Fedorčákova, A. M., Bachledová, E., Niederland, T. R., and Bózner, A. (1968). *Exp. Geront.* **3**, 63.

Fels, I. G. (1969). *Gerontologia* **15**, 308.

Fenger, F. (1915). *J. Biol. Chem.* **20**, 115.

Fisher, A. M., and Scott, D. A. (1934). *J. Biol. Chem.* **106**, 305.

Fletcher, M. J., and Sanadi, D. R. (1961). *J. Geront.* **16**, 255.

Flexner, L. B., and Flexner, J. B. (1949). *J. Cell Comp. Physiol.* **34**, 115.

Flexner, L. B., and Flexner, J. B. (1950a). *Anat. Rec.* **106**, 413.

Flexner, J. B., and Flexner, L. B. (1950b). *J. Cell Comp. Physiol.* **36**, 351.

Flückiger, E., and Verzár, F. (1955). *J. Geront.* **10**, 306.

Franks, L. M., and Chesterman, F. C. (1964). *Nature (London)* **202**, 821.

Frantz, A. G., and Rabkin, M. T. (1964). *New England J. Med.* **13**, 75.

Friedfeld, L. (1949). *Geriatrics* **4**, 211.

Friedman, S. M., Hinke, J. A. M., and Friedman, C. L. (1956). *J. Geront.* **11**, 286.

Friedman, S. M., Stretér, F. A., and Friedman, C. L. (1963). *Gerontologia* **7**, 65.

Friedman, S. M., Nakashima, N., and Friedman, C. L. (1965). *Gerontologia* **11**, 129.

Friedman, M., Green, M. F., and Sharland, D. E. (1969). *J. Geront.* **24**, 292.

Frolkis, V. V. (1966). *J. Geront.* **21**, 161.

Fujita, T., Ohata, M., Orimo, H., and Yoshikawa, M. (1971). *J. Geront.* **26**, 20.

Gala, R. R., and Westphal, U. (1965). *Endocrinology* **76**, 1079.

Gartner, L. M., and Arias, I. M. (1963). *Amer. J. Physiol.* **205**, 663.

Giarnieri, D., and Lumia, V. (1961). *Clin. Chim. Acta* **6**, 144.

Gold, P. H., Gee, M. V., and Strehler, B. L. (1968). *J. Geront.* **23**, 509.

Goodman, M., Rosenblatt, M., Gottlieb, J. S., Miller, J. S., and Chen, C. H. (1963). *Arch. Gen. Psychiat.* **8**, 518

Gordon, H. A., Bruckner-Kardoss, E., and Wostmann, B. S. (1967). *J. Geront.* **21**, 380.

Grad, B. (1969). *J. Geront.* **24**, 5.

Grad, B., and Khalid, R. (1968). *J. Geront.* **23**, 522.

Grad, B., and Kral, V. A. (1956). *J. Geront.* **12**, 172.

Grad, B., Kral, V. A., Payne, R. C., and Berenson, J. (1967). *J. Geront.* **22**, 66.

Greep, R. O., and Chester-Jones, I. (1950). *Recent Progr. Horm. Res.* **5**, 197.

Gregerman, R. I., Gaffney, G. W., Shock, N. W., and Crowder S. E. (1962). *J. Clin. Invest.* **41**, 2065.

Griffiths, M. (1941). *J. Physiol. London* **100**, 104.

Griswold, R. L., and Pace, N. (1965). *J. Geront.* **12**, 150.

Grossie, J., and Turner, C. W. (1965). *Proc. Soc. Exp. Biol. Med.* **118**, 28.

Gsell, D., von Hahn, H. P., and Schaub, M. C. (1964). *Gerontologia* **9**, 36.

Gutmann, E. (1970). *Exp. Geront.* **5**, 357.

Gutmann, E., Hanzlíková, V., and Jakovbek, B. (1968). *Exp. Geront.* **3**, 141.

Haensly, W. E., and Getty, R. (1965). *J. Geront.* **20**, 544.

Haensly, W. E., Jermier, J. A., and Getty, R. (1964). *J. Geront.* **19**, 54.

Haining J. L., and Legan J. S. (1973). *Exp. Geront.* **8**, 85.

Hall, D. A., Keech, M. K., Reed, R., Saxl, H., Tunbridge, R. E., and Wood, M. J. (1955). *J. Geront.* **10**, 388.

Halliday, R., and Kekwick, R. A. (1957). *Proc. Roy. Soc. B.* **146**, 431.

Hamilton, J. B. (1965). *J. Geront.* **20**, 96.

Hamilton, J. B., and Mestler, G. E. (1969). *J. Geront.* **24**, 395.

Hamilton, J. B., Hamilton, R. S., and Mestler, G. E. (1969). *J. Geront.* **24**, 427.

Harding, H. R., Rosen, F., and Nichol, C. A. (1961). *Amer. J. Physiol.* **201**, 271.

Harman, D. (1961). *Lancet* **1**, 200.

Harman, D., and Piette, L. H. (1968). *J. Geront.* **21**, 560.

Harrison, B. J., and Holliday, R. (1967). *Nature (London)* **213**, 990.

Hartog, M., Gaafer, M. A., and Fraser, R. (1964). *Lancet* **ii**, 376.

Hastings, A. B., Blumgart, H. L., Lowry, O. H., and Gilligan, D. R. (1939). *Trans. Ass. Amer. Phys.* **54**, 237.

Hayflick, L. (1965). *Exp. Cell Res.* **37**, 614.
Hayflick, L., and Moorhead, P. S. (1961). *Exp. Cell Res.* **25**, 585.
Hellon, R. F., and Lind, A. R. (1958). *J. Physiol. London* **141**, 262.
Henry, J. P., Meehan, J. P., Stephens, P., and Santisbeban, G. A. (1965). *J. Geront.* **20**, 239.
Herbeuval, R., Duheille, J., Cuay, G., and Haagen, A. (1967). *Presse Méd* **75**, 731.
Hines, H. M., and Knowlton, G. C. (1939). *Proc. Soc. Exp. Biol. Med.* **42**, 133.
Hochschild, R. (1971). *Exp. Geront.* **6**, 133.
Hodgkins, J. (1962). *J. Geront.* **17**, 385.
Hollander, J., and Barrows, C. H. (1968). *J. Geront.* **23**, 174.
Holliday, R. (1969). *Nature (London)* **221**, 1224.
Horton, D. L. (1967). *J. Geront.* **22**, 43.
Horvath, S. M. (1946). *J. Geront.* **1**, 213.
Hrůza, Z. (1970). *Exp. Geront.* **6**, 199.
Hrůza, Z. (1971). *Exp. Geront.* **6**, 103.
Hubble, D. (1957). *Lancet* ii, 301.
Inay, M., and Thompson, K. W. (1938). *Amer. J. Physiol.* **123**, 106.
Ingle, D. J., Morley, E. H., and Nezamis, J. E. (1952). *Endocrinology* **51**, 487.
Iob, V., and Swanson, W. W. (1934). *Amer. J. Dis. Child.* **47**, 302.
Irvine, W. J., Davies, S. H., Teitebaum, S., Delamore, I. W., and Williams, A. W. (1965). *Ann. N.Y. Acad. Sci.* **124**, 657.
Jakubczak, L. F. (1967). *J. Geront.* **22**, 421.
Jayne, E. P. (1963). *J. Geront.* **18**, 227.
Jelínková, M. (1964). *Physiol. Bohemoslov.* **13**, 327.
Jelínková, M., Stuchlíková, E., and Smrz, M. (1970). *Exp. Geront.* **5**, 257.
Jennings, H. S., and Lynch, R. S. (1928). *J. Exp. Zool.* **50**, 345.
Jones, D. C., Osborn, G. K., and Kimeldorf, D. J. (1967). *Gerontologia* **13**, 211.
Jost, A. (1951). *Ciba Foundation Colloq. Ageing* **2**, 18.
Kallmann, F. J. (1961). Genetic factors in ageing. *In* "Psychopathology of Ageing" (P. H. Hoch and J. Zubin, eds.), p. 227. Grune and Stratton, New York.
Kanabrocki, E. L., Fells, I. G., Decker, C. F., and Kaplan, E. (1963). *J. Geront.* **18**, 18.
Kemény, V., Kemény, A., and Vecsei, P. (1964). *Acta Physiol. Hung.* **25**, 31.
Kibler, H. H., and Johnson, H. D. (1966). *J. Geront.* **21**, 52.
Kiessling, K. H. (1962). *Exp. Cell Res.* **28**, 145.
Kirk, J. E. (1969). "Enzymes of the Arterial Wall." Academic Press, New York.
Kirk, J. E., and Dyrbye, M. (1956). *J. Geront.* **11**, 273.
Kohn, R. R. (1963). *J. Geront.* **18**, 14.
Korenchevsky, V. (1942). *J. Pathol. Bact.* **54**, 13.
Korenchevsky, V. (1961). "Physiological and Pathological Ageing." Karger, New York.
Korenchevsky, V., Paris, S. K., and Benjamin, B. (1953). *J. Geront.* **8**, 415.
Krag, C. L., and Kountz, W. B. (1950). *J. Geront.* **5**, 227.
Krag, C. L., and Kountz, W. B. (1952). *J. Geront.* **7**, 61.
Krecek, J., Dlouha, H., Jelinek, J., Kreckova, J., and Vacek, Z. (1958). *Ciba Foundation Colloq. Ageing* **4**, 165.
Kurnick, N. B., and Kernen, R. L. (1962). *J. Geront.* **17**, 245.
Lansing, A. I. (1942). *J. Exp. Zool.* **91**, 195.
Lansing, A. I. (1947). *J. Geront.* **2**, 228.

Lansing, A. I. (1948). *Proc. Nat. Acad. Sci. U.S.* **34**, 304.

Lansing, A. I., Rosenthal, T. B., and Alex, M. (1950). *J. Geront.* **5**, 211.

Lesser, G. T., Deutsch, S., and Markofsky, J. (1970). *J. Geront.* **25**, 108.

Lewis, W. H., and Alving, A. S. (1938). *Amer. J. Physiol.* **123**, 500.

Longcope, C. (1971). *Amer. J. Obstet. Gynec.* **111**, 778.

Lowry, O. H., Hastings, A. B., Hull, T. Z., and Brown, A. N. (1942). *J. Biol. Chem.* **143**, 271.

Lowry, O. H., Hastings, A. B., McCay, C. M., and Brown, A. N. (1946). *J. Geront.* **1**, 345.

Lunsford, W. R., McCay, C. M., Lupien, P. J., Pope, F. E., and Sperling, G. (1963). *Gerontologia* **7**, 1.

MacNider, W., de B. (1917). *J. Exp. Med.* **26**, 1.

MacNider, W., de B. (1943). *Proc. Soc. Exp. Biol. Med.* **53**, 1.

MacNider, W., de B. (1946). *J. Geront.* **1**, 189.

Magladery, J. W., Teasdall, R. D., and Norris, A. H. (1958). *J. Geront.* **13**, 282.

Mainwaring, W. I. (1967). *Gerontologia* **13**, 177.

Marine, D., Manley, O. T., and Baumann, E. J. (1924). *J. Exp. Med.* **40**, 429.

Markovsky, J., and Perlmutter, A. (1973). *Exp. Geront.* **8**, 65.

Mason, H. L., and Engstrom, W. W. (1950). *Physiol. Rev.* **30**, 321.

McCay, C. M. (1942). *In* "Cowdrey's 'Problems of Ageing,' " 2nd ed. Williams and Wilkins, Baltimore, Maryland.

McCay, C. M. (1952). *In* "Cowdry's Problems of Ageing" Williams and Wilkins, Baltimore Maryland.

McFarland, R. A., and Fisher, M. B. (1955). *J. Geront.* **10**, 424.

McHale, J. S., Mouton, M. L., and McHale, J. T. (1971). *Exp. Geront.* **6**, 89.

McKay, L. L., McKay, E. M., and Addis, T. (1924). *J. Clin Invest.* **1**, 576.

Meites, J., Hopkins, T. F., and Deuben, R. (1962). *Fed. Proc. Fed. Amer. Soc. Exp. Biol.* **21**, 196.

Meyer, R. K., Rao, M. A., and Aspinall, R. L. (1959). *Endocrinology* **64**, 890.

Miller, J. H., McDonald, R. K., and Shock, N. W. (1951). *J. Geront.* **6**, 213.

Miller, J. H., McDonald, R. K., and Shock, N. W. (1952). *J. Geront.* **7**, 196.

Moon, H. D., Koneff, A. A., Li, C. H., and Simpson, M. E. (1956). *Proc. Soc. Exp. Biol. Med.* **93**, 74.

Moore, C. R. (1945). *Amer. J. Anat.* **76**, 1.

Morrison, A. B., and Staroscik, R. N. (1964). *Gerontologia* **9**, 65.

Murray, H. A. (1926). *J. Gen. Physiol.* **9**, 789.

Myers, V. C., and Lang, W. C. (1946). *J. Geront.* **1**, 441.

Nagorny, A. V., and Golubitzkaia, R. I. (1947). *Sci. Rec. Kharkov Univ. Ukr.* **25**, 149.

Needham, A. E. (1949). *J. Exp. Zool.* **112**, 49.

Needham, A. E. (1959). *Quart. Rev. Biol.* **34**, 189.

Needham, J. (1931). "Chemical Embryology." Cambridge Univ. Press, London and New York.

Norris, A. H., Shock, N. W., and Yiengst, M. J. (1955). *J. Geront.* **10**, 145.

Novak, L. P. (1973). *J. Geront.* **27**, 438.

Oberman, J. W., Gregory, K. O., Burke, F. G., Ross, S., and Rice, E. C. (1956). *New England J. Med.* **255**, 743.

Orgel, L. E. (1963). *Proc. Nat. Acad. Sci. U.S.* **49**, 517.

Palmer, B. G. (1966). *J. Endocrinol.* **36**, 73.

Pauker, J. D., Kheim, T., Mensh, J. N., and Kountz, W. B. (1958). *J. Geront.* 13, 389.

Pecile, A., Muller, E., Falconi, G., and Martini, L. (1965). *Endocrinology* 77, 241.

Peters, V. B., and Flexner, L. B. (1950). *Amer. J. Anat.* 86, 133.

Pierson, W. R., and Montoye, H. J. (1958). *J. Geront.* 13, 418.

Pincus, G., Romanoff, L. P., and Carlo, J. (1954). *J. Geront.* 9, 113.

Porkrajac, N., Rabadija, L., Vranic, M., and Allegretti, N. (1959). *Naturwissenschaften* 46, 338.

Praetorius, E. (1951). *J. Geront.* 6, 135.

Rahman, Y. E., and Peraino, C. (1973). *Exp. Geront.* 8, 93.

Raisz, L. G., O'Brien, J. E., and Au, W. Y. W. (1965). *Proc. Soc. Exp. Biol. Med.* 119, 1048.

Rafsky, H. A., Newman, B., and Krieger, C. I. (1949). *Amer. J. Med. Sci.* 217, 206.

Ram, J. S. (1967). *J. Geront.* 22, 92.

Rapport, R. L., and Curtis, G. M. (1950). *J. Clin. Endocrinol. Metab.* 10, 735.

Reichel, W. (1968). *J. Geront.* 23, 145.

Reiter, R. J., McCreight, C. E., and Sulkin, N. M. (1964). *J. Geront.* 19, 485.

Riegle, G. D., and Nellor, J. E. (1966). *J. Geront.* 21, 435.

Riegle, G. D., Przekop, F., and Nellor, J. E. (1968). *J. Geront.* 23, 187.

Rigby, M. K., Soule, S. D., Barber, W., and Rothman, D. (1964). *J. Geront.* 19, 313.

Ring, G. C., Dupuch, G. H., and Emeric, D. (1964). *J. Geront.* 19, 215.

Ringle, D. A., and Dellenback, R. J. (1963). *Amer. J. Physiol.* 204, 275.

Robertson, T. B. (1916). *J. Biol. Chem.* 24, 363.

Rockstein, M., and Brandt, K. (1962). *Nature (London)* 196, 142.

Rockstein, M., Gray, F. H., and Berberian, P. A. (1971). *Exp. Geront.* 6, 211.

Rodeck, H. (1966). *Proc. Int. Cong. Geront., 7th* p. 153. Egermann, Vienna.

Rodeck, H., Lederis, K., and Heller, H. (1960). *J. Endocrinol.* 21, 225.

Rogers, J. B. (1951). *J. Geront.* 6, 13.

Roman, L., and Strong, L. C. (1961). *J. Geront.* 17, 37.

Romanoff, L. P., Morris, C. W., Welch, P., Grace, M. P., and Pincus, G. (1963). *J. Clin. Endocrinol. Metab.* 23, 286.

Romanoff, L. P., Thomas, A. W., and Baxter, M. N. (1971). *J. Geront.* 25, 98.

Root, A. W., and Oski, F. A. (1969). *J. Geront.* 24, 97.

Rudzinska, M. A. (1951). *Science* 113, 10.

Ruger, H. A., and Stoessiger, B. (1927). *Ann Eugen.* 2, 76.

Samuels, L. T. (1956). "Hormones and the Ageing Process." Academic Press, New York.

Samson, F. E., Balfour, W. M., and Dahl, N. A. (1958). *J. Geront.* 13, 248.

Schaub, M. C. (1963). *Gerontologia* 8, 114.

Schaus, R., and Kirk, J. E. (1956). *J. Geront.* 11, 147.

Schaus, R., Kirk, J. E., and Laursen, T. J. S. (1955). *J. Geront.* 10, 170.

Schmukler, M., and Barrows, C. H. (1967). *J. Geront.* 22, 1.

Schneeberg, N. G., and Finestone, I. (1952). *J. Geront.* 7, 54.

Schol, A. T. (1939). "Mineral Metabolism," Amer. Chem. Soc. Monograph. Van Nostrand Reinhold, Princeton, New Jersey.

Schumacher, S. S., and Premachandra, B. N. (1968). *J. Geront.* 23, 311.

Scott, D. A., and Fisher, A. M. (1938). *J. Clin. Invest.* 17, 725.

Serio, M., Piolanti, P., Romano, S., de Magistris, L., and Guisti, G. (1970). *J. Geront.* **25**, 95.

Shamoto, M. (1968). *J. Geront.* **23**, 1.

Shattuck, G. C., and Hilferty, A. (1932). *Amer. J. Trop. Med.* **12**, 223.

Shaw, K. E. (1952). *Amer. J. Vet. Res.* **23**, 1217.

Shock, N. W. (1952). "Cowdry's Problems of Ageing." Williams and Wilkins Baltimore, Maryland.

Shock, N. W., and Yiengst, M. J. (1950). *J. Geront.* **5**, 1.

Shock, N. W., and Yiengst, M. J. (1955). *J. Geront.* **10**, 31.

Shock, N. W., Watkin, D. M., Yiengst, M. J., Norris, A. H., Gaffney, G. W., Gregerman, R. I., and Falzone, J. A. (1963). *J. Geront.* **18**, 1.

Silverstone, F. A., Brand fon Brener, M., Shock, N. W., and Yiengst, M. J. (1957). *J. Clin. Invest.* **36**, 504.

Simms, H. S. (1942). *J. Gen. Physiol.* **26**, 169.

Sobel, H., and Hewlett, M. J. (1967). *J. Geront.* **22**, 196.

Sobel, H., Hrubant, H. E., and Hewlett, M. J. (1968). *J. Geront.* **23**, 387.

Solderwall, A. L., Kent, H. A., Turbyfill, C. L., and Britenbaker, A. L. (1960). *J. Geront.* **15**, 246.

Soltz, W. B., Horonick, A., and Chow, B. F. (1963). *J. Geront.* **18**, 151.

Sonneborn, T. M. (1930). *J. Exp. Zool.* **57**, 57.

Sonneborn, T. M., and Schneller, M. (1960a). *Biol. Ageing* **6**, 283.

Sonneborn, T. M., and Schneller, M. (1960b). *Biol. Ageing* **6**, 286.

Sonneborn, T. M., and Schneller, M. (1960c). *Biol. Ageing* **6**, 290.

Starr, P. (1955). *J. Amer. Geriat. Soc.* **3**, 217.

Stave, U. (1964). *Biol. Neonat.* **6**, 128.

Stearns, G. (1939). *Physiol. Rev.* **19**, 415.

Steward, V. W., and Brandes, D. (1961). "Structural Aspects of Ageing." Pitman, London.

Storer, J. B. (1967). *J. Geront.* **21**, 404.

Streicher, E. (1958). *J. Geront.* **13**, 356.

Streicher, E., and Garbus, J. (1955). *J. Geront.* **10**, 441.

Strobel, T. (1939). *Z. Ges. Neurol. Psychiat.* **166**, 161.

Sulkin, N. M., and Sulkin, D. F. (1967). *J. Geront.* **22**, 485.

Sulkin, N. M., and Sulkin, D. F. (1968). *J. Geront.* **22**, 485.

Swanson, P., Leverton, R., Gram, M. R., Roberts, H., and Pesek, I. (1955). *J. Geront.* **10**, 41.

Szilard, L. (1959). *Proc. Nat. Acad. Sci. U.S.* **45**, 30.

Tait, J. F., Rosemberg, E., and Pincus, G. (1966). *Proc. Int. Cong. Geront., 7th.*

Talbot, N. B., and Richie, R. (1958). *Ciba Foundation Colloq. Ageing* **4**, 139.

Tannenbaum, A. (1947). *Ann. N.Y. Acad. Sci.* **49**, 6.

Tanner, J. M. (1959). *Ciba Foundation Colloq. Ageing* **5**, 88.

Tauchi, H., and Sato, T. (1962). *J. Geront.* **17**, 254.

Tauchi, H., and Sato, T. (1968). *J. Geront.* **23**, 454.

Templeton, H. A., and Ershoff, B. A. (1949). *Amer. J. Physiol.* **159**, 33.

Thrasher, J. D. (1971). *Exp. Geront.* **6**, 19.

Tonna, E. A. (1973). *Exp. Geront.* **8**, 9.

Tonelli, G., Partridge, R., and Ringler, I. (1965). *Proc. Soc. Exp. Biol. Med.* **119**, 136.

Troup, G. M., and Walford, R. L. (1969). *J. Geront.* **24**, 444.

Trout, E. C., Arnett, W., Hume, E. B., and McGavack, T. H. (1970). *J. Geront.* **25**, 229.

Tuttle, W. W., Horvath, S. M., Preson, L. F., and Daum, K. (1952). *J. Appl. Physiol.* **5**, 631.

Tyler, F. H., Eik-Nes, K., Sandberg, A. A., and Florentin, A. A. (1955). *J. Amer. Geront. Soc.* **3**, 79.

Van Putten, L. M., Van Bekkum, D. M., and Querido, A. (1963). *Acta Endocrinol. Copenhagen* **12**, 159.

Verzár, F. (1957). *Gerontologia* **1**, 363.

Verzár, F., and Spichtin, H. (1966). *Gerontologia* **12**, 48.

Von Hahn, H. P. (1966). *J. Geront.* **21**, 291.

Vranic, M., and Pokrajac, N. (1961). *J. Geront.* **16**, 110.

Walford, R. L. (1967). *Advan. Gerontol. Res.* **2**, 159.

Walford, R. L. (1969). "The Immunological Theory of Aging." Munksgaard Copenhagen.

Walker, D. G., Simpson, M. E., Asling, C. W., and Evans, H. M. (1950). *Anat. Rec.* **106**, 539.

Weaver, J. A. (1955). *J. Pathol. Bact.* **69**, 133.

White, A., and Dougherty, T. F. (1947). *Endocrinology* **41**, 230.

Whiteford, R., and Getty, R. (1966). *J. Geront.* **21**, 31.

Widdowson, E. M., and Kennedy, G. C. (1962). *Proc. Roy. Soc. B* **156**, 96.

Winter, C. A., Sibler, R. H., and Stoerk, H. C. (1950). *Endocrinology* **47**, 60.

Withrow, C. D., Woodbury, D. M., and Wilcox, W. D. (1964). *Amer. J. Physiol.* **206**, 521.

Wulff, V. J., Piekielniak, M., and Wayner, M. J. (1963). *J. Geront.* **18**, 322.

Wulff, V. J., Quastler, H., and Sherman, F. G. (1964). *J. Geront.* **19**, 294.

Yannet, H., and Darrow, D. C. (1938). *J. Biol. Chem.* **123**, 295.

Yiengst, M. J., Barrows, C. H., and Shock, N. W. (1959). *J. Geront.* **14**, 400.

Yu, S. Y., and Blumenthal, H. T. (1963). *J. Geront.* **18**, 119.

Yuan, G. C., Chang, R. S., Little, J. B., and Cornil, G. (1967). *J. Geront.* **22**, 174.

Zorzoli, A., and Li, J. B. (1967). *J. Geront.* **22**, 151.

10

GENERAL SUMMARY

Joseph LoBue and Albert S. Gordon

In the first chapter of this volume, Dr. Maslennikova briefly reviews humoral regulation of insect morphogenesis as a prelude to his interesting analysis of that most intriguing phenomenon, insect diapause and its control. We are reminded that insect ontogenesis is regulated by three neurohormones: activation hormone (AH), molting hormone (MH, or ecdysone), and juvenile hormone (JH). The role of AH is an indirect one since it functions to stimulate prothoracic gland (a source of MH) and corpora allata (a source of JH) secretory activity, with regulation of growth and development depending upon the MH:JH ratio achieved. Special neurosecretory cells in the brain produce AH, the glycoprotein D_1 component of "brain hormone." MH by its well-known gene activation function controls molting and metamorphosis and chemically is a $\alpha-\beta$, unsaturated ketosteroid. JH has diverse actions, but the overall effect of this humoral principle is maintenance of larval characteristics.

Insect diapause is a temporary suspension of the hormonally regulated processes of growth, differentiation, and reproduction. It represents an adaptation for survival under adverse environmental conditions and may occur at the egg (embryonic), larval, pupal, and imaginal stages of insect development. Diapause is *obligate* in some species (univoltine species) of insects, in which case environmental factors effect its duration but not its onset, and *facultative* in other species (polyvoltine species). In these latter insects alterations in temperature, food supply, and photoperiod can act as the stimuli that induce (via the endocrine and nervous system) the onset of diapause. That is, as Dr. Maslennikova states it, environmental information "is translated into hormonal language," which has profound effects upon insect metabolism, thus allowing these animals to resist the otherwise lethal effects of extremes in environmental conditions.

Egg (embryonic) diapause appears to be regulated by a specific humoral principle the "diapause hormone," which has been extracted from subesophageal glands. Its ultimate effect is to reduce metabolism. Secretion of diapause hormone is controlled by neurosecretions from the brain and corpora allata. In addition, corpora allata secretions stimulate metabolism. Thus, whether or not eggs become determined to diapause depends upon the balance that exists between diapause hormone levels and the corpora allata antagonistic neurohumor. The target cells for diapause hormone are oöcytes, and these are most sensitive during the second half of the pupal period. Hence, we find that egg diapause is determined by a set of humoral conditions acting within maternal generations.

Larval, pupal, and imaginal diapause is believed, by some, to be due to hormonal deficiency. However, to what extent such a deficiency is a "consequence" of diapause rather than a cause has not been settled. Specific agents, however, seem to be involved. For example, larval diapause may be controlled through the action of a humor, "proctodon," produced by proctadeal cells. Proctodon stimulates brain hormone secretion and thereby "breaks" diapause. Finally, inhibitors such as "diapause factor," by their inhibitory effect on hormone secretion, may initiate and maintain pupal diapause in facultative insects.

In Chapter 2, Dr. Bode has no difficulty in convincing us that hydra is an extremely useful model with which to study control of growth and differentiation. This is because of this organism's relative simplicity, the occurrence of primitive cells in the adult (interstitial cells), the fact that growth and differentiation go on continuously, and the remarkable regenerative powers that hydra possesses.

Growth of tissue layers in hydra is exponential, and the dynamic steady-state is achieved by exfoliation and bud formation. It is noteworthy, particularly to those interested in problems of cell proliferation, that 90% of all epithelial cells in hydra are in active cell cycle. Since the proliferation rate is not altered after tissue removal, there does not appear to be any need to postulate a humoral regulatory mechanism involving stimulators or inhibitors (i.e., chalone-like agents). Thus, the situation is considerably different than that which obtains, for example, in mammalian liver regeneration.

The mechanism whereby specific cell-type ratios and their gradients are maintained constant over many generations remains to be determined. In some cases a differentiation "sink" may supply a partial answer to the problem. However, at present, there is no evidence for any humoral regulation of this phenomenon.

Regeneration in hydra is polarized, and some evidence has been presented suggesting that diffusible "humoral" regulators may indeed be involved. For example, an agent extracted from neurosecretory vesicles seems to possess hypostome-forming activity, whereas specific inhibitors of hydranth formation have been found to be present in marine hydroids. Thus, it may turn out that regeneration polarity arises from the unequal distribution of stimulatory and inhibitory diffusible regulators.

The interstitial cell is the multipotential stem cell in hydra. This cell is required to differentiate into somatic cell types and gametes, and, when hydra passes over from asexual to sexual reproduction and vice versa, mechanisms must exist to alter interstitial cell differentiation accordingly. Humoral control here is an obviously acceptable concept, and some evidence that diffusible humoral substances are the responsible agents indeed exists.

Hydra is unique in that the cells of the body column are constantly undergoing morphogenic movements. Thus the position of a stem cell determines its ultimate differentiation. Moreover, an abrupt alteration in the location of a cell due to extirpation of adjacent parts above or below that cell has a substantial effect upon the ultimate outcome of differentiation. Unfortunately, it is not known whether humors or gradients of diffusable substances act as causative factors in the manifestation of position effects in this animal.

Sponges, the most primitive of all metazoan animals, present an excellent group in which to study selective cell reaggregation and cell–cell interaction. We are grateful to Dr. Kuhns and his associates for their most interesting and readable account of these phenomena and their humoral control.

Reaggregation in sponges is mediated by aggregation factor (AF). In bioassay, aggregation effects of AF are definitely dose dependent. Chemically, AF is a glycoprotein, the primary sugars of which appear to be galactose, fucose, glucose, and hexosamine. The carbohydrate:protein ratio varies depending upon the species, but protein and intact disulfide groups as well as glucuronic acid are important for its bioactivity. Electron microscopy has indicated that AF is a 20–25 Å spheroid that functionally appears to be arranged into a "supramolecular, sunburst configuration."

AF is effective in aggregating dispersed sponge cells within a given order. Hence, species must be considerably diverse to demonstrate AF taxonomic specificity.

Immunological studies have strongly suggested that an antigen–antibody type of complementarity may account for AF-directed cell reaggregation. Thus, a binding site ("baseplate") has been isolated by hypotonic treatment of sponge cells and baseplate–AF interaction is inhibited by glucuronic acid. This clearly indicates that binding site recognition of AF occurs, in part at least, via this component of the AF molecule. Additional studies have shown that although plant lectins do not agglutinate sponge cells, lectin aggregation of erythrocytes is inhibited by AF. This would suggest the presence of AF-like binding sites on red cell membranes. Similarly, AF-mediated sponge cell reaggregation is inhibited by some lectin preparations. In particular, studies with these plant lectins have indicated complementarity to AF sugars, especially α-methyl-D-mannose and 3-o-methyl-D-glucose.

The cellular slime molds are a truly remarkable group of organisms that undergo an unusual life cycle of growth and differentiation. In Chapter 4, Dr. Bonner engrosses us with a lucid study of this curious developmental story and its humoral regulation.

The spores of *Dictyostelium discordeum* each contain a single uninucleate amoeba. Under suitable environmental conditions these amoebae emerge and begin feeding on soil bacteria. When the food supply becomes exhausted, these organisms enter their "social phase" by humorally controlled aggregation into cell masses. These masses migrate for a time and then transform into sessile, stalked "fruiting bodies" possessing an apical mass of cellulose encapsulated spores. Growth and differentiation is complex and spatially separate, and, as Dr. Bonner indicates, ". . . many of the steps in development . . . are hormone controlled . . . . These factors are essential for morphogenetic movements, pattern formation and differentiation."

The most important humoral agents involved may be loosely classed as morphogenetic movement hormones, pattern hormones, and differentiation hormones. Among the group of agents regulating morphogenetic movements, slime molds seem to secrete cell repulsion hormones that play a role in feeding dispersion and cell attraction hormones ("acrasins") that are important in some species for aggregation during their "social phase" of development. The acrasins are cyclic AMP in some but apparently not all social amoebae. Another movement hormone is a repellent molecule (a cell mass repulsion factor) that induces repulsion of adjacent fruiting bodies during the sessile stage of development. These agents appear to be volatile substances that insure optimal spore dispersal. The chemical nature of these factors is presently unknown.

The pattern hormones are spacing agents that prevent centers of aggregation from developing and thus also ultimately have the effect of encouraging spore dispersal. It is in fact possible that these "spacing substances" may be identical to cell mass repulsion factors.

In some species of slime molds, amoebae occasionally bypass aggregation and fruiting body stages to encyst directly to form microspores. This special developmental behavior is initiated by changes in the environment and is apparently regulated by a differentiating principle, the microcyst-inducing hormone. This is a volatile substance that might actually be NH_3. Cyclic AMP, in addition to acting as an acrasin, may also promote stalk cell differentiation. It is interesting that social amoebae, which utilize cyclic AMP as an acrasin and a differentiating principle, develop slowly but start early; whereas, those species using cyclic AMP solely as a promotor of differentiation develop rapidly but begin to realize their developmental potentials only relatively late in their life history.

Doctor Darden has treated us to a most illuminating description of the phenomenon of sexual differentiation and its control within the genus *Volvox*.

As a preface to his detailed evaluation of *Volvox*, Dr. Darden has reviewed items relating to humoral regulation of sexual development in algae, generally. We thus learn that throughout this group agents serving to attract gametes and mating types of the opposite set are secreted. Many chemical substances such as carbon monoxide, acetylene, and hydrogen sulfide will attract gametes. Natural "gamete attractants" are volatile, and, recently, *n*-hexane, a potent attractant, has been extracted from female tips of *Fucus*. Also, "mating-type agglutinins," which cause flagellar agglutination of compatible gametes, have been isolated from flagella. These are glycoproteins, and male and female substances are chemically different. One cannot help but wonder as to their sugar composition relative to other glycoprotein aggregation agents.

In addition to attractants and agglutinins, certain algae secrete hormones that actually initiate sexual differentiation. This is particularly true in *Volvox*. This organism is a haploid colonial flagellate. The spherical colonies (coenobia) consist of a single layer of chlamydomonad-like cells imbedded within a gelatinous matrix. Reproduction is both asexual and sexual with true fertilization producing a diploid zygospore that returns to the haploid condition meiotically during germination. *Volvox* is a valuable experimental tool because it can be grown axenically in well-defined media and manifests true cellular differentiation. One advantage that *Volvox* possesses is its remarkable simplicity—it develops only two basic cell types: reproductive cells and somatic cells. The reproductive cells are usually able to differentiate into sperm or egg cells, and chemical inducers controlling both male and female gamete formation have been extracted from sexual cultures. It is interesting that asexual colonies exposed to these inducers in fact produce gametes. Gamete inducers are inactivated by heat and proteolytic enzymes but are unaffected by nucleases. Sephadex chromatography and sucrose gradient centrifugation have suggested a heterogeneity of molecular weight of anywhere from 10,000 to 2,000,000. Such a size variation could indicate the occurrence of different species of inducers, membrane binding, or subunit interaction.

The number and kind of reproductive cells that differentiate vary with the species of *Volvox*. In *Volvox carteri*, for example, which is a heterothallic species, unequal divisions occur leading to the production of large gonidial initials and small somatic initials. In both males and females, inducer delays the onset of these asymmetrical divisions. Eventually the large daughter progeny become gamates. *Volvox aureus* M5 is homothallic and dioecious. It does not differentiate special egg cells; rather, the undivided gonidia assume the role of female gametes. Inducer substances in young vegetative colonies of this species cause division of gonidia and formation of male colonies with subsequent sperm packet formation. Hormone treatment here actually results in an initial inhibition of gonidial differentiation since androgonidia are morphologically indistinguishable from somatic cells. *Volvox rousseletii* and *V. gigas* are heterothallic species possessing three types of cells within their vegetative colonies: large and small reproductives and somatic cells. The large reproductive cells usually form new colonies; whereas, the small ones develop into somatic cells. In female strains, inducer substance causes direct gamete formation from reproductive cells; cell division is inhibited, and both large and small reproductives differentiate into eggs. In male strains, inducer has an opposite effect initiating, as it does, division of large reproductive cells and their subsequent differentiation into sperm-producing elements. Small reproductives are also stimulated to divide and produce sperm packets, but it is interesting that this will not happen until after division and differentiation of large reproductives have taken place.

In Chapter 6, Dr. Mertz presents us with a comprehensive treatment of a difficult subject, namely, the humoral control of root growth.

The root is a heterotrophic organ. Most are terrestrial and nonphotosynthetic, and, while supplying micro- and macronutrients to the plant generally, they in turn must be "fed" by their more aerial component parts. Determination of root nutritional rquirements have been technically difficult; however, most roots show defects in the synthesis of vitamins and other growth factors. The root's terminal history is uncomplicated by formation of "lateral appendages"; moreover, the "stages of differentiation are aligned in a linear sequence from apex to base." An interesting feature of the root apical meristem is the fact that a major portion of the central zone of cells is not actively proliferating but rather makes up a so-called quiescent center. In *zea mays,* for example, this quiescent center is composed of roughly 600 cells. Of these, less than 50% undergo mitosis, and half of such cells have a generation time of as long as 40 hours. Contrast this to the meristem peripheral to the quiescent center in which 80% of the cells undergo mitosis, and the generation time is reduced to 14 hours. This quiescent zone, although relatively sluggish metabolically, apparently plays an important role in the production of growth regulators and in root morphogenesis.

Many directed movements in the aerial parts of plants ("tropisms") result from the asymmetrical distribution of growth-regulating hormones. This would seem also to be true for root tropisms. Extirpation of the root tip eliminates the geotropic response, and this is a very old and well-established observation. Recent evidence indicates that the columella cells are the receptors for geotropic stimuli with their contained cytoplasmic amyloplasts acting as the actual "geosensors." Auxins appear to play an important role in this bending of roots toward gravity by stimulating the production of an inhibitory agent (ethylene). Thus, as Dr. Mertz indicates, certain investigators "have concluded that the geotropic curvature is related to an IAA-dependent ethylene production on the lower side of the root." The sources of auxin supply are not totally clear since data from auxin translocation studies have been conflicting. Another complication is the evidence pointing to the existence of a growth inhibitor elaborated by the root cap cells. Just how this factor fits into the scheme of tropic curvature remains to be established.

There is no question that auxins, gibberellins, and cytokinins are essential for root growth and development. Each agent is synthesized in the root and in the aerial parts, and, as indicated earlier, the dominance of source depends upon the direction and efficacy of translocation. Moreover, whether these humoral influences are inhibitory or stimulatory is a function of concentration. Auxins, in particular, have a well-established effect upon cell elongation. They increase both the elasticity and plasticity of the cell wall. Plastic expansion is irreversible since this results from an auxin-induced rupture of the crosslinkages between cellulose microfibrils. Actual elongation is due to inhibition of water and addition of new cellulose fibrils among the old.

Plant hormones have a substantial influence upon root vascular differentiation. Both cytokinins and auxins are necessary for xylogenesis in primary root vascularization. In secondary vascularization, the auxin:gibberellic acid ratio is important. Thus, when the auxin to gibberellic acid ratio is greater than one, xylem formation is favored; otherwise, phloem differentiation predominates.

Sensory receptors in animals are extremely sensitive to exogenous air- and waterborne chemical signals. In Chapter 7, Dr. Happ discusses these "allomones," "kairomones," and "pheromones" in a most illuminating and pleasing fashion.

We learn that chemical signals ("semiochemicals") are usually exocrine secretions. When these are primarily concerned with intraspecific communication, the semiochemicals are referred to as "pheromones"; otherwise, they are "alleochemics". Alleochemics benefiting the organism secreting them are known as "allomones"; those advantageous to the recipient are "kairomones." These agents have been studied most extensively in invertebrates, especially insects; however, a large and rapidly growing literature also exists in the field of vertebrate semiochemicals.

Recipient organisms only respond to chemical signals when they are presented to them in an amount exceeding threshold concentration. Physical space bearing at least this concentration is designated "active space." The shape of active space is affected by many factors (e.g., emitter position and behavior, air and water currents, etc.), and mathematical relationships have been developed to allow computation of active space volume and prediction of molecular size constraints for potential chemical signals.

Allomones by broad definition would include such diverse secretions as antibodies, venoms, and escape substances. For brevity, Dr. Happ has limited his treatment to defense (repellent) and symbiotic allomones. A classic example of the group of small repellent molecules that serve as defense allomones is the formic acid repellent released by certain ants. This is a volatile cytotoxin that is effective in the liquid and gaseous state. To act, cytotoxins of this type must often penetrate the protective coverings of potential predators, and many interesting mechanisms have been evolved to insure this happening. For example, in some species, cytotoxic irritants are mixed with substances that lyse the protective armor of the attacker. Larger repellent molecules include steroids, alkaloids, and apparently even some nucleoproteins. These are generally pharmacologically active, especially the alkaloids, many of which are well-known neurotoxins. An intriguing question, when considering toxic agents, relates to autointoxication. That is, what prevents the repellent from causing injury to the host that synthesizes and presumably stores these dangerous agents? Apparently one answer lies in the existence of a reactor gland system in which a relatively nontoxic precursor is actually converted to the active toxin only upon its release into a more

reproductive potential. This is invariably lower in cell cultures initiated from older animals. A similar diminution of proliferative activity is observed *in vivo* when cells are serially transplanted from increasingly older animals. Such "mortal" behavior is in contrast to that seen in cell "lines" derived from normal cell strains. Cell lines often behave as neoplastic cells when inoculated into recipients and seem to be "capable of indefinite proliferation *in vitro*." Hence, it would appear that these abnormal cell lines do not, in point of fact, age relative to their cytokinetic potential. Dr. Bellamy suggests that one key to this "immortality" may be that these cells have developed a mechanism of "perfect duplication and self-maintenance" possibly by means of "the unmasking of a self-correcting code which is normally only expressed in germ cells."

Turning our attention to postmitotic cells, several theories of aging have been advanced. These include the following:

1. Accumulation of inert metabolites. Materials such as lipofucin granules ("age pigment") and "amyloid deposits" are known to accumulate in certain cells with age. These substances are believed by some to represent metabolic end products formed by the interaction of macromolecules with free radicals. Such collections of inert metabolites could themselves have serious deleterious side effects or may represent evidence of the extent of an aged cell's exposure to the harmful effects of these highly reactive free radicals.

2. Somatic mutation theory. Cellular aging may be the result of a progressive increase in somatic cell mutations. These could lead either to a gradual deterioration of cell function or immediate cell death. In either case the functional state of the tissue would be reduced with time, thereby lowering resistance to death.

3. Gene suppression theory. Aging might decrease a cell's ability to synthesize RNA and protein because of an age-dependent "irreversible binding of repressors to corresponding structural or operator genes." Thus, taken from this vantage point, aging would be the result of a gradual turning off of genes until eventually their numbers became such as to be incompatible with life.

4. Incorrect translation theory. In this potential explanation, aging is considered to be due to faulty protein synthesis, which leads to a state of reduced cellular biochemical efficiency incapable of sustaining life.

5. Epigenetic theory. Highly specialized postmitotic cells are really "surviving" or dying cells. In many instances, such as for skin epidermis, aging and dying are an integral part of the differentiation process. Aging in these cells then may be taken as a genetically programmed physiological by-product of cytodifferentiation.

AUTHOR INDEX

Numbers in italics refer to the pages on which the complete references are listed.

A

Aasheim, T., 129, *143, 144*
Abbo, F. E., 240, *271*
Ackermann, P. G., 227, 234, 248, *271*
Adamo, N. J., 198, *216*
Addicott, F. T., 126, *143*
Addis, T., 231, *276*
Adelman, R. C., 237, *271*
Aetter, E. G., 123, *145*
Akert, K., 207, *216*
Akintobi, T., 103, *118,* 178,*186*
Albert, A., 241, *271*
Albone, E. S., 164, *181*
Alex, M., 228, *276*
Alexander, N., 16, 17, 27, 28, *29*
Alkufaishi, H., 228, 247, 248, *272*
Allegretti, N., 242, *277*
Allen, T. H., 228, *271*
Almestrand, A., 126, *143*
Alpin, R. T., 169, *181*
Altman, N., 237, *272*
Alving, A. S., 239, *276*
Amoore, J. E., 177, 179, *181*
Andersen, D. H., 246, *271*
Anderson, E. C., 228, *271*
Andrewartha, H. G., 3, 8, 10, *29*
Andrews, G. F., 62, *77*
Aneshansley, D., 159, *181*
Anfinsen, C. B., 69, *77*
Angrist, A., 228, *272*
Angervall, L., 235, *271*
Arias, I. M., 234, *274*
Arimura, A., 208, *215*
Arndt, A., 89, *97*
Arnett, W., 240, *278*
Asling, C. W., 237, *279*
Aspinall, D., 135, *145*

Aspinall, R. L., 236, *276*
Atlan, H., 263, *271*
Au, W. Y. W., 242, *277*
Audus, L. J., 128, 130, 132, *144, 145, 146*
August, C. J., 150, *181*
Axelrod, J., 210, *217*
Axelrod, L. R., 241, *271*
Azmitia, E. C., 203, *215*

B

Baard, G., *33*
Baca, Z. V., 242, *271*
Bachledová, E., 231, *273*
Baker, D. H., 210, *216*
Baker, J. R., 158, *181*
Balázs, R., 195, *213*
Balfour, W. M., 267, *277*
Ballantine, J. A., 172, *183*
Barber, W., 248, *277*
Bardach, J. E., 164, 171, *181*
Barkley, D. S., 84, 85, 89, *97*
Barksdale, A. W., 101, *117*
Barlow, G., 67, 68, *78*
Barlow, P. W., 131, 134, *143, 144*
Barnabei, O., 237, *271*
Barnes, R. D., 60, *77*
Barraclough, C. A., 197, 198, 199, *214*
Barras, S. J., 163, 174, *181, 184*
Barrett, L., 208, *215*
Barrows, C. H., 227, 228, 231, 239, 258, 267, 268, *272, 273, 275, 277, 279*
Barth, L., 53, *55*
Barth, R. H., 180, *181*
Bartley, M. A., 126, *145*
Bashey, R. I., 228, *272*
Batali, M., 242, *272*

Baumann, E. J., 247, *276*
Baxter, M. N., 240, *277*
Beauchene, R. E., 239, 267, *272*
Beck, S. D., 3, 9, 16, 17, 27, 28, *29*
Beery, S. J., 12, 14, 27, 28, *31*
Beierwaltes, W. H., 193, *214*
Belasco, I. J., 236, *272*
Bellamy, D., 226, 228, 234, 238, 241, 244, 246, 247, 248, 251, 268, *272*, *273*
Belozerov, V. N., 12, *29*
Bender, A. D., 252, *272*
Benjamin, B., 247, *275*
Berberian, P. A., 244, *277*
Berenson, J., 240, *274*
Berg, B. N., 229, *272*
Bergen, J. R., 205, *214*
Berger, H., 5, *30*, 210, *214*
Bergmann, E. D., 173, *185*
Bergström, G., 171, *185*
Berking, S., 35, 37, 38, 39, 40, 43, 50, 51, 52, *55*, *56*
Bern, H. A., 150, *189*
Bernick, S., 251, *272*
Beroza, M., 170, 171, 177, *181*, *183*
Berry, R. J., 223, *272*
Berthold, G., 102, *117*
Beslow, D. T., 141, *143*
Bettus, J. S., 193, *214*
Bhattacharya, S. K., *273*
Bierl, B. A., 170, 171, *183*
Binnard, R., 263, *271*
Birch, L. C., 8, 10, *29*
Birch, M. C., 169, *181*
Birren, J. E., 232, *272*
Bitonto, G. di, 237, *271*
Bjerke, J., 7, *32*
Bjorksten, J., 238, 263, *272*
Blankenbaker, D., 38, *57*
Blickenstaff, C. C., 21, 27, 28, *29*
Blum, M. S., 157, 159, 164, 165, 166, 168, 177, 179, *181*
Blumenthal, H. T., 228, 241, 242, *272*, *279*
Blumgart, H. L., 245, *274*
Boch, R., 168, 177, *182*
Bodansky, M., 236, *272*
Bode, H., 35, 37, 38, 39, 40, 43, 50, 51, 52, *55*, *56*
Boer, T. A. de, 18, 19, 21, 27, *33*

Bogart, B. I., 268, *272*
Bogdonoff, M. D., 236, *273*
Bohus, B., 208, *214*
Boll, W. G., 126, *143*
Bonner, H., 127, *143*
Bonner, J., 126, 127, 133, *143*, *145*
Bonner, J. T., 82, 84, 85, 86, 87, 88, 89, 91, 92, 94, 97, *98*, 101, *117*
Bonnett, H. T., 130, *143*
Bonsall, R. W., 172, *183*, *186*
Boot, L. M., 174, 176, *185*
Bossert, W. H., 46, *57*, 151, 152, 153, 163, 165, 179, *182*, *190*
Botwinick, J., 232, *272*
Bourne, G. H., 242, *272*, *273*
Bovaird, J., 38, 48, 53, *56*
Bowers, W. S., 21, 27, 28, *29*
Boyd, E., 246, *272*
Boyd, W. C., 71, *77*
Boysen-Jensen, P., 128, *143*
Bózner, A., 231, *273*
Brady, U. E., 169, *182*
Brandes, D., 242, *278*
Brand fon Brener, M., *278*
Brandham, P. E., 102, *117*
Brandt, K., 239, *277*
Brauner, L., 131, *143*
Braunitzer, G., 103, *118*
Brefeld, O., 81, *97*
Brian, P. W., 134, *143*
Briant, R. E., 130, *145*
Brien, P., 36, 37, 46, 47, 60, *56*, *77*
Briggs, F. N., 201, *215*
Britenbaker, A. L., 230, *278*
Britton, V. J., 268, *272*
Broadbent, D., 137, *143*
Brody, S., 226, *272*
Bronson, F. H., 163, 174, 175, 176, 180, *182*, *186*, *190*
Brower, J. V. Z., 169, *182*
Brower, L. P., 169, *182*
Brown, A. N., 227, 245, *276*
Brown, J. B., 240, *272*
Brown, R., 137, *144*
Brown, W. L., Jr., 150, *182*
Browne, E., 46, *56*
Brownlee, R. G., 167, 169, 173, *182*, *189*
Broza, M., 20, *29*
Bruce, H. M., 174, *182*

Bruckner-Kardoss, E., 252, *274*
Bucher, N. L. R., 231, *272*
Buckley, E., 160, *182*
Bücherl, W., 160, *182*
Brumfield, R. T., 124, *144*
Bullock, M., 6, *29*
Bullough, W. S., 38, *56*, 265, *272*
Bulos, B., 237, *272*
Bunge, R. P., 194, *215*
Bünning, E., 9, *29*
Burdette, W., 6, *29*
Burg, S. P., 131, 133, *144*
Burge, K. G., 200, *214*
Burger, M. M., 60, 64, 65, 67, 68, 69, 77, *78*, *79*
Burghardt, G. M., 164, 172, *182*
Burkart, J. F., 194, 195, *215*
Burke, A. W., 161, 162, *183*
Burke, F. G., 241, *276*
Burke, W. E., 232, 238, *272*
Burkholder, W. E., 169, *187*, *189*
Burnett, A. L., 36, 37, 41, 42, 43, 47, 49, 50, 52, 53, *56*, *57*
Burnett, D., 132, *144*
Burokas, S., 67, 68, *78*
Burrows, H., 235, *272*
Burstrom, H., 133, 134, 135, 136, *144*
Burszstyn, H., 202, *217*
Butcher, D. N., 134, 136, *144*
Butenandt, A., 6, *29*, 150, *185*
Butler, C. G., 164, 168, 173, 174, *182*, *183*
Byrd, J. B., 159, *182*
Byrne, J. M., 125, *144*

C

Calam, D. H., 168, *182*
Caldwell, B. M., 248, *273*
Callow, R. K., 174, *182*
Cameron, D. E., 210, *215*
Cameron, G. R., 251, *273*
Cameron, I. L., 258, *273*
Campbell, R. D., 36, 37, 39, 40, *56*
Cane, A. R., 129, *146*
Cardé, R. T., 170, *187*
Carlo, J., 240, *277*
Carlson, A. J., 226, *273*
Carlson, D. A., 170, 171, *183*
Carr, E. A., Jr., 193, *214*

Casper, R., 199, *214*
Caviezel, F., 191, 199, *214*
Cavill, G. W. K., 156, *183*
Cayer, A., 212, *214*, *216*
Ceccarini, C., 92, 97
Celand, R., 133, *144*
Chadwick, A. V., 131, 133, *144*
Chang, R. S., 252, 258, *279*
Chang, Y. P., 133, *145*
Chang, Y. Y., 84, 85, 89, 97
Charriper, H. A., 242, *273*
Chauchard, P., 210, *214*
Chayes, Z. W., 247, *273*
Chefurka, W., 10, *29*
Chen, C. H., 259, *274*
Chester-Jones, I., 234, 241, 246, *272*, *273*, *274*
Chesterman, F. C., 244, *274*
Chi, Y. Y., 86, 97
Chiang, S., 123, *145*
Child, C. M., 40, 41, *56*, 57
Chino, H., 25, *29*
Chiodi, H., 242, *271*
Chiquoine, A. D., 92, 97
Cholodny, N. G., 128, *144*
Chow, B. F., 238, *278*
Christianson, M., 201, *214*
Chu, E. W., 210, *217*
Church, N. S., 16, 17, 27, *29*
Chvapil, M., 268, *273*
Ciesielski, T., 130, *144*
Claret, J., 12, *29*
Clarke, M. R. B., 45, 57
Clarkson, S. G., 37, 38, 44, *56*
Clayton, R. B., 179, *183*
Cleveland, L. R., 161, 162, *183*
Clever, U., 6, *29*
Cloutier, E. J., 17, 28, *29*
Clowes, F. A. L., 124, 125, *144*
Cocks, W. A., 195, *213*
Cohen, L. W., 178, *189*
Cohen, M. H., 92, 94, 97
Coleman, A. W., 103, 104, *117*
Collins, D. D., 231, *273*
Colvin, I. B., 17, *29*
Comeau, A., 169, 177, 178, *187*
Comfort, A., 172, *183*
Conrad, R. A., 232, *273*
Cook, A. H., 102, *117*
Coombe, B., 135, *145*

Coppel, H. C., 167, 177, *186, 189*
Corbet, S. A., 180, *183*
Cornil, G., 252, 258, *279*
Correll, W. W., 237, *273*
Cotter, D. A., 92, *97*
Cranston, F. P., 169, *182*
Cross, B. A., 204, *214, 215*
Crossley, A. C., 159, *183*
Crowder, S. E., 240, *274*
Cuatrecasas, P., 69, *77*
Cuay, G., 260, *275*
Curley, A., 157, *181*
Curtis, A. S. G., *77*
Curtis, G. M., 240, *277*
Curtis, R. F., 172, *183*

D

Dafney, N., 207, *214*
Dahl, N. A., 267, *277*
Dahm, K. H., 7, *32*
Dalderup, L. M., 226, 228, *273*
Daly, J., 160, *189*
Daniel, C. W., 244, *273*
Danilevsky, A. S., 3, 4, 8, 9, 10, 12, *29*
D'Arcy, P. F., 251, *273*
Darden, W. H., 104, 105, 106, 107, 108, 110, 111, 113, *117, 118, 119*
Darimont, E., 136, *144*
Darjo, A., 20, *29*
Darrow, D. C., 244, 245, *279*
Das, B. C., 227, 234, 241, 255, *273*
Dateo, G. P., 171, *188*
Daum, K., 267, *279*
Davenport, V. D., 210, *217*
David, C. N., 35, 37, 38, 39, 40, 43, 50, 51, 52, *55, 56*
Davidson, J. M., 206, *214*
Davies, S. H., 259, *275*
Davis, L. V., 44, *56*
Deason, T. R., 105, *118*
Decker, C. F., 227, *275*
Defendi, V., 60, *78*
Delamore, I. W., 259, *275*
Dellenback, R. J., 241, *277*
DeMaggio, A. E., 140, *144*
de Magistris, L., 240, *278*
De Ome, K. B., 244, *273*
Desaulles, P. A., 237, 243, *273*

Dethier, V. G., 159, *183*
Detwiler, T. C., 267, *273*
Deuben, R., 241, *276*
Deulofeu, V., *182*
Deutsch, S., 249, *276*
de Vellis, J., 202, *214*
De Wied, D., 208, *214*
Dezell, H. E., 174, *182*
Diamond, M. C., 202, *214*
DiBella, F., 86, *97*
Dicker, S. E., 239, *273*
Diehl, N. A., 47, 49, *56*
Diemer, R., 131, *143*
Dierschke, D. J., 200, *215*
Digby, J., 139, *144*
Diwald, K., 104, *118*
Dlouha, H., 243, *275*
Dodd, M. R., 87, 88, *97*
Dodd, R. Y., 63, *78*
Dodson, E. O., 136, *146*
Dodson, V. N., 193, *214*
Doecke, F., 198, *214*
Doerner, G., 198, *214*
Doniach, D., 259, *273*
Donike, M., 103, *118*, 178, *186*
Dontas, A. S., 250, *273*
Doolitle, R. E., 177, *181*
Doskočil, J., 26, *29*
Dougherty, T. F., 247, *279*
Downing, R. D., *145*
Drabkina, A. A., 7, *32*
Draper, H. H., 267, *273*
Drew, W., 177, *186*
Duff, V. B., 236, *272*
Duheille, J., 260, *275*
Duinter, C. S., *33*
Dullaart, J., 132, *144*
Dunihue, F. W., 242, 249, *273*
Dupuch, G. H., 267, *277*
Duvall, L. K., 250, *273*
Dyer, R. G., 204, *214*
Dyrbye, M., 228, *275*

E

Eakin, R. E., 42, 44, *56*
Eayrs, J. T., 194, 195, 197, 196, *213, 214*
Edwards, D. A., 200, *214, 217*

Eibl-Eibesfeldt, I., 164, *183*
Eik-Nes, K., 240, *279*
Eisdorfer, C., 236, *273*
Eisenbraun, E., 177, *186*
Eisner, H. E., 156, 159, *183*
Eisner, T., 150, 156, 157, 158, 159, 160, 169, *181, 182, 183, 186, 187, 188*
Eleftheriou, B. E., 174, *182*
Elson, G. W., 134, *143*
Elvidge, J. A., 102, *117*
Ely, S., 105, *118*
Ely, T. H., 107, *118*
Emeric, D., 267, *277*
Emme, A. M., 9, *29*
Emmerich, H., 19, 21, *29*
Emslie, A. G., 154, *184*
Endo, K., 20, *30*
Endröczi, E., 209, *215*
Enesco, A. E., 268, *273*
Engelman, F., 19, 21, *29*
Engstrom, W. W., 240, *276*
Enzmann, E. V., 267, *273*
Erickson, R. O., 101, *118*
Erlanger, M., 268, *273*
Ershoff, B. A., 226, 250, 251, 272, *273, 278*
Erslev, A. J., 46, 48, *56*
Esau, K., 122, 123, 124, 137, *144*
Evans, H. M., 237, *279*
Everitt, A. V., 228, 236, 248, 249, 250, *273*
Ewen, A. B., 26, *29*
Eymann, H., 45, *57*

F

Fairey, E. M., 174, *182*
Falconi, G., 241, *277*
Faludi, G., 247, *273*
Falzone, J. A., 227, 229, 231, 258, 267, *272, 273*
Fanestil, D. D., 239, 267, *272*
Fawcett, D. W., 46, *57*
Fedorčákova, A. M., 231, *273*
Feeny, P. P., 150, 155, *190*
Feit, I., 89, *97*
Feldman, S., 205, 207, 208, *214, 216*
Fell, P., 62, *78*

Fells, I. G., 227, *275*
Fels, I. G., 223, *273*
Fenger, F., 246, *273*
Finestone, I., 232, *277*
Fisher, A. M., 241, 242, *273, 277*
Fisher, F., 5, *30*
Fisher, M. B., 232, *276*
Fisher, S., 156, *183*
Fitzgerald, D. C., 42, *56*
Fletcher, M. J., 267, *273*
Flexner, J. B., 245, *273*
Flexner, L. B., 194, 195, *215*, 245, *273, 277*
Flick, K., 35, *56*
Florentin, A. A., 240, *279*
Florini, J. R., 268, *272*
Flückiger, E., 231, *274*
Förster, H., 103, *118*
Ford, D. H., 194, *216*
Foskett, D. E., 141, 142, *144, 146*
Fox, M. W., 164, *181*
Frame, B., 210, *215*
Francis, D. W., 86, 88, *97*
Francke-Grosmann, H., 163, *184*
Franks, L. M., 244, *274*
Frantz, A. G., 251, *274*
Fraser, R., 251, *274*
French, A., 131, *145*
Friedebold, G., 211, *216*
Friedfeld, L., 230, *274*
Friedman, C. L., 248, 249, *274*
Friedman, L., 175, *184*
Friedman, M., 236, *274*
Friedman, S. M., 248, 249, *274*
Frisch, K. von, 164, 168, 176, *184*
Frolkis, V. V., 233, 253, *274*
Frost, J., 177, *186*
Fry, W. G., 60, *78*
Fujita, T., 238, *274*
Fukaya, M., 16, 17, 28, 29, *31*
Fukuda, S., 4, 20, 22, 24, 27, 28, 29, *30*
Fulton, C., 45, *56*

G

Gaafer, M. A., 251, *274*
Gabba, A., 164, 165, 166, *184*

Gaffney, G. W., 228, 231, 240, 267, *274, 278*
Gala, R. R., 238, 240, 247, *274*
Galanti, N., 68, *78*
Galey, F., 202, *214*
Galson, E. C., 136, *145*
Galtsoff, P., 61, 63, *78*
Gandolfi, G., 171, *184*
Garber, B., 60, 64, *78*
Garbus, J., 231, *278*
Garciá Argiz, C. A., 195, *214, 216*
Gartner, L. M., 234, *274*
Gary, N. E., 171, *184*
Gasie, G., 68, *78*
Gaspar, T., 136, *144*
Gaston, L. K., 180, *186, 189*
Gee, M. V., 267, *274*
Geel, S., 194, 195, *214*
Gehlbach, F. R., 159, *182, 184, 189*
Geispitz, K. F., 10, 27, 28, *30*
Geitler, L., 103, *118*
Geldiay, S., 20, *30*
Gerall, A. A., 200, *216*
Gerisch, G., 86, *97*
Gersch, M., 4, 5, 6, 7, 10, *30*
Gershon, D., 268, *273*
Gessner, P. K., *214*
Gesteland, R. C., 175, *184*
Getty, R., 242, 262, *274, 279*
Ghent, R. L., 157, 158, 159, 165, *183, 184*
Giarnieri, D., 236, *274*
Gibbons, G. S. B., 130, *144*
Gierer, A., 35, 37, 38, 39, 40, 43, 44, 50, 51, 52, *55, 56, 57*
Gifford, E. M., 123, *145*
Gilbert, L. I., 4, 5, 6, 7, 8, *30*
Gilby, A. R., 158, *184*
Gilligan, D. R., 245, *274*
Gilmore, D. P., 205, *215*
Gitlin, D., 202, *214*
Glaser, G. H., 205, *214*
Gleason, K. K., 172, 173, *184*
Glinos, A. D., 231, *272*
Gold, P. H., 267, *274*
Goldschneider, I., 64, *78*
Goldsmith, M. H. M., 129, *144*
Goldstein, M., 102, *118*
Golubitzkaia, R. I., 237, *276*
Gomez, C. J., 195, *214, 216*

Goodman, M., 259, *274*
Goodwin, B. C., 94, *97*
Gordon, H. A., 252, *274*
Gormon, J. E., 169, *187, 189*
Goryshin, N. I., 4, 5, 9, 18, *29, 30*
Gorzalka, B. B., 197, 200, 205, *217*
Goss, R., 38, 46, 48, *56*
Gottlieb, J. S., 259, *274*
Gottlieb, R., 173, *185*
Goy, R. W., 200, *214, 216*
Grace, M. P., 238, *277*
Grad, B., 230, 238, 240, *274*
Gram, M. R., 228, *278*
Gray, F. H., 244, *277*
Green, M. F., 236, *274*
Green, P. B., 101, *118*
Greenstein, J. A., *190*
Greep, R. O., 246, *274*
Gregerman, R. I., 228, 231, 240, 267, *274*
Gregg, J. H., 87, *97*
Gregory, E., 176, *187*
Gregory, K. O., 241, *276*
Griffiths, M., 242, *274*
Grison, P., 18, 21, 27, *30*
Griswold, R. L., *274*
Grodzinska, L., 210, *215*
Gross, J., 49, *56*
Grossie, J., 251, *274*
Grossman, I. W., 237, *272*
Groves, S., 130, *145*
Gsell, D., 234, *274*
Guardagni, D. G., 175, *185*
Gueldner, R. C., 170, *189*
Guillemin, R., 208, *215*
Guisti, G., 240, *278*
Gutmann, E., 233, 246, *274*

H

Haagen, A., 260, *275*
Habermehl, G., 160, *184*
Hackney, A. L., 87, *97*
Haensly, W. E., 242, *274*
Hainer, R. M., 154, *184*
Haining, J. L., 237, 267, 273, *274*
Hall, D. A., 228, *274*
Hall, E. M., 84, 85, 86, *97*
Halliday, R., 241, *274*
Ham, R. G., 42, 44, *56*

Hamburg, D. A., 205, *214*
Hamburgh, M., 193, 194, 195, *214, 215*
Hamilton, J. B., 226, 248, 250, *274*
Hamilton, R. S., 226, 248, 250, *274*
Hamilton, S., 37, 38, *57*
Hangartner, W., 167, *184*
Hansmann, G., 35, *56*
Hanstein, 124, *144*
Hanzliková, V., 233, *274*
Happ, C. M., 163, 174, *184*
Happ, G. M., 150, 155, 159, 163, 174, 180, *184*
Hardee, D. D., 170, *189*
Harding, H. R., 237, *274*
Harman, D., 263, *274*
Harmiston, C. R., 229, *272*
Harner, R., *215*
Harper, R. A., 89, *97*
Harris, J. H., 175, *185*
Harrison, B. J., 264, *274*
Hartmann, M., 102, *118*
Hartog, M., 251, *274*
Harvey, W. R., 4, 10, *30*
Hasegawa, K., 22, 24, *30, 33*
Hastings, A. B., 227, 245, *274, 276*
Hatotani, N., 196, *215*
Hawk, R. E., 157, *181*
Hawker, L. E., 128, *144*
Hayflick, L., 258, 259, *275*
Hedin, P. A., 170, *189*
Heilborn, J., 102, *117*
Heim, L. M., 199, *215*
Heimsch, C., 123, 125, *144*
Heller, H., 249, *277*
Hellon, R. F., 230, *274*
Hemming, H. G., 134, *143*
Henderson, L. J., 179, *185*
Hendry, L. B., 157, *183*
Henkin, R. I., 206, *215*
Henry, J. P., 229, *275*
Henry, S. M., 161, *185*
Herbeuval, R., 260, *275*
Herman, S. W., 5, *30*
Hertel, R., 128, *144*
Hervey, A., 132, *145*
Hewlett, M. J., 228, *278*
Heyne, A. N. J., 133, *144*
Hicklin, J., 41, 45, 52, *57*
Highnam, K. C., 9, 20, *30*
Hilferty, A., 230, *278*

Hillman, S. K., 129, *144*
Hills, J. I., 175, *188*
Hines, H. M., 244, 248, *275*
Hinke, J. A. M., 249, *274*
Hinks, C. F., 5, *30*
Hinton, H. E., 4, *30*
Hinz, G., 198, *214*
Hlubucek, J. R., 102, *118*
Hoagland, H., 210, *215*
Hochschild, R., 252, *275*
Hodges, J. R., 211, *215*
Hodgkins, J., 232, *275*
Hölldobler, B., 161, *185*
Hoelzel, F. J., 226, *273*
Hoffman, L., 102, *118*
Hoffman, M. E., 89, *97*
Hofinger, M., 136, *144*
Holaday, J. W., 107, *118*
Hollander, J., 267, *275*
Holliday, R., 264, *274, 275*
Holoubek, H., 159, *188*
Holthans, L. M., 7, *32*
Hopkins, T. F., 241, *276*
Hora, J., 102, *118*
Hornbruch, A., 41, 45, 52, *57*
Horonick, A., 238, *278*
Horton, D. L., 244, *275*
Horvath, S. M., 227, 267, *275, 279*
Howard, E. M., 251, *273*
Hrbacek, I., 164, *185*
Hrubant, H. E., 228, *278*
Hrůza, Z., 236, 237, 244, *275*
Hsiao, C., 16, *30*
Hsiao, T. H., 16, *30*
Hubble, D., 236, *275*
Hull, T. Z., 245, *276*
Hume, E. B., 240, *278*
Hummel, H., 167, *185*
Humphreys, S., 68, 75, *78*
Humphreys, T., 60, 63, 67, *78*
Hurst, J. J., 156, 159, *183*
Hutt, W., 108, *118*
Hyde, J. E., 206, *216*
Hymen, L. N., 60, *78*

I

Ibuki, T., 204, *215*
Iijima, T., 23, 25, *31*

Ikan, R., 173, *185*
Ikeda, S., 23, 25, *31*
Inay, M., 246, *275*
Ingham, C., 202, *214*
Ingle, D. J., 251, *275*
Inglish, D., 202, *214*
Iob, V., 244, *275*
Irvine, W. J., 259, *275*
Iversen, T., 129, 130, *143, 144*

J

Jacobs, W. P., 129, 130, 140, *144, 145*
Jacobson, A., 154, *184*
Jaenicke, L., 103, *118*, 178, *186*
Jakobsen, M. E., 223, *272*
Jakovbek, B., 233, *274*
Jakubczak, L. F., 231, *275*
James, J. D., 171, *183*
Janczewski, E. von, 124, *144*
Janney, C. D., 232, 238, *272*
Janssens, P. A., 247, *272*
Jayne, E. P., 242, *272, 275*
Jeff, R. A., 141, *144*
Jelinek, J., 243, *275*
Jelinková, M., 236, 239, *275*
Jennings, H. S., 225, *275*
Jensen, W. A., 125, 128, 137, *144*
Jermier, J. A., 242, *274*
Joerrens, G., 9, *29*
Johnson, C. E., 241, *271*
Johnson, H. D., 250, *275*
Johnson, R. E., 202, *214*
Johnston, J. W., Jr., 151, *185*
Joly, P., 20, *30*
Jones, B. M., 25, 26, *30*
Jones, C. I., 201, *214*
Jones, D. C., 229, *275*
Jones, R. L., 135, *144*
Jones, W. C., 60, *78*
Jost, A., 235, *275*
Juniper, B. E., 130, 131, *144*

K

Kabitza, W., 5, *30*
Kafatos, F. C., 46, *57,* 156, 159, *183*
Kahn, A., 88, *97*
Kaissling, K. E., 149, 176, 177, *185*

Kallmann, F. J., 222, *275*
Kanabrocki, E. L., 227, *275*
Kanaev, I. I., 52, *56*
Kaplan, E., 227, *275*
Kaplun, B., 195, *214, 216*
Karlson, P., 4, 6, *29, 30,* 150, 161, 162, 167, *183, 185*
Karpati, G., 210, *215*
Karsch, F. J., 200, *215*
Kastin, A. J., 208, *215*
Kato, J., 199, *215*
Kato, Y., 134, *145*
Kavalijian, L. G., 125, 137, *144*
Kawakami, M., 203, 204, 205, 207, *215, 216*
Kay, H., 232, *272*
Keeble, F., 128, *145*
Keech, M. K., 228, *274*
Kekawa, N., 25, *32*
Kekwick, R. A., 241, *274*
Keller, G. H. M., 228, *273*
Kemény, A., 241, *275*
Kemény, V., 241, *275*
Kende, H., 136, *145*
Kennedy, G. C., 251, *279*
Kent, H. A., 230, *278*
Kephart, J., 134, *146*
Kernen, R. L., 267, *275*
Kerr, W. E., 166, 167, *185, 186*
Keverne, E. B., 172, 176, 180, *183, 186*
Khalid, R., 240, *274*
Khan, A. A., *145*
Kheim, T., 227, 234, 248, *271, 277*
Kibler, H. H., 250, *275*
Kidman, A. D., 213, *217*
Kiessling, K. H., 239, *275*
Kikuchi, S., 23, 25, *31*
Kimeldorf, D. J., 229, *275*
Kind, T. V., 5, 15, 16, 17, 26, *30, 31*
Kingsbury, J. M., 102, *119*
Kirimura, J., 5, *31*
Kirk, J. E., 228, 268, *275, 277*
Kirk, S. C., 129, 130, *145*
Klee, C. B., 203, *215, 216*
Kline, E. S., 44, *56*
Knight, R. P., 205, *214*
Knobil, E., 200, *215*
Knowlton, G. C., 244, 248, *275*
Kobayashi, K., 5, *31*
Kobayashi, M., 5, 25, *32, 33*

Kochert, G., 105, 107, 108, 109, 110, 111, 113, 114, 115, 117, *118*
Kogure, M., 22, *31*
Kohn, R. R., 238, 275
Kolderie, M. Q., 92, 97
Koneff, A. A., 236, 248, *276*
Konijn, T. M., 84, 85, 89, 97
Konings, H., 130, *145*
Koob, K., 162, *186, 188*
Korányi, L., 209, *215*
Korenchevsky, V., 245, 247, 250, *275*
Kormendy, C. G., 252, 272
Kornfield, D. S., 205, *214*
Kort, C. A. de, 33
Kostowski, W., *215*
Kountz, W. B., 230, 248, *275, 277*
Kovács, S., 195, *213*
Kozanchikov, I. V., 10, *31*
Krag, C. L., 230, 275
Kral, V. A., 230, 240, *274*
Krecek, J., 243, 275
Kreckova, J., 243, 275
Krieger, C. I., 241, 277
Krivanek, J. O., 87, 97
Krivoy, W. A., 208, *215*
Kroeger, H., 6, *31*
Kroll, J. C., 159, *189*
Kubo, K., 203, *215*
Kuhns, W. J., 65, 78
Kullenberg, B., 171, *185*
Kumate, J., 202, *214*
Kunieda, H., 102, *118*
Kurland, G., 10, *31*
Kurnick, N. B., 267, 275

L

Lahiri, A. N., 132, *145*
Laidlan, J., 205, *215*
Lang, A., 135, *145*
Lang, W. C., 227, *276*
Langerham, W. H., 228, *271*
Lansing, A. I., 225, 228, 249, *275, 276*
Lauffer, M., 206, *216*
Laurence, E. B., 38, *56*
Laursen, T. J. S., 227, *277*
Law, J. H., 150, 151, 165, 170, *185*
Lederer, E., 151, 157, 170, 172, *185*
Lederis, K., 249, 277
Lee, S., van der, 174, 176, *185*

Lees, A. D., 3, 4, 8, 9, 10, 27, 28, *31*
Legan, J. S., 267, *274*
Lehn, H., 51, *56*
Leitereg, T. J., 175, *185*
Leith, A., 60, 64, 75, *78*
Le Magnen, J., 172, 180, *185*
Lemon, S. M., 77
Lenhoff, H. M., 38, 48, 53, *56, 57*
Lenicque, P. M., 41, 42, 44, *56*
Lentz, T. L., 41, 43, 46, 50, *56*
Leonard, R. A., 238, 241, 247, *272*
Leopold, A. C., 128, *144*
Lerche, W., 103, *118*
Lesh, G., 41, 42, 43, 53, *56*
Lesser, G. T., 249, *276*
Lester, E. J., 210, *217*
Lettvin, J. Y., 175, *184*
Leverton, R., 228, *278*
Levi, C., 60, *78*
Levine, S., 201, *215, 216*
Lewis, W. H., 239, *276*
Li, C. H., 236, 248, *276*
Li, J. B., 267, *279*
Libertun, C., 198, *216*
Lilien, J. E., 60, 64, *78*
Lilley, J. J., 207, *216*
Lincoln, D. W., 204, *215*
Lind, A. R., 230, *275*
Lindauer, M., 167, *185*
Ling, C. M., 210, *215*
Lis, H., 71, *79*
Lisk, R. D., 199, 204, *215, 216*
Little, J. B., 252, 258, *279*
Lockhart, J. A., 133, *145*
Löfqvist, J., 165, *181*
Logothetis, J., *215*
Loher, W., 174, *185*
Longcope, C., 240, *276*
Lonski, J., 87, 90, 97
Loof, A. de, 19, 21, 28, *31, 33*
Loomis, R. S., 139, 140, *145, 146*
Loomis, W. F., 38, 49, 53, *56, 57*
Losey, G. S. Jr., 171, *185*
Lowry, O. H., 227, 245, *274, 276*
Lozina-Lozinsky, L. K., 10, *31*
Lüscher, M., 150, 174, *185, 186*
Lumia, V., 236, *274*
Lundblad, M., 41, 42, 44, *56*
Lundin, P. M., 235, *271*
Lunsford, W. R., 244, *276*

Lupien, P. J., 244, *276*
Luttage, W. G., 197, 200, 205, *217*
Lynch, R. S., 225, *275*
Lynn, E., 194, *215*

M

McCay, C. M., 226, 227, 244, 245, 249, *276*
McClary, J. E., 126, *145*
McClay, D., 64, *78*
McComb, A. J., 135, *145*
McComb, J. A., 135, *145*
McCracken, M. D., 105, 106, 107, 110, 111, *118*
McCready, C. C., 127, *145*
McCreight, C. E., 231, 239, *277*
McDaniel, C. N., 12, 14, 27, 28, *31*
McDonald, P. G., 205, *215*
McDonald, R. K., 232, *276*
McEwan, B. S., 199, 200, 203, 207, *215*
McFarland, R. A., 232, *276*
McGavack, T. H., 240, *278*
McGuire, J. L., 204, *215, 216*
Mc Gurk, D. J., 177, *186*
McHale, J. S., *276*
McHale, J. T., *276*
McHenry, F., 159, *183*
Machlis, L., 102, 104, *118*
McIsaac, W. M., *214*
McKay, E. M., 231, *276*
McKay, L. L., 231, *276*
Macklin, M., 53, *57*
MacLennan, A. P., 63, 64, 67, 68, 69, *78*
McLeod, D. G. R., 17, 28, *29*
MacNider, W., de B., 231, 232, *276*
McShan, W. H., 7, *32*
MacWilliams, H. K., 46, *57*
Magladery, J. W., 233, *275*
Mainwaring, W. I., 269, *276*
Mainx, F., 102, *118*
Mallory, T. E., 123, *145*
Malmos, B. K., 250, *273*
Manery, J. F., 76, *78*
Manley, O. T., 247, *276*
Marczynski, T. J., 210, *215, 216*
Margoliasch, E., 67, 68, *78*
Marine, D., 247, *276*
Markofsky, J., 226, 249, *276*

Marler, P., 166, *186*
Marsden, H. M., 174, 176, *182, 186*
Martin, M. M., 162, *186*
Martini, L., 191, 199, *214,* 241, *277*
Maschwitz, E., 159, 164, *188*
Maschwitz, U., 159, 160, 162, 166, *186, 188*
Maslennikova, V. A., 12, 13, 27, *31*
Mason, H. L., 240, *276*
Mason, J. W., 85, 86, *97*
Mathew, G. D., 240, *272*
Mathur, S. N., 140, *145*
Matsumura, F., 167, 177, *186, 189*
Mayer, M. S., 170, 171, *183*
Mazzocchi, P. H., 169, *186*
Medina, D., 244, *273*
Medoza, L. A., 194, 195, *215*
Meehan, J. P., 229, *275*
Meinwald, J., 156, 158, 159, 160, 169, *183, 186*
Meinwald, Y. C., 157, 160, 169, *186*
Meites, J., 241, *276*
Menon, M., 180, *186*
Mensh, J. N., 248, *277*
Mertz, D., 135, *145*
Mestler, G. E., 226, 248, 250, *274*
Metcalf, C., 62, *79*
Metz, C. B., 103, *119*
Meyer, A., 7, *31*
Meyer, K., 160, *186*
Meyer, R. K., 236, *276*
Meyerson, B., 207, *216*
Michael, R. P., 172, 176, 180, *183, 186*
Miller, J. G., 175, *184*
Miller, J. H., 232, *276*
Miller, J. S., 259, *274*
Miller, R. E., 208, *216*
Millett, K., 200, *216*
Mills, L. C., 247, *273*
Minyard, J. P., 170, *189*
Miquel, J., 263, *271*
Misztal, S., 210, *216*
Mitsuhashi, J., 16, 17, 28, 29, *31*
Möller, J., 20, *32*
Moguilevsky, J. A., 198, *216*
Money, J., 203, *216*
Montoye, H. J., 232, *277*
Mook, L., *33*
Moon, H. D., 236, 248, *276*
Moore, B. P., 167, *186*

Moore, C. R., 235, *276*
Mooré, D. J., 133, *145*
Moore, R. E., 103, *118*
Moorhead, P. S., 259, *274*
Morales, C., 202, *214*
Morgan, W. T. J., 78
Morin, K. L., 180, *189*
Morley, E. H., 251, *275*
Morohoshi, S., 4, 21, 22, 23, 24, 25, *31*
Morrall, S., 102, *118*
Morrell, F., *215*
Morris, C. W., 238, *277*
Morris, D. A., 130, *145*
Morrison, A. B., 249, *276*
Moscona, A. A., 60, 63, 64, 67, 68, *78*
Moser, J., 164, 166, 167, *186, 189*
Mon, T. R., 175, *185*
Moulton, D. G., 151, *185*
Moure, J. S., 166, *186*
Mouton, M. L., *276*
Müller, D. G., 103, *118, 178, 186*
Müller, W. A., 41, 42, 43, *57*
Müller-Velten, H., 164, *186*
Mugford, R. A., 173, *186*
Muller, E., 241, *277*
Müller, H. J., 4, 10, *31*
Muller-Schwarze, D., 173, *182, 186*
Munro, A., 157, 158, 159, *183*
Murlin, J. R., 236, *272*
Murphy, J. V., 208, *216*
Murray, H. A., 244, *276*
Muscatine, L., 38, *57*
Myant, N. B., 193, 196, *216*
Myers, V. C., 227, *276*
Mykytowycz, R., 163, 172, *186*

N

Nagao, M., 128, *145*
Nagorny, A. V., 237, *276*
Nakashima, N., 248, 249, *274*
Nayer, K. K., 21, *31*
Neales, T. F., 126, *145*
Necco, A., 53, *56*
Needham, A. E., 223, 270, *276*
Needham, J., 244, *276*
Nellor, J. E., 228, 236, *277*
Nelson, M. G., 128, *145*
Newman, B., 241, *277*

Nezamis, J. E., 251, *275*
Nichol, C. A., 237, *274*
Nicholls, P., 135, *145*
Nicoll, R. A., 175, 176, *187*
Niederland, T. R., 231, *273*
Nieman, E. A., 202, *216*
Nissl, D., 133, *145*
Nogueira-Neto, P., 166, *186*
Noller, S., 86, *97*
Norgard, D. W., 7, *32*
Norris, A. H., 228, 231, 232, 233, 267, *276, 278*
Northcote, D. H., 141, *144*
Novak, L. P., 270, *276*
Novák, V. J. A., 4, 5, 6, 7, 10, 16, *31*
Nowell, N. W., 173, *186*
Nunn, J., 239, *273*

O

Oberman, J. W., 241, *276*
O'Brien, J. E., 242, *277*
Odhnoff, C., 140, *145*
Ogawa, N., 208, *216*
Ohata, M., 238, *274*
Ohwaki, Y., 128, *145*
O'Keefe, G., III, 85, *97*
Olausson, B., 133, *144*
Oleson, F. B. Jr., 86, *97*
Olewine, D. A., 268, *272*
Opdam-Stockmann, H., 226, *273*
Orgel, L. E., 264, *276*
Orimo, H., 238, *274*
Orlovskaya, E. I., 10, 27, 28, *30*
Ortmeyer, A. B., 38, 49, *57*
Osborn, G. K., 229, *275*
Oski, F. A., 236, *277*
Overbeek, J. van, 132, *145*

P

Pace, N., *274*
Padovani, F., 157, *181*
Page, I. H., *214*
Paleg, L., 135, *145*
Palmer, B. G., 237, 251, *276*
Palmieri, G., 177, *181*
Panov, A. A., 5, *31*
Papanayiotou, P., 250, *273*
Papanicolaou, N. T., 250, *273*
Paris, S. K., 247, *275*

Park, H. D., 38, 49, *57*
Park, K. E., 22, 25, *31*
Partridge, R., 247, *278*
Pascher, A., 102, *118*
Pasquini, J. M., 195, *214, 216*
Pastan, I., 95, *98*
Pasteels, J. M., 161, *186*
Pauker, J. D., 248, *277*
Pavan, M., 164, 165, 166, *184*
Payne, R. C., 240, *274*
Payne, T. L., 177, 180, *187*
Peakall, D. B., 157, *183*
Pearlstein, A., 242, *273*
Pecile, A., 241, *277*
Pener, M. P., 20, *29*
Peraino, C., 237, *277*
Percy, J. E., 150, 156, 157, *189*
Perlman, R., 95, *98*
Perlmutter, A., 226, *276*
Perry, T., 163, *181*
Pesek, I., 228, *278*
Pessac, B., 60, *78*
Peters, V. B., 245, *277*
Pettus, J. A., 103, *118*
Pfaff, D. W., 176, *187*, 199, 200, 208, *215, 216*
Pfaffmann, C., 176, *187*
Pfeiffer, W., 164, *187*
Pflügfelder, O., 7, *32*
Phillips, H. L., 125, 135, *145*
Phillips, I. D. J., 129, 140, *144, 146*
Phillips, J. G., 241, *272*
Phoenix, C. H., 200, *216*
Piechocki, T., *215*
Piekielniak, M., 227, *279*
Pierson, W. R., 232, *277*
Piette, L. H., 263, *274*
Pilet, P. E., 129, 136, *145*
Pincus, G., 238, 240, 267, *273, 277, 278*
Pintner, I. J., 104, *118*
Piolanti, P., 240, *278*
Pitts, W. H., 175, *184*
Pliske, T. E., 169, *187*
Pokrajac, N., 234, 243, *279*
Pope, F. E., 244, *276*
Porkrajac, N., 242, *277*
Porter, R. W., 208, *214*
Powell, A. H., 236, *273*
Powell, R., 252, *272*
Powers, J. A., 45, *57*

Powers, J. H., 105, *118*
Praetorius, E., 227, *277*
Premachandra, B. N., 232, *277*
Preson, L. F., 267, *279*
Pridantzeva, E. A., 7, *32*
Priesner, E., 149, 176, 177, *185*
Provasoli, L., 104, *118*
Przekop, F., 236, *277*

Q

Quastler, H., 268, *279*
Querido, A., 251, *279*

R

Raabe, M., 5, *32*
Rabadija, L., 242, *277*
Rabkin, M. T., 251, *274*
Radius, R., 77
Radley, M., 134, *143*
Rafsky, H. A., 241, *277*
Raghavan, V., 135, *145*
Rahman, Y. E., 237, *277*
Raisz, L. G., 242, *277*
Ralls, K., 172, *187*
Ralph, C. L., 24, *32*
Ram, J. S., 259, *277*
Raman, G., 193, *214*
Ramussen, H., 77, *78*
Randall, R. V., 241, *271*
Rao, M. A., 236, *276*
Raper, J. R., 178, *187*
Raper, K. B., 81, 82, 88, 89, 92, 97, *98*
Rapport, R. L., 240, *277*
Rasmont, R., 61, 62, *78*
Rasmussen, H., 86, *97*
Ratner, A., 198, *216*
Rawitscher-Kunkel, E., 102, 104, *118*
Ray, P. M., 133, *145*
Rechsteiner-de Vos, H., 226, *273*
Reed, R., 228, *274*
Regnier, F. E., 150, 151, 165, 166, 170, *185, 187, 190*
Rehm, M., 16, *32*
Reichel, W., 228, 262, *277*
Reiners-Decoen, M., 36, 37, *56*
Reinhard, E., 137, *145*
Reiter, R. J., 231, 239, *277*
Remold, H., 160, *187*

Reno, H. W., 159, *184*
Rewerski, W., *215*
Reyniersé, J. H., 172, 173, *184*
Rhines, R., 194, *216*
Rice, S., 241, *276*
Rich, F., 51, 52, 54, 57
Richardson, S. D., 135, *145*
Richie, R., 243, *278*
Richmond, P. A., 101, *118*
Richter, K., 5, *30*
Richter, W., 67, 68, *78*
Riddiford, L. M., 9, 32, 169, 178, 180, *187*
Riegle, G. D., 228, 236, *277*
Rier, J. P., 141, *143*, *146*
Rigby, M. K., 164, *189*, 248, *277*
Ring, G. C., 267, *277*
Ringle, D. A., 241, *277*
Ringler, I., 247, *278*
Robbins, W. J., 126, 132, *145*
Roberts, A. B., 86, 97
Roberts, E. H., 126, *145*
Roberts, H., 228, *278*
Roberts, L. W., 140, *145*
Robertson, A. D. J., 94, 97
Robertson, P. L., 156, *183*
Robertson, R. T., 200, *217*
Robertson, T. B., 242, 246, *277*
Rock, J., 180, *189*
Rockstein, M., 239, 244, *277*
Rodeck, H., 249, *277*
Rodin, J. O., 169, 170, *187*, *189*
Roeder, L. M., 231, 267, 268, *272*
Roelofs, W. L., 169, 170, 177, 178, *187*
Röller, H., 7, *32*
Rogers, J. B., 227, 242, 272, *277*
Roitt, I. M., 259, *273*
Roman, L., 230, *277*
Romano, B., 237, *271*
Romano, S., 240, *278*
Romanoff, L. P., 238, 240, *277*
Root, A. W., 236, *277*
Ropartz, P., 174, *187*, *188*
Rorke, J., 86, *98*
Rose, S. M., 45, 57
Rosemberg, E., 240, *278*
Rosen, F., 237, *274*
Rosenblatt, M., 259, *274*
Rosenthal, G., 86, *98*
Rosenthal, T. B., 228, *276*

Ross, S., 241, *276*
Rossi, A. C., 171, *188*
Roth, L. M., 150, 156, 157, 171, *188*
Roth, S. A., 64, 79
Roth, Z., 268, *273*
Rothman, D., 248, *277*
Rubin, M. A., 210, *215*
Rudinsky, J. A., 163, *188*
Rudzinska, M. A., 226, *277*
Ruesink, A. W., 133
Ruf, K., 205, 207, *216*
Ruffing, F. E., 53, *56*
Ruger, H. A., 232, *277*
Rulon, O., 41, 57
Russell, G. F., 175, *188*
Russell, G. K., 92, *98*
Ruth, J. M., 167, *189*
Ryugo, K., 135, *146*

S

Sachs, R. M., 135, *146*
Sachs, T., 142, *145*
Sachsenmaier, W., 84, 97
Sacktor, B., 237, *272*
Salpeter, M. M., 159, *183*
Samson, F. E., 267, *277*
Samuel, E. W., 84, *98*
Samuels, L. T., 240, *277*
Sanadi, D. R., 267, *273*
Sandberg, A. A., 240, *279*
Sansing, N., 107, *118*
Santisbeban, G. A., 229, *275*
Sapognikova, F. D., 27, 28, *30*
Sato, T., 256, 267, *278*
Saunders, F. J., 199, *216*
Sawyer, C. H., 205, 207, *215*, *216*
Saxl, H., 228, *274*
Sayers, E. R., 104, 110, *118*
Scacchi, P., 198, *215*
Schachar, B. L., 130, *145*
Schaller, H., 35, 37, 38, 39, 40, 41, 42, 43, 44, 45, 50, 51, 52, *55*, *56*, 57
Schally, A. V., 208, *215*
Scharer, B., 21,̇ *32*
Schaub, M. C., 228, 234, *274*, *277*
Schaus, R., 227, *277*
Schenck, H. M., 67, 68, *78*
Schiaffini, O., 198, *216*
Schildknecht, H., 150, 156, 157, 158,

159, 160, 162, *186, 188*
Schmidt, M. B., 126, *145*
Schmukler, M., 267, *277*
Schneck, L., 194, *216*
Schneeberg, N. G., 232, *277*
Schneider, D., 176, 177, *188*
Schneider, I. A., 163, 179, *188*
Schneider, V., 51, 54, 57
Schneiderman, H. A., 4, 7, 10, 11, *30, 31, 32*
Schneller, M., 260, *278*
Schol, A. T., 244, *277*
Schooneveld, H., 19, *32*
Schroeder, M. E., 150, *184*
Schumacher, S. S., 232, *277*
Schwartz, L. S., 207, *215*
Schwinck, I., 168, *188*
Scott, D. A., 241, 242, *273, 277*
Scott, T. K., 129, *145, 146*
Seboek, T. A., 151, *188*
Seelet, B., 201, *216*
Selye, H., 205, *216*
Sereni, F., 237, *271*
Serio, M., 240, *278*
Sexton, W. A., 158, *189*
Shaffer, B. M., 86, 89, *98*
Shamoto, M., 239, 256, *278*
Shane, I. L., 16, 17, 27, *29*
Shapiro, S., 201, *216*
Shapleigh, E., 71, *77*
Sharland, D. E., 236, *274*
Sharon, N., 71, *79*
Sharpless, N. E., 49, *57*
Shattuck, G. C., 230, *278*
Shaw, K. E., 236, *278*
Shearer, D. A., 168, 177, *182*
Sherman, F. G., 268, 272, *279*
Shigemura, Y., 139, *146*
Shock, N. W., 227, 228, 231, 232, 236, 240, 258, 267, *272, 273, 274, 276, 278, 279*
Shoemaker, D. W., 103, *119*
Shorey, H. H., 180, *187, 189*
Sibler, R. H., 250, 251, *279*
Siegel, R. W., 178, *189*
Siew, Y. C., 20, *32*
Silhacek, D. L., 17, 28, 29, 170, 171, *183*
Silva, M. T. A., 208, *216*

Silverman, G., 236, *273*
Silverstein, R. M., 167, 169, 170, 173, *182, 187, 189*
Silverstone, F. A., 236, *278*
Simeone, J. B., 150, 151, *189*
Simms, H. S., 256, *278*
Simpson, J., 173, *183*
Simpson, M. E., 236, 237, 248, *276, 279*
Simpson, T. L., 62, *79*
Singer, A. G., 173, *182*
Sláma, K., 10, 16, 20, *32*
Slautterback, D. B., 46, *57*
Slusher, M. A., 206, *216*
Smith, G. M., 104, *118*
Smith, R. A., 241, *271*
Smrz, M., 239, *275*
Snow, R., 128, *145*
Sobel, H., 228, *278*
Sokoloff, L., 203, *215, 216*
Solderwall, A. L., 230, *278*
Soltz, W. B., 238, *278*
Sondheimer, E., 136, *145,* 150, 151, *189*
Sonneborn, T. M., 260, 262, *278*
Sorokin, H. P., 140, *145*
Soule, S. D., 248, *277*
Sperling, G., 244, *276*
Spichtin, H., 250, *279*
Spiegel, M., 62, 65, *79*
Spindler, K. D., 41, 42, *57*
Sproul, M., 201, *216*
Staal, G. B., *33*
Ställberg-Stenhagen, S., 171, *185*
Stambaugh, R. A., 193, *214*
Staroscik, R. N., 249, *276*
Starr, P., 248, *278*
Starr, R. C., 104, 105, 106, 107, 108, 110, 111, 112, 113, 114, 115, 116, *118,* 178, *189*
Staudt, J., 198, *214*
Stave, U., 239, *278*
Stearns, G., 244, *278*
Steinberg, M., 60, 64, 75, *78*
Steinberg, M. S., 45, *57*
Steiner, F. A., 205, 207, 206, *216*
Stephens, P., 229, *275*
Steward, V. W., 242, *278*
Stiven, A. E., 38, *57*
Stoboy, H., 211, 212, *216*
Stoerk, H. C., 250, 251, *279*
Stoessiger, B., 232, *277*

Stone, B., 202, *214*
Storer, J. B., 252, *278*
Strand, F. L., 211, 212, *214, 216*
Street, H. E., 125, 126, 132, 134, 136, *144, 145, 146*
Strehler, B. L., 267, *274*
Streicher, E., 228, 231, *278*
Streifler, M., 205, *216*
Stretér, F. A., 249, *274*
Strobel, T., 245, *278*
Strong, L. C., 230, *277*
Stroo, M. M., 228, *273*
Stuart, A. M., 164, 166, *189*
Stuchliková, E., 239, *275*
Sulkin, D. F., 231, 256, *278*
Sulkin, N. M., 231, 239, 256, 277, *278*
Supnieski, J., 210, *216*
Suto, S., 102, *118*
Suzuki, M., 25, *32*
Swanson, P., 228, *278*
Swanson, W. W., 244, *275*
Sweely, C. C., 7, *32*
Szilard, L., 221, *278*

T

Tacheci, H., 160, *188*
Tai, A., 167, 177, *186, 189*
Tait, J. F., 240, *278*
Takami, T., 25, *32*
Takeuchi, S., 22, *30*
Talbot, N. B., 243, *278*
Tamarina, I. A., 4, 7, *32*
Tanaka, Y., 22, *32*
Tannenbaum, A., 226, *278*
Tanner, J. M., 250, *278*
Tanton, J., 193, *214*
Taranetz, M. N., 27, 28, *30*
Tardent, P., 45, 48, 49, 51, 52, 54, *57*
Tauchi, H., 256, 267, *278*
Taylor, S. H., 194, *214*
Teasdall, R. D., 233, *276*
Teichgraeber, P., 195, *213*
Teitebaum, S., 259, *275*
Templeton, H. A., 226, *278*
Teranishi, R., 175, *185*
Terasaw, E., 204, *215*
Thimann, K. V., 127, 129, 132, 140, *144, 145, 146*
Thoman, E. B., 201, *216*

Thomas, A. W., 240, *277*
Thomasi, V., 237, *271*
Thompson, A. G., 170, *189*
Thompson, C. W., 232, 238, *272*
Thompson, K. W., 246, *275*
Thomsen, E., 20, *32*
Thomsen, M., 20, *32*
Thomson, P. G., 130, *145*
Thorn, G. S., 205, *216*
Thrasher, J. D., 230, *278*
Tietz, A., 136, *146*
Timiras, P. S., 194, 195, 196, 197, 199, 203, 205, 210, *214, 215, 216, 217*
Todd, J. H., 164, 171, *181*
Todt, J. C., 208, *214*
Tokuyama, T., 160, *189*
Tonelli, G., 247, *278*
Tonna, E. A., 258, *278*
Torii, S., 228, *272*
Torres, F., *215*
Torrey, J. G., 122, 123, 125, 130, 131, 135, 137, 138, 139, 140, 141, 142, *143, 144, 145, 146*
Toube, T. P., 102, *118*
Travis, J. R., 159, *182*
Trenkner, E., 35, 37, 38, 39, 40, 43, 50, 51, 52, *55, 56*
Tripp, K., 36, 37, *57*
Trost, B. H., 7, *32*
Troup, G. M., 232, *278*
Trout, E. C., 240, *278*
Truman, J. W., 9, *32*
Tschermak-Woess, E., 102, *119*
Tschinkel, W., 150, *189*
Tsubo, Y., 102, *119*
Tucker, R. G., 108, *118*
Tumlinson, J. H., 167, 169, 170, *182, 189*
Tunbridge, R. E., 228, *274*
Turbyfill, C. L., 230, *278*
Turk, A., 151, *185*
Turner, C. W., 251, *274*
Turner, M. D., 237, *273*
Turner, R. S., 60, 64, 65, 67, 68, *79*
Tuttle, W. W., 232, 238, 267, 272, *279*
Tweedall, K. S., 45, *57*
Twitty, V. C., 84, 98, 171, *189*
Tyler, A., 65, *79*
Tyler, F. H., 240, *279*

Tyshchenko, V. P., 4, 5, 9, 28, 29, 30, 32

U

Uhrstrom, I., 133, 144
Umeya, Y., 22, 32
Under, H., 5, 30
Ushatinskaya, R. S., 8, 10, 18, 32

V

Vacek, Z., 243, 275
Valcana, T., 197, 199, 216
Valenta, J. G., 164, 189
Valentine, J. M., 150, 189
Van Bekkum, D. M., 251, 279
Vande Berg, W. L., 105, 106, 107, 110, 111, 113, 119
Van de Meene, J. G. C., 84, 85, 97
Van Hof-Van Durin, J., 211, 216
Van Marthens, E., 202, 217
Van Putten, L. M., 251, 279
Vecsei, P., 241, 275
Vernadakis, A., 199, 202, 214, 216
Vernikos, J., 211, 215
Verzár, F., 231, 250, 257, 274, 279
Vick, K., 177, 186
Vierling, J. S., 180, 189
Villee, C. A., 199, 215
Vinogradova, E. B., 26, 32
Viosca, S., 208, 215
Vite, J. P., 170, 189
Von Hahn, H. P., 234, 264, 279
Vranic, M., 234, 242, 243, 277, 279

W

Waku, J., 16, 17, 32
Walford, R. L., 232, 259, 278, 279
Walker, D. G., 237, 279
Walker, P. B., 84, 97
Walker, S. M., 210, 217
Walton, D. C., 136, 145
Wang, J. C. H., 150, 184
Wangermann, E., 139, 144
Wanke, E., 177, 181
Ward, I., 201, 217
Wareing, P. E., 140, 146
Waterhouse, D. F., 158, 159, 183, 184

Watkin, D. M., 228, 231, 267, 278
Watkins, J. F., II, 159, 182, 184, 189
Watkins, W., 79
Watson, R., 248, 273
Wayner, M. J., 227, 279
Weatherston, J., 150, 156, 157, 189
Weaver, J. A., 247, 279
Webb, C., 228, 273
Webb, D., 60, 79
Weber, N. A., 162, 189
Weber, R. J., 232, 238, 272
Webster, G., 37, 38, 40, 44, 45, 46, 52, 54, 57
Weedon, B. C. L., 102, 118
Weil, F., 194, 195, 215
Weinbaum, G., 64, 65, 67, 68, 69, 79
Weis, F. H., 160, 188
Weiss, B., 213, 217
Weiss, E. P., 194, 195, 215
Weiss, J. M., 203, 207, 208, 215, 216
Weiss, P., 65, 79
Welch, P., 238, 277
Wenneis, W. F., 160, 188
Went, F. W., 127, 146
Weston, V. A., 64, 79
Westphal, U., 238, 240, 247, 274
Wetmore, R. H., 141, 146
Whalen, R. E., 49, 57, 197, 200, 205, 217
Whaley, W. G., 134, 146
Wheeler, J. W., 150, 155, 184
White, A., 247, 279
White, P. R., 126, 146
Whiteford, R., 262, 279
Whitmoyer, D. I., 207, 216
Whittaker, R. H., 150, 155, 182, 190
Whitten, W. K., 174, 175, 176, 180, 190
Widdowson, E. M., 251, 279
Widom, B., 159, 181
Widom, J. M., 159, 181
Wiese, L., 102, 103, 104, 118, 119
Wiese, L. S., 103, 119
Wigglesworth, V. B., 4, 6, 7, 8, 32, 33
Wightman, F., 137, 144
Wilby, O. K., 45, 57
Wilcox, W. D., 234, 279
Wilde, J. de, 4, 9, 18, 19, 21, 27, 28, 31, 33
Wilkins, M. B., 127, 129, 130, 144, 145, 146

Williams, A. W., 259, *275*
Williams, C. M., 4, 7, 10, 11, 27, *29, 32, 33,* 180, *187*
Willson, C., 150, *189*
Wilson, E. O., 150, 151, 152, 153, 154, 161, 163, 164, 165, 166, 167, 173, 174, 175, 178, 179, 180, *182, 187, 190*
Wilson, H. V., 61, 62, *79*
Wimersma Greidanus, Tj., B., van, 208, *217*
Winter, C. A., 250, 251, *279*
Withrow, C. D., 234, *279*
Witkop, B., 160, *189*
Wolfe, P. B., 85, *97*
Wolpert, L., 37, 38, 41, 45, 52, 54, *56, 57*
Wood, D. L., 170, *189*
Wood, M. J., 228, *274*
Woodbury, D. M., 197, 202, 203, 206, 210, *216, 217,* 234, *279*
Woodward, R. B., 160, *190*
Woolley, D. E., 205, *217*
Wostmann, B. S., 252, *274*
Wright, E. B., 210, *217*
Wright, R. H., 177, *190*
Wulff, V. J., 227, 268, *279*
Wurtman, R. J., 210, *217*
Wylie, A. W., 135, *146*

X

Xhaufflaire, A., 136, *144*

Y

Yamada, M., 177, *190*
Yamaguchi, N., 210, *215*
Yamashita, O., 24, *33*
Yamazaki, M., 5, *33*
Yang, D., 136, *146*
Yannet, H., 244, 245, *279*
Yarbrough, J. D., 107, *118*
Yates, I., 113, *118*
Yeomans, L. M., 128, *146*
Yiengst, M. J., 227, 228, 231, 232, 236, 267, *272, 276, 278, 279*
Yoshikawa, M., 238, *274*
Yoshitake, N., 22, 25, *31*
Young, J., 177, *186*
Young, L. J. T., 244, *273*
Young, W. C., 200, 204, *216, 217*
Yu, S. Y., 228, *279*
Yuan, G. C., 252, 258, *279*

Z

Zamenhof, S., 202, *217*
Zeigler, J. R., 102, *119*
Zenk, M. H., 133, *145*
Zigmond, R. E., 200, 203, *215, 216*
Zimmerman, W., 106, *119*
Zinsmeister, H. D., 132, *144*
Zongker, J., 53, *56*
Zorzoli, A., 267, *279*
Zumstein, A., 48, 54, *57*
Zysin, U. S., 7, *32*

SUBJECT INDEX

A

Abscisic acid (ABA), 136
"Acrasin", 85, 86, 91
"Acrasinase", 86
Activation hormone
  brain hormone component, 5
  effect on
    corpora allata, 5
    metabolism, 5
    prothoracic gland, 5
  neurohormone, 5
  source, 5
"Adoption gland", 161
Age involution
  thymus
    corticosteroids in, 247
    sex hormones in, 246
    thyroid in, 247
Age pigment
  chemistry, 262
Aggregating factor (AF), 62–66
  anti-AF antibodies, 62
  assay, 67
  cell receptor sites, 69–71
  chemistry, 67–69
  effect on plant agglutinins, 71–73
  effect of sugars, 73
  glucuronic acid and bioactivity, 68
  preparation, 66
  properties, 66–69
  specificity, 64
Aging
  "accumulated waste" theory, 262–263
  acid–base balance, 227
  age involution, 228–229
  "age pigment", 228
  "amyloid deposits" in, 262
  autoimmunity in, 259
  sex differences, 259–260
  cardiovascular effects, 229

cell cycle in, 230
cell proteins, 238
cellular, 257–269
  aerobic metabolism, 267
  anaerobic metabolism, 267
  biochemical constancy, 266
  "clonal selection" theory, 259
characteristics, 224
cholesterol levels, 228
chromosomal abnormalities, 263
connective tissue degeneration, 257
defined, 220
dehydration in, 244
development, 220
drug toxicity, 231
effect of
  castration, 250
  corticosteroids, 251–252
  diet, 249
  growth hormone, 249
  hypophysectomy, 250
  prednisolone, 251
  temperature, 250
  thyroidectomy, 250
effect on
  biological variability, 255–256
  cell cultures, 258
  effector organs, 254
  extracellular substances, 256
  macromolecules, 256
  mitotic cells, 258–261
  postmitotic cells, 261–269
endocrine system in, 233–253
  changes in endocrine gland activity,
    239–243
  changes in endocrine gland size, 242
  functional changes, 243
  histological changes, 242
  response of target tissues, 236–239
    ACTH, 236
    growth hormone, 236

hormone metabolism, 238
hormone receptors, 238
hormone uptake, 238
insulin, 236
noradrenalin, 236
steroids, 237
thyroid hormone, 236
thyrotropin, 236
vasopressin, 236
enzyme induction, 237
enzymes, 238–239, 266–269
epigenetic theory, 265–269
evolution, 269–270
faulty translation in, 264–265
gene suppression theory, 264
graft-versus-host reaction, 232
homeostasis in, 226–256
hormonal control, 244–253
parabiosis experiments, 244
tissue involution, 244–248
hormone deficiency in, 248–249
hormones and pharmacological control, 249–252
"hydration" in, 245
inert metabolites in, 262
leukocyte count, 228
lipofuscin granules, 262
maturation rate in, 226
metabolic rate, 230
multifactorial theory of, 224
muscular atrophy, 245
nature, 223–224
neurophysiological adjustment in, 232
nuclear DNA, 269
nuclear RNA, 267–268
organic constituents, 227, 228
physiological, 226–233
protozoa, 260
regeneration in, 223
renal physiology, 232
reproduction, 229
rotifers
egg development, 225
environmental factors, 225
somatic mutation theory, 263–264
somatic mutations, 263
susceptibility to infection, 232
temperature stress, 230
tissue regeneration, 231
tissues, 227, 256–257

traumatic injury, 231
unitary hypothesis, 224
vascular changes, 254
water and mineral balance, 227, 228
Aging factor, 225
"Aging genes", 265
Aging rate
relation to development rate, 249
Algae
gamete attractants, 102
mating-type agglutinins
location, 103
physicochemical characteristics, 103
sexual differentiation, 102–104
Alleochemic
defined, 150
Allomones, 155–163
appeasement substances, 161
defensive, 156–160
defined, 150, 155
ecdysone, 161
nucleoprotein, 160
promoting symbiosis, 161–163
small repellent molecules, 156–160
ants, 156
penetration of integumen, 158
prevention of self-intoxication, 158
"reactor glands", 159
steroids and alkaloids, 160
Androgens
hypothalamic maturation, 197
Anemotaxis, 168
Antiandrogenic drugs, *see* cyproterone acetate
Archeocytes, 61
Auxins, 127, 132
cell elongation, 133
cell wall effects, 133

B

Barban, 134
Bioassay
gamete inducers in algae, 107
Bivoltine insects, 21
photoperiod effects, 22, 23
temperature effects, 22, 23
Bombykol, 176
Brain development
critical period, 195

sex hormone binding sites, 199–200
Brain hormone
  chemistry, 5
Bruce effect, 174

C

Calyptrogen, 124
Castration
  effect on patterns of sexual activity,
    197
Cell
  aggregation, 62–66
Central nervous system
  effect of
    ACTH, 205–208
    adrenal cortical hormones, 205–208
    cortisol, 206
    deoxycorticosterone, 206
    glucocorticoids, 206
    insulin, 208–210
    melanocyte-stimulating hormones,
      208
    melatonin, 210
    parathormone, 210
    sex hormones, 203–205
    thyroid deficiency, 202
    thyroxin, 202–203
Chalones, 38
Chemical ecology, 150, 180
Chemical signals
  "active space", 152
    maximum value, 153
    optimum value, 153
  characteristics, 151–154
  environmental noise, 153
  mathematical model of signal trans-
    mission, 151–154
  methods of study, 154–155
  volatility, 153
Coenobia, 105
"Copulin", 172
Corpora allata, 5, 6
  extirpation, 5
  implantation, 7
Corpora cardiaci, 5
Cyclic AMP
  as acrasin, 85, 94
  as cell attractant, 85
  cell permeability effects in slime mold,
    86

role in slime mold differentiation, 90
  as "second messenger", 95–96
  significance in evolution of hormonal
    control, 95–96
  spacing patterns in social amoebae,
    89
Cyproterone acetate, 197
Cytokinins, 136

D

Dermatogen, 124
Dexamethasone
  effects on central nervous system, 206
Diapause
  characteristics, 8
  definition, 3
  ecological and physiological features,
    8–11
  photoperiodic regulation, 9
  resistance to environmental extremes,
    10
Diapause factor, 14
Diapause hormonal control, 26, *see also*
    Pupal diapause, Imaginal diapause,
    Larval diapause, Embryonic
    diapause
  bifactorial principle, 26–29
Diapause-inducing hormone
  source, 22
Diapausing and nondiapausing eggs
  biochemistry, 25
DNA synthesis
  developing brain
    thyroid hormone, 195
  hydroxyurea, 134
  kinetin, 141
  root meristem, 125

E

Ecdysone
  α and β forms, 6
  as allomone, 161
  chemistry, 6
  mechanism of action, 6
  source, 6
  target tissues, 6
Electroantennogram, 176

Electroencephalogram
  effect of
    insulin, 208
    melatonin, 210
    steroids, 205
Embryonic diapause
  hormonal control, 21–26
  maturation genes, 23
  moltinism genes, 23
  voltinism genes, 23
Endocrineurology, 192
Environment and life span, 224–226
Enzyme induction
  effect of
    aging, 237
    steroids, 237
Estrogens
  effect on
    hypothalamic maturation, 197
    hypothalamus, 204
    lordosis reflex, 204
Exocrines versus endocrines, 178–180

F

Feedback control
  ACTH secretion, 207–208
  nematocyst differentiation in hydra, 48
"Female protein", 19
Founder cell, 86

G

Gamete chemotactic attractants
  algae, 102–103
  chemistry, 102–103
  female, 102
  male, 103
Gamones, 178
Generation time
  digestive cell (hydra), 39
  effect of aging, 230
  epithelial muscle cell (hydra), 39
  interstitial cells (hydra), 39
  root meristem, 125
  slime mold amoebae, 82
  volvox, 104
Geotropism in roots
  auxins in, 131
  geosensors in, 130

inhibitor substances in, 130–131
receptor cells in, 130
Gibberellins, 134
Gonadotrophin secretion
  centers controlling, 198
  sex hormone effects, 198
Gonidia, 105
Ground meristem, 123
Growth
  hormonal control in insects, 4

H

"Heteropolarity factor"
  interstitial cell differentiation, 53
Histogen concept, 124
Homeostasis during aging 226–256
Hydra
  growth, 36–40
  growth patterns of tissue layers, 36
  growth rate
    environmental factors, 38
  growth zone, 36
    evidence against, 37
  "head" regeneration, 40–46
  hypostome formation
    inhibitors, 44–46
    promotion substances, 41–44
  inhibition of sexual reproduction in,
    49
  interstitial cell differentiation, 46–50
  interstitial cells as stem cells, 46
    humoral control of differentiation,
      47–50
    multipotential nature, 47
  migration of tissue layer cells, 36
  model system for study of growth and
    differentiation, 35
  position dependent cell differentiation,
    50–53
  tissue layer cell ratios, 38–40
  wounding and mitotic rate, 38

I

IAA, 132
  cell elongation, 133
  cell proliferation, 134
  transport in roots, 128–130
Imaginal diapause

effect of
  allatectomy, 18
  implantation of brain, 18
  implantation of corpora allata, 18
  hormonal control, 18–21
  inhibitors, 21
  photoperiod effects, 18
  syndrome of hormonal deficiency, and
    arguments against, 20
Immunology of sponge cell–cell inter-
    action, 71–75
  antigen-antibody forms of comple-
    mentarity in AF-receptor site ad-
    hesion, 74
Indole-3-acetic acid, *see* IAA

J

Juvenile hormone, 6–8
  action, 6
  chemistry, 7
  effect on
    gonads, 7
    metabolism, 7
    prothoracic gland, 7
  source, 6

K

Kairomones
  defined, 150
Kinetin, 136

L

Larval diapause
  effect of
    α-ecdysone, ceasterone, ecdysterone,
      and inocosterone, 16
    brain implantation, 16
    parabiosis with postdiapause
      partner, 16
    prothoracic gland implantation, 17
    removal of corpora allata, 17
    removal of corpora cardiaca, 17
  hormonal control, 16–18
  inhibitors, 17
  termination, 16
Lee-Boot effect, 174
Life span

city versus rural living, 222
genetic control, 221, 223
identical twins, 222
sex chromatin in, 223
sex differences in, 222
Life tables, 221–223
  parameters required for construction,
    222
Liver regeneration
  aging effects, 231

M

Meristem
  apical initials, 122, 124
  quiescent center of, 124–125
Mitotic activity
  effect of
    IAA, 134
    kinetin, 141
    root meristem, 125
    xylogenesis, 142
Mitotic cells, 257
Monovoltine insects, *see* Univoltine
  insects
Morphogenesis
  hormonal control in insects, 4
Morphogenetic movements, 84–88
  cell attraction, social amoebae, 85–86
    bacterial substances in, 84
    cyclic AMP in, 85
  cell repulsion, social amoebae, 84–85
  hormones affecting, 84–88
  hydra, 36
  repulsion of cell masses, social amoe-
    bae, 86–88
    hormones, 87

N

Neurendocrine centers in insects, 17
Neurogenesis
  criteria for hormonal involvement in,
    192–193
  effect of
    ACTH, 201–202
    adrenal cortical hormones, 201–202
    adrenalectomy, 201
    growth hormone, 202
    phenylthiouracil, 194

sex hormones, 197–200
thyroidectomy, 194
thyroxine, 193–197
hormonal involvement, 192–202
Neurosecretory cells
control of aging, 244
hydra, source of promotor substances
in hypostone formation, 43
insects, 5, 19
Neurotoxins
as allomones, 160
amphibia, 160

O

Olfaction, 175
stereochemical theory, 177
vibrational theory, 177
Ontogenesis
in holometabolous insects, 4
Operator genes, 264

P

Parasites, "social", 161
Parthenospores, 110
Periblem, 124
Peripheral nerves
effect of
ACTH, 210–212
adrenal cortical hormones, 210–212
Pheromones, 163–173
defined, 150, 163
alarm substances ("Schreckstoff"),
164–166
chemistry, 165
interspecific, 166
"colony odors" of hymenoptera, 172
lower plants, 178
perception, 175–178
airborne pheromones, 175
receptor protein, 178
primers, 163, 173–175
in caste development in social insects, 174
in mammals, 174
in reproduction, 173
"propaganda substances", 166
recognition scents in vertebrates, 173
recruitment substances, 166–168

stationary scent marks, 166
trail substances, 166
releasers, 163
sex substances, 168–172
ants, 168
aphrodisiacs, 169, 171–172
signal ambiguity, 169
territoriality scents, 172
Phialopore, 105
Phosphodiesterase
role in chemotaxis, 86
Plant hormones, 127–137
Cholodny-Went theory, 127–130
Darwin and, 127
geotropic response, 130–131
quiescent center in production of, 137
root elongation, 131–137
effects of
abscisic acid, 136–137
auxin, 132–134
cytokinins, 136
gibberellins, 134–135
vascular differentiation, 137–142
pattern, 137–139
primary vascularization, 140–142
effect of sugars, 140–141
secondary vascularization, 139–140
Plerome, 124
Polyvoltine insects, 9
Postmitotic cells, 258
Primary meristems, 123
Procambium, 122
Proctodon, 17
Progesterone
anesthetic effect, 205
hypothalamic maturation, 199
lordosis reflex, 204
Promeristem, 122
Protein synthesis
developing brain
thyroid hormone, 194
thyroidectomy, 194
Prothoracic gland, 5, 6
effect of extirpation, 6
implantation, 11
Protocerebrum, 5
Protoderm, 122
Protozoa
aging in, 260

differential gene expression, 261
Pupal diapause
  effect of extirpation of
    brain, 12
    subesophageal ganglion, 12
  effect of implantation of
    brain, 11
    corpora allata, 12
    gonad, 12
    prothoracic gland, 11
    subesophageal ganglion, 12
    thoracic ganglia, 12
  endocrine hypofunction as cause, 12
  hormonal control, 11–16
  parabiosis experiments, 11
  specific inhibitors in, 12–15
  temperature effects, 11–12

R

Receptor cells
  odor generalists, 176
  odor specialists, 176
Regeneration
  effect of aging, 223
  hydra, 40–46
    gradients of diffusible substances,
      41
    polarity, 40
Repellent hormone in social amoebae,
  85
Repressors, 264
RNA synthesis
  effect of ecdysone, 6
Root nutritional requirements, 125–127
Root organization, 122–125
  apical meristem, 123
Root tip
  synthesis of
    auxins, 122
    cytokinins, 122
    gibberellins, 122
Ropartz effect, 174

S

"Schreckstoff", *see* Pheromones
"Seducin", 171
Selective reaggregation, 60
Semiochemicals

as distinguished from hormones, 150
Senescence
  defined, 220
Senile purpura, 257
Sex steroids
  life-shortening effect, 248
Sexual behavior
  effect of sex hormones in development
    of, 200
Sexual differentiation
  algae, 102–104
    histones, 108
    hormones in, 103
    inducers, 106–107
Slime mold
  advantages for studies of development,
    82
  life cycle (*D. discoideum*), 82
  spore germination, 82
  summary of humoral agents in, 92–93
Slime mold differentiation
  cyclic AMP, 90
  hormones in, 90–92
    microcyst inducing hormone, 90
  sequence of hormone action in, 93–94
Somatic mutations, 263
Spacing hormone, in social amoebae,
  88–90
Spacing patterns in social amoebae
  acrasins in, 88
  cyclic AMP in, 89
  factors involved, 88–89
Sponges
  cell aggregation
    autoradiographic analysis, 64
    species specificity, 63
    temperature effects, 63
  differentiation and development, 60–62
  gemmules, 61
  general characteristics, 59
Spore germination
  inhibitory substances in, 92
Stem cells, 46
  archeocytes in sponge, 61
Stenotele, 48
  position dependent differentiation,
    50–51
Steroids
  anesthetic effects, 205
  central nervous system effects, 203–208

Structural genes, 264
Subesophageal ganglion
  source of diapause-inducing hormone,
    22
Symbiosis
  allomones in, 161

T

Thyroid deficiency
  brain development in, 193
Thyroid hormone
  effect on
    amino acid incorporation in brain,
      203
    brain mitochondria, 203
    cultured neurons, 194
    electrical activity of brain, 196–197
    fetal rat, 194
    neurogenesis, 193
Thyroxin
  in treatment of athyreotic fetuses, 193
Tritiated thymidine autoradiography
  root growth, 124–125

U

Univoltine insects, 9, 11

V

Volvox

"androembryo", 116
asexual colonies, 105
asexual reproduction, effect of inducers, 114
biochemistry of development, 107–108
cell types, 104, 110
  genetic equivalence of, 113
control of differentiation in, 104–117
dedifferentiation in, 114
description, 105
developmental patterns, 108–111
effect of inducers in
  Volvox aureus, 110
  Volvox carteri, 108
  Volvox gigas, 110
  Volvox Rousseletii, 110
female induction, 113–116
  effect of cell proliferation, 113
  inhibition of cell division, 113
generation time, 104
male induction, 116–117
sexual reproduction, 105–106
"spontaneous female", 115
susceptibility to gamete induction,
  111–112

W

Whitten effect, 174

A 4
B 5
C 6
D 7
E 8
F 9
G 0
H 1
I 2
J 3